THE TEAL

Published 2014 by T & AD Poyser, an imprint of
Bloomsbury Publishing Plc, 50 Bedford Square, London WC1B 3DP

Copyright © Matthieu Guillemain and Johan Elmberg

The moral right of the authors has been asserted

No part of this publication may be reproduced or used in any form or by any means –
photographic, electronic or mechanical, including photocopying, recording, taping
or information storage or retrieval systems – without permission of the publishers.

www.bloomsbury.com

Bloomsbury is a trademark of Bloomsbury Publishing Plc

Bloomsbury Publishing: London, New Delhi, New York and Sydney

A CIP catalogue record for this book is available from the British Library

ISBN (print) 978-1-4729-0850-6
ISBN (epub) 978-1-4729-0851-3

10 9 8 7 6 5 4 3 2 1

Commissioning Editor: Jim Martin

Design by Mark Heslington Ltd, Scarborough
Illustrations by Federico Gemma

Printed and bound in Great Britain by Berforts Information Press Ltd.

THE TEAL

Matthieu Guillemain

and

Johan Elmberg

T & AD POYSER

London

Contents

	List of figures	7
	List of tables	12
	List of boxes included in main text	13
	Foreword	14
	Preface	16
	Acknowledgements	18
1	Names and relatives	21
2	General appearance and identification	27
	Distinguishing Teal from other ducks	27
	Age and sex determination	30
	Measurements	35
3	Distribution and numbers	38
	Subpopulations and long-term trends in numbers	38
	Conservation status	48
4	Movements and habitat use	49
	The annual cycle	49
	The migrations	50
	Habitat use during the annual cycle	69
5	Feeding ecology	82
	Energy requirements	82
	Daily foraging time	85
	Foraging methods and postures	90
	General diet and seasonal changes	101
6	Breeding ecology	115
	Pair formation	115
	From nest establishment to hatching of ducklings	127
	From hatching to fledging	140
7	Mortality and limiting factors	153
	Diseases	153
	Parasites	157
	Pollution	161
	Predation	166
	Hunting	170
	Human disturbance	176
	Adverse weather	179
	Habitat loss and change	183

8	Demography	186
	Teal survival rates	186
	Fecundity and age structure	193
	Life cycle and population trends	195
	Demographic features and resilience to harvest	197
9	Management, harvest and conservation	198
	Teal popularity with humans	198
	Teal on the table, then and now	199
	Teal in captivity	204
	Harvest management	205
	Habitat management	207
10	The Teal's future	213
	General numbers and trends	213
	Black boxes: the Teal's big remaining secrets	214
	Limiting factors for Teal populations	216
	A proactive approach to Common Teal population management in Eurasia	217

Appendices	219
References	265
Index	315

List of figures

(CS = colour section)

FIGURE 1.1.	*Illustration of a Teal in Gesner (1555)*	22
FIGURE 1.2.	*Phylogeny of the tribe Anatini (dabbling ducks) after Gonzales et al. (2009)*	23
FIGURE 1.3.	*Common and Green-winged Teal males*	CS
FIGURE 1.4.	*Polar view of the globe illustrating nuclear DNA gene flows between Common and Green-winged Teal*	26
FIGURE 2.1.	*Male Common and Green-winged Teal*	CS
FIGURE 2.2.	*The white horizontal line on the side in Common Teal is formed by the scapulars. Only Common Teal bear such white colour on the scapulars, while scapulars of Green-winged Teal are vermiculated grey*	28
FIGURE 2.3.	*Female Common Teal*	CS
FIGURE 2.4.	*Male Common Teal in active moult to breeding (alternate) plumage*	CS
FIGURE 2.5.	*Female and male wing speculums. While female speculums generally have fewer than four metallic green feathers, males have four or more such feathers*	CS
FIGURE 2.6.	*(Colour section) Sexing of Teal by the most distal tertial of females and males*	CS
FIGURE 2.7.	*Tails of Teal: a juvenile female with all feathers (rectrices) notched, a moulting juvenile male and an adult female with only pointed feathers*	31
FIGURE 2.8.	*Standard terminology used to age Teal in Europe and in the Americas*	32
FIGURE 2.9.	*Age determination by examination of the cloaca*	33
FIGURE 2.10.	*Abdominal manipulation to sex Teal by the sound produced*	34
FIGURE 2.11.	*Preening female Green-winged Teal*	35
FIGURE 2.12.	*Mean winter body mass and wing length of males and females of the species in the dabbling duck guilds of North America and Europe*	36
FIGURE 2.13.	*Mean body mass (g) and wing length (mm) for Common Teal by sex and age class*	37
FIGURE 3.1.	*Common Teal (Eurasia) and Green-winged Teal (North America) geographic ranges*	39
FIGURE 3.2.	*Extent of flyways in Common and Green-winged Teal*	40
FIGURE 3.3.	*Monitoring of waterbirds as part of the 1974 mid-January count of the Bureau International de Recherche sur la Sauvagine, south of Tan-Tan, Moroccan Atlantic coast*	CS

FIGURE 3.4. *Trends in numbers of Green-winged Teal (based on spring pair surveys) and of Common Teal (based on winter counts)*   43
FIGURE 3.5. *Ratification of the Ramsar Convention in Iran, 1971*   48
FIGURE 4.1. *Recoveries in Europe and adjacent areas of birds initially captured in Tour du Valat, southern France*   53
FIGURE 4.2. *Footage from a film showing ring recoveries in western Europe of Common Teal initially ringed in Tour du Valat, southern France*   54
FIGURE 4.3. *Mean proportion of Teal recoveries from areas outside the Camargue to the north and east*   56
FIGURE 4.4. *Teal may have to wait for their breeding lakes to become ice-free when they reach the breeding grounds in spring. Sometimes they have to endure icy and snowy conditions for a couple of weeks before they can settle on their breeding ponds*   57
FIGURE 4.5. *Marcel Klaassen measuring Common Teal flight characteristics of a live bird in a wind tunnel at the Ecology Building of Lund University in Sweden*   58
FIGURE 4.6. *Distribution of ring-recovery distances for Green-winged Teal*   59
FIGURE 4.7. *Teal ring recoveries in western Europe during mild and cold winters*   63
FIGURE 4.8. *Changes in adult male Common Teal body mass over time during the February 1956 cold spell in the Camargue, inferred from live birds weighed during ringing operations*   64
FIGURE 4.9. *Ringing of a Common Teal in Tour du Valat, the Camargue, in the 1950s*   66
FIGURE 4.10. *Annual number of Common Teal ringed in the United Kingdom and Ireland from 1909 to 2011*   66
FIGURE 4.11. *Common Teal with a nasal saddle*   67
FIGURE 4.12. *Common Teal with a radio tag*   68
FIGURE 4.13. *GPS tag of a type that can be fitted to a Teal*   68
FIGURE 4.14. *Eleven lakes in northern Sweden were used in an experiment to study the effect of Pike presence on breeding performance in Teal*   67
FIGURE 4.15. *A preferred breeding and moulting habitat of Common Teal: grassy shores, patches of open water and floating vegetation*   CS
FIGURE 4.16. *Typical beaver pond selected by breeding Common Teal in the Finnish boreal forest*   73
FIGURE 4.17. *Teal do not only rely on standing waters but can also be observed in river habitats, as here in Kamo River, Japan ('Duck river' in Japanese)*   75
FIGURE 4.18. *Very little is still known about the ecology of spring-staging Teal, especially their whereabouts and habits during the last few weeks before settling on their breeding wetlands. River deltas and estuaries in the boreal biome may have a crucial role in the annual cycle by providing foraging opportunities at this time*   76
FIGURE 4.19. *Thousands of Common Teal at a winter day-roost at Tour du Valat, Camargue, southern France*   77
FIGURE 4.20. *Marsh Harrier forcing roosting Common Teal to take flight*   79

FIGURE 4.21. *The functional unit of wintering Teal* 80
FIGURE 5.1. *Relationship between species-specific Resting Metabolic Rate (RMR) and species-specific body mass among dabbling ducks* 83
FIGURE 5.2. *Change in mean body mass in adult male Teal through winter* 84
FIGURE 5.3. *Mean proportion of Common Teal foraging during daylight hours over a tidal cycle in the Netherlands* 87
FIGURE 5.4. *Mean proportion of time spent foraging by wintering seed-eating dabbling ducks (most of these being Teal) in relation to mean January temperature* 88
FIGURE 5.5. *Mean percentage of time spent foraging during daylight hours per month by Green-winged Teal wintering in Louisiana* 89
FIGURE 5.6. *Mean proportion of time spent foraging by Common Teal in Europe during daylight hours (08:00–18:00) and during the night (18:00–08:00) per period of the annual cycle* 90
FIGURE 5.7. *A 24-hour perspective on foraging activity and foraging methods in Common Teal in relation to prey availability in a boreal Swedish lake* 91
FIGURE 5.8. *The foraging repertoire of Teal, allowing them to reach food at various depths* 92
FIGURE 5.9. *Typical posture of a foraging Teal in winter* 94
FIGURE 5.10. *Mean foraging depth of Common Teal and frequency of raptor fly-overs at study sites from early winter to the end of summer moult* 95
FIGURE 5.11. *Feeding behaviour of Common Teal on a Swedish breeding lake and availability of invertebrate food in the vegetation and on the lake bottom* 96
FIGURE 5.12. *Mean foraging depth of Common Teal over the morning hours in Finland in June and July* 97
FIGURE 5.13. *Proportion of time spent vigilant while foraging in Common Teal depending on whether the eyes are above or under water* 98
FIGURE 5.14. *Close-up view of a Teal head, showing the bill lamellae used to filter water and soft sediment for food items* 99
FIGURE 5.15. *Schematic functioning of a duck bill, showing water and food item fluxes within the apparatus* 100
FIGURE 5.16. *Mean ± SE calorific density (kJ/cm3) of invertebrates and seeds per sampling period* 102
FIGURE 5.17. *Dissection of Teal digestive tract* 107
FIGURE 5.18. *Mean number of breeding Common Teal pairs per lake depending on whether wing-clipped Mallard were previously experimentally introduced to the lakes or not* 110
FIGURE 5.19. *Divergent foraging strategies of Teal and Mallard* 112
FIGURE 5.20. *Teal often forage in muddy environments, and may then become dispersers ('vectors') of seeds, eggs and other propagules stuck on their bill and/or feet* 114
FIGURE 6.1. *Percentage of Teal engaged in courtship activity during scan samples performed in the Camargue during winters 2002–2003 to 2004–2005* 120
FIGURE 6.2. *Annual variation in percentage of female ducks being paired at Ismaning lakes, Germany* 121

FIGURE 6.3.   *Copulation by a pair of Green-winged Teal*   CS
FIGURE 6.4.   *Grunt-whistling male Common Teal*   CS
FIGURE 6.5.   *Male Teal in Down-up position*   CS
FIGURE 6.6.   *Penis of a first-year Teal in February*   CS
FIGURE 6.7.   *Forced copulation is a common reproductive strategy of Teal males*   125
FIGURE 6.8.   *Pre-breeding Common Teal in a mixed-species flock in an estuarine habitat in the boreal region (Västerbotten, Sweden)*   128
FIGURE 6.9.   *Common Teal nest and eggs concealed in vegetation, Iceland*   129
FIGURE 6.10.   *Percentage of time spent foraging by the two members of Green-winged Teal pairs during pre-Rapid Follicular Growth (pre-RFG, at least a week before laying), Rapid Follicular Growth (RFG, the week before laying) and incubation*   132
FIGURE 6.11.   *Common Teal egg beside a Mallard egg*   134
FIGURE 6.12.   *Probability of presence on the nest across the day for four Green-winged Teal females*   136
FIGURE 6.13.   *Common Teal female with brood (southern Sweden)*   CS
FIGURE 6.14.   *Downy Common Teal duckling of the youngest (Ia) age class, in other words being at the most six days old*   CS
FIGURE 6.15.   *Variability of facial patterns in Common Teal ducklings*   141
FIGURE 6.16.   *Time needed to attain fledging mass in Teal, Shoveler, Gadwall and Mallard, expressed as proportion of final fledging body mass*   146
FIGURE 6.17.   *Mean brood size of ducklings in each age group class, based on data available in the literature*   148
FIGURE 6.18.   *Semi-natural nest of the type used by Elmberg et al. (2009) to compare the predation rate over time in southern France versus northern Sweden*   150
FIGURE 6.19.   *Mean (± SE) predation rate of semi-natural duck nests over time in northern Sweden and southern France*   151
FIGURE 7.1.   *Modelled potential spread of AIV across western Europe and Russia by Common Teal wintering in the Camargue, southern France, depending on the duration of viral excretion duration*   CS
FIGURE 7.2.   *Green-winged Teal poisoned by botulism*   157
FIGURE 7.3.   *X-ray photography of a Teal gizzard content in a Petri dish*   165
FIGURE 7.4.   *Lead-poisoned Common Teal*   166
FIGURE 7.5.   *Peregrine Falcon, Gadwall and Green-winged Teal*   168
FIGURE 7.6.   *Estimated Green-winged Teal harvest per state in the USA during the 2011–2012 hunting season*   171
FIGURE 7.7.   *Common Teal fitted with both a standard numbered ring and a reward ring*   174
FIGURE 7.8.   *Relationship between the proportion of juvenile birds in the autumn catch of Common Teal in the Camargue, used as an index of breeding success, and the body mass of individuals wintering in the same area by the end of the previous winter*   180
FIGURE 7.9.   *Teal sometimes have to endure periods of adverse weather during winter. Besides the physiological stress this may cause, cold periods also quickly lead to the freezing of shallow waterbodies and snow cover on the ground, making food inaccessible*   181

FIGURE 7.10. *Annual survival rates of male and female Common Teal ringed at Tour du Valat, Camargue, France* — 182

FIGURE 7.11. *Present and predicted distribution of Common Teal under current global climate change scenarios* — 184

FIGURE 8.1. *Percentage of first-year birds among Common Teal in the bags of hunters throughout the western European flyway* — 191

FIGURE 8.2. *Proportion of the total number of ring recoveries obtained per number of winters elapsed since ringing* — 191

FIGURE 8.3. *Total number of Common Teal of each age and sex class among the 56,604 individuals ringed in the Camargue between 1952 and 1978* — 192

FIGURE 8.4. *Annual investment in reproduction (expressed as the ratio between mean clutch mass and mean female body mass) plotted against mean adult survival rate in Anatidae* — 193

FIGURE 8.5. *Formal model of Teal life cycle* — 195

FIGURE 9.1. *One of the very few representations of Teal in ancient art: mosaic from the House of the Faun in Pompeii, currently presented in the Naples Museum of Archaeology* — CS

FIGURE 9.2. *Green-winged Teal was selected as the emblem species for the youth education programme of Ducks Unlimited, Inc. and Ducks Unlimited Canada* — 199

FIGURE 9.3. *A duck decoy in operation* — 201

FIGURE 9.4. *Label of a can of Teal meat from the Island of Föhr, Germany* — 202

FIGURE 9.5. *Commercial harvest of Teal by means of duck decoys has been discontinued. However, a few are still on display or even operated to catch ducks for ringing purposes* — 202

FIGURE 9.6. *Common Teal and other live birds for sale in a market, Damietta, Egypt, January 1980* — 203

FIGURE 9.7. *Amateur and professional breeders have produced a range of Teal varieties in captivity, which mainly differ in their colouration* — CS

FIGURE 9.8. *Opening and closing dates of Common Teal hunting season in the Camargue, southern France, over the period 1955–2012* — 206

FIGURE 9.9. *European countries in which lead shot is completely banned for hunting waterfowl or hunting in wetlands, countries with a partial ban in some wetlands and countries where the use of lead shot is still legal in all wetlands* — 208

FIGURE 9.10. *Changes in Common Teal distribution within a winter quarter* — 209

FIGURE 9.11. *Artificial damming of a creek in Evo, southern Finland, to imitate Beaver activity* — 211

List of tables

TABLE 1.1. *Names for Common Teal in various languages* — 23
TABLE 3.1. *Breeding Green-winged Teal population estimates for 2012* — 39
TABLE 3.2. *Mean Teal numbers in European countries during winter and breeding seasons* — 44
TABLE 3.3. *Numbers of wintering Common Teal recorded during the Asian Waterbird Census, 1997–2001* — 45
TABLE 4.1. *Some record distances between Common Teal ringing and recovery locations* — 58
TABLE 4.2. *Teal brood density (broods per km pond shoreline) and invertebrate abundance index (mean ±SE) on Beaver and non-Beaver ponds of the Finnish boreal forest* — 72
TABLE 5.1. *Main factors leading to nocturnal foraging in Teal* — 86
TABLE 5.2. *Maximum foraging depth associated with different foraging methods of Teal* — 93
TABLE 7.1. *Summary table of diseases known to occur in Common and Green-winged Teal* — 155
TABLE 7.2. *Summary of parasites known to occur in Common and Green-winged Teal, and whose prevalence was quantified* — 158
TABLE 7.3. *Concentrations of contaminants recorded in tissue from Common and Green-winged Teal* — 163
TABLE 7.4. *Reported Teal predators in the Holarctic* — 169
TABLE 7.5. *Reported mortality causes of ringed Teal later recovered, expressed as percentages of total number of recoveries* — 173
TABLE 8.1. *Annual survival rate estimates for Common Teal and Green-winged Teal based on analysis of ring recoveries* — 188

List of boxes included in main text

BOX 1.1.	*How many Teal species are there? The conventions used in this book*	24
BOX 2.1.	*Teal plumage: waterproof and in need of care*	33
BOX 3.1.	*History of waterfowl surveys*	41
BOX 3.2.	*Population sizes and trends*	45
BOX 3.3.	*Teal conservation status by listing authorities*	46
BOX 4.1.	*Annual movements of Teal wintering in the Camargue, southern France*	52
BOX 4.2.	*Individual histories: marking techniques for Teal and other ducks*	64
BOX 4.3.	*An unexpected ecological relationship: facilitation of Teal by Beaver*	72
BOX 4.4.	*Daily habitat-use strategies by wintering Teal: the 'Functional Units'*	78
BOX 5.1.	*Dabbling ducks possess an efficient filtering tool: how the bill works*	98
BOX 5.2.	*Methods of assessing the diets of Common and Green-winged Teal*	105
BOX 5.3.	*Teal in the dabbling duck community*	108
BOX 5.4.	*Teal and Mallard use different strategies when facing food depletion*	110
BOX 5.5.	*Grit and lead poisoning*	112
BOX 5.6.	*Teal give wings to plants and flightless invertebrates*	113
BOX 6.1.	*Duckling description and ageing of Teal broods*	141
BOX 6.2.	*Why migrate to breed thousands of kilometres from the wintering grounds? An experiment at both ends of the flyway*	149
BOX 7.1.	*Lead poisoning*	164
BOX 8.1.	*Survival of juvenile Teal during the first months after fledging*	187
BOX 9.1.	*Duck decoys: from game trade to scientific duck ringing*	200
BOX 9.2.	*Lead poisoning and the ban on toxic hunting shot*	207
BOX 10.1.	*A wish list for Teal research, monitoring and conservation in the near future*	218

Foreword

This is a wonderful book about a wonderful little duck! It is a masterful work that will be relished as much by those of us in search of new insights into the natural history of waterbirds as by students of evolutionary ecology who are curious about the various trade-offs that shape avian life histories. With the authors, we travel with the Common Teal and Green-winged Teal as they move between their respective breeding and wintering regions in Eurasia and North America, and learn how Teal ply their trade – finding food while evading predators during migration, coping with the challenges of cold winter weather and selecting high-quality breeding habitat. These and many other examples form a central core of this book, and collectively reveal a series of remarkable balancing acts throughout the Teal's annual life cycle.

We are told that – while being small bodied by duck standards – Teal more than compensate for their diminutive body size by possessing great beauty, both in the finery of the males in breeding plumage and in the gorgeously intricate yet highly cryptic females, and by the 'serious attitude' of males defending mates. Widespread and relatively common on both continents, Teal migrate north to remote breeding areas, virtually disappearing for several summer months, only to re-appear in autumn as they migrate to their winter quarters. Those of you who have stood in the field behind binoculars or a shotgun have no doubt been impressed by (and perhaps cursed) the speed and agility of the Teal as they literally rocket past you. This perceived speed on the wing potentially reflects their general pace of life, as Teal invest more heavily in reproduction and have lower annual survival rates than most members of their genus, *Anas*. Indeed, comparisons with congeners such as Mallard provide a rich context for a better understanding of the behavioural and ecological strategies adopted by Teal.

We also gain insights from learning about differences in approaches to conservation and harvest management between Eurasia and North America. Although the book highlights massive data gaps in regions like Central and eastern Asia, where very little is known about Common Teal, deficiencies also arise in North America and even in western Europe, regions of the world where most duck species are relatively well studied. For instance, whatever the reason, details about Teal spring staging and moulting in particular have eluded us. It remains a sobering thought that very few biologists have ever found nests of Common Teal in the western Palearctic and certainly not in numbers helpful to really understand their breeding biology. Hence, despite the wealth of new information and synthesis presented here, this book also abounds with fresh research ideas and ongoing challenges to effective management. Fortunately for us, and the authors, mysteries remain unsolved, creating rich opportunities for future field studies.

This synthesis represents the culmination of years of collaborative research, combining thoughtful and careful observation with thorough investigation and experimentation carried out by two very different biologists, following very different approaches, at opposite ends of the range of the Common Teal. We are fortunate to have the views of two outstanding scientists with differing perspectives on the same organism: Johan Elmberg's view of the breeding grounds of a solitary and tantalisingly shy duck of beaver ponds and peat bogs in the boreal forest of Finland and his native Sweden, in contrast to Matt Guillemain's perspective from the winter quarters in Mediterranean France, of a social and flocking species, torn between acquiring food resources while avoiding being eaten. It is these contrasting insights and deep knowledge that both authors contribute to this book. There can be few waterbirds about which so much has been discovered in very recent years, and our guides are outstanding in the way they peel back the layers of knowledge to reveal the biological basis and explanation for much of what we enjoy when watching these birds.

And what an achievement this has been. For the Common Teal, the authors have had to overcome diverse barriers to complete their treatise – funding shortfalls, political boundaries and the sheer distances between themselves and their respective study areas along the western European flyway. So how did they succeed? By virtue of their undiminished curiosity, their remarkable abilities to coordinate their efforts with those of close collaborators, a staunch commitment to learning and strong science. Herein lies the future, because by their shining example it is evident that the next step involves galvanising coordinated efforts to fill in the blanks for Teal in both hemispheres. More importantly, through their 'road map' for Teal, the authors have provided a means to extend the template to other species, perhaps especially to those with more urgent conservation needs. We congratulate Matt and Johan on their great foresight and innovation – and success! No doubt, the authors of this book and their compatriots are already designing new studies to address the remaining gaps. We wish them well on their new paths to exciting discoveries, made all the easier by the contents of this book.

**Bob Clark, Environment Canada, Saskatoon,
and Tony Fox, Aarhus University, Denmark**

Preface

More than any other species, Teal is the very incarnation of a 'wild duck' in popular imagery: an emblematic migrant that is present for only a part of the year in the wetlands of any geographic area. Although most Teal live no longer than a year or two, this delicate and petite duck travels widely during its life, being equally at home in clear-water tarns in the boreal forest and on estuaries close to our largest cities. Teal is also a generally well-known bird, owing to its abundance and widespread distribution, as well as to the male's beautiful bright breeding plumage that makes it easy to identify. Because of its abundance, flight speed and also its tasty meat, Teal has been a highly regarded gamebird since antiquity. This led John James Audubon (1860b) to write, for Green-winged Teal, that 'Nothing can be more pleasing to an American sportsman, than the arrival of this beautiful little Duck in our Southern or Western States'. As birdwatchers and scientists living in the present day we can only agree. Teal are special. We have always been humbled and inspired by their appearance and actions, in our work as well as in our leisure time. For almost 20 years we have been working on Teal ecology, together for more than a decade. We are delighted and honoured to have had the privilege to write this book. As you will see, Teal are well-studied birds and the long list of references at the end of the book amply illustrates how many colleagues we are indebted to for producing this collective wealth of knowledge.

On purpose we have chosen to be rather factual in our writing style, and not to speculate too much about what interesting secrets lurk in the large knowledge gaps that still exist. Yes, it is true, and a blessing for waterfowl biologists to follow, that an amazing amount remains to be discovered about Teal. Unlike its name in our own mother tongues, 'Teal' is a convenient word in English as it is both singular and plural. However, already here in the foreword we feel obliged to underline that this book is actually about two birds, the Eurasian Common Teal and the North American Green-winged Teal. They have sometimes been regarded as subspecies of the same species, sometimes as full species. The latter view prevails today and is the treatment we have chosen to adopt. Depending on continent of residence, readers with their own close relationship with either species may in places find our treatment of 'his/her' Teal superficial, and the language perhaps a bit dry. We regret if this is the case, but the reason is again that we have chosen to publish a text based mainly on written sources, and some topics are well researched in one but not in the other species. Accordingly, the two species are not equally prominent in all parts of the text, and each tends to feature more in topics where it has been studied more than the other.

Now to do like the Teal: jumpstart into the air and fly into the book on light wings!

Matthieu Guillemain and Johan Elmberg,
Tour du Valat, France, and Kristianstad, Sweden, July 2014

Acknowledgements

The writing of this book has been all about friendship, not only between the authors but also with tens of students, colleagues, birdwatchers and hunters on both sides of the Atlantic. It has been a story of early inspiration during misty autumn mornings along the mighty Mississippi, mosquito nights at Beaver ponds in the boreal forest and windy Provence mornings full of duck wings. Writing this book has also been a story of critical questions, data sharing, active help and continuous support by many, whom we are pleased to acknowledge here.

We would first like to thank Jim Martin and Bloomsbury Publishing for believing in this project and supporting us wholeheartedly in making this book a reality. We are also delighted to thank our reviewers, especially Olivier Devineau (Hedmark University College, Norway) for Chapter 8, Jean-Baptiste Mouronval and Jean-Yves Mondain-Monval (Office National de la Chasse et de la Faune Sauvage, France) for Chapters 2 and 7, Hannu Pöysä (Finnish Game & Fisheries Research Institute) for Chapter 6, and Erland Björklund and Ola Svahn (Kristianstad University, Sweden) for Chapter 7. We are particularly grateful to Géraldine Simon (Office National de la Chasse et de la Faune Sauvage), Bob Clark (Environment Canada) and Tony Fox (Aarhus University, Denmark), who agreed to review the entire manuscript; there was of course much more to do and much less time to do it than we initially told them. All these reviewers did a fantastic job, and the errors that remain have surely been added to the manuscript after they performed their review.

We extend a very special '*grazie*' to Federico Gemma for producing the wonderful artwork adorning the covers and many other places in this book.

Many people and organisations contributed to this project by providing original, sometimes unpublished data. For this we acknowledge the valuable help of Simon Delany, Tom Langendoen and Szabolcs Nagy at Wetlands International (the Netherlands), who provided Teal numbers from the International Waterbird Censuses, as well as Bernard Deceuninck and the Ligue pour la Protection des Oiseaux (France) for the French count data specifically. Alain Tamisier (Centre National de la Recherche Scientifique, France) and Michel Gauthier-Clerc (Centre de Recherche de la Tour du Valat, France) kindly provided the datasets from their monthly Camargue duck counts over the 1964–2013 period. A lot of the information in this book comes from ringing/banding data. We are of course thankful to the countless fieldworkers who ringed the birds and the many who sent the rings back. We thank the Bird Banding Laboratory at the United States Geological Survey Patuxent Wildlife Research Center (USA), the Centre de Recherche de la Tour du Valat, and the Wildfowl and Wetlands Trust (England) for letting us use their ringing databases. We thank Danny Bystrak (USGS Bird Banding Laboratory), Jacquie Clark (British Trust for Ornithology, UK) and David Rodrigues (University of Coimbra, Portugal) for their help with ring recovery data. Rob Clay, John Cornell and Lincoln Fishpool at BirdLife International provided invaluable information and data on Important Bird Areas. Jorge Koppen from the U.S. Fish and Wildlife Service (USA) helped us greatly in finding comparable information for North America. Véronique Cherguy (Mairie d'Arles, France) is acknowledged for her help while searching the archives for old hunting season rules.

We were lucky to have been advised by a lot of people who helped us with technicalities in ecology, taxonomy and legislation in their specific countries, along with various questions they never imagined anyone would ever ask them: Preben Clausen and Thomas Kjaer Christensen (Aarhus University), Kathryn M. Dickson and Michel Gendron (Environment Canada), Kathy Fleming and Kathy Pearse (U.S. Fish and Wildlife Service), Ghislain Fuster, Cy Griffin (Federation of Associations for Hunting and Conservation of the European Union – FACE), Richard Hearn and Kerry Mackie (Wildfowl and Wetlands Trust), John Harradine (British Association for Shooting and Conservation, UK), Alan Johnson and Marion Vittecoq (Tour du Valat), Alain Kaps, Marcel Klaassen (Deakin University, Australia), Wim C. Mullié (Dakar, Senegal, formerly Foundation for Ornithological Research in Egypt), Petri Nummi (University of Helsinki, Finland) and Steven Portugal (University of London, UK).

Most of the illustrations in this book we got from others. We first thank the very numerous amateur and professional photographers who let us use their wonderful pictures: Jean-Pierre Artel, Jerome Beziat, Romain Blanc, Trent K. Bollinger (Canadian Cooperative Wildlife Health Centre), Didier Buysse, Michel Collard, Don Delaney, Danièle Delvart, Magnus Elander, Dominique Gest, Tom Grey, John Houston (Abbotsbury Swannery, Dorset, UK), Christian Hovette, Kenneth Johansson, Alan Johnson, Johannes B. Jonsson, Marcel Klaassen, Serge Lardos, David Lédan, Christelle Lucas, Charles McDonald, Jean-Baptiste Mouronval, Staffan Müller, Wim C. Mullié, Marie-Lan Nguyen (Wikimedia Commons), Mathieu Remacle, Özden Sağlam, Kjell Sjöberg, Pär Söderquist and Marcus Wikman. Sharing your photographs on the internet may result in you getting an email request from someone on the other side of the planet one day.

Jon Fjeldså kindly let us reprint his very nice duckling drawings, while Clémence Deschamps (Tour du Valat) and Jean-Paul Rodrigue (Hofstra University, USA) helped us greatly with maps. Merebeth Switzer (Ducks Unlimited Canada) kindly provided the Greenwing program logo, and Matt Kaminski (Ducks Unlimited, Inc., USA) sent Green-winged Teal feathers. We also thank Adeline Rouilly for access to the collections of the Museum of

Nîmes (France), where we could photograph Teal and Mallard eggs, and Jutta Kollbaum-Weber (Dr Carl Häberlin Friesen Museum, Germany) for providing the photograph of a Teal can label from the Isle of Föhr. Dafila Scott kindly gave us permission to reproduce the line drawing of a duck decoy by her father, Sir Peter Scott.

Some illustrations were reprinted from other publications. For this courtesy we thank Wildlife Biology, Lynx Edicions, John Wiley & Sons, Inc., the Ornithologischer Anzeiger, Boreal Environment Research, the Netherlands Ornithologists' Union, and Springer as publishers. We also thank Brian Huntley (Durham University, UK), Rhys Green (University of Cambridge, UK), David Gibbons (Royal Society for the Protection of Birds, UK), Dietlind Willer and Bärbel Mund (SUB Göttingen, Germany), Rob G. Bijlsma (Editor of Ardea), Olivier Devineau (for a figure in his PhD thesis), Kjell Sjöberg and Kjell Danell (Swedish University of Agricultural Sciences), Robert Pfeifer and Einhard Bezzel for letting us reuse figures they authored or for helping us to access difficult material.

Librarians have played a key role during the writing process. We received invaluable help from Peter Bengtsson (Kristianstad University), Jacqueline Crivelli and Gwenael Wasse (Tour du Valat library) and Marie-Solange Landry (ONCFS library). Mark Drever and Pierre Legagneux provided important additional references. For help with translation we are indebted to Mathilde Balas (Chinese), Anis Guelmami and Wed Abdou Abdelatif Ibrahim (Arabic), Hélène Guillemain (Japanese), Christiane Jakob (German), Frédéric Joly (Russian) and Jean Saint-Zéby (Latin).

In addition to studies published by others, this book summarises Teal work we have ourselves been involved in over the last 20 years. It is a pleasure to use this opportunity to here thank all the people who contributed to these studies over that time. We would first like to thank our many duck students, especially the MSc and PhD students, who always pushed our limits (both in the field and in front of the computer): Frédéric Albespy, Céline Arzel, Fabricio Basilio de Almeida, Romain Blanc, Anne-Laure Brochet, Jocelyn Champagnon, Lisa Dessborn, Olivier Devineau, Karin Folkesson, Gunnar Gunnarsson, Perrine Lair, Claire Pernollet, Maud Poisbleau, Pär Söderquist and Emilien Weissenbacher. It is also a pleasure to thank our closest duck colleagues, in particular Jean-Marie Boutin, François Cavallo, Bob Clark, Patrick Duncan, Tony Fox, Hervé Fritz, Jonathan Fuster, Michel Gauthier-Clerc, Andy Green, Noël Guillon, Jean-Dominique Lebreton, Pierre Legagneux, Michel Lepley, Grégoire Massez, Jean-Baptiste Mouronval, Petri Nummi, Hannu Pöysä, Vincent Schricke and Kjell Sjöberg.

Matthieu Guillemain wishes to thank Géraldine, Jules and Léa for their patience and continuous support over the years.

Johan Elmberg thanks family, friends and foes for never ceasing to ask the difficult questions about priorities in life.

We dedicate this book to Luc Hoffmann and to the late Guy A. Baldassarre, who did so much for Teal on the two sides of the Atlantic.

CHAPTER 1

Names and relatives

The Common Teal *Anas crecca*, or Eurasian Teal, was given its scientific name in 1758 in Linnaeus's monumental *Systema Naturae*. In present-day systematics it belongs to the Anatidae family, like all ducks, geese and swans. Teal is placed in the genus *Anas* ('duck' in Latin), along with most other dabbling ducks (such as Mallard *A. platyrhynchos* and Northern Pintail *A. acuta*). 'Dabbling' ducks generally forage from the water's surface, as opposed to 'diving' ducks like the *Aythya* species (for example, Ring-necked Duck *Aythya collaris*, Tufted Duck *A. fuligula*, Redhead *A. americana* and Common Pochard *Aythya ferina*). The Teal's scientific name is onomatopoeic, in other words describing its sound, in this case the flight call. For the same reason its vernacular name is *Krickente* in German, *Kricka* in Swedish, *Krikkand* in Norwegian, and similar names in some dialects in countries where the national name is different (for example *Criquet* or *Crac* in some French regions versus the common French name *Sarcelle d'hiver*, Ternier 1922) (Table 1.1). The literal English translation of the Teal's Russian name (*Chirok-svistunok*) is 'whistler-teal' (Dement'ev & Gladkov 1952).

Some of the common national names also bear witness to the fact that Teal is a well-known and abundant bird. Accordingly, its English name is *Common Teal*, its Spanish name *Cerceta común* and its Portuguese name *Marrequinha-comum*. Other vernacular names refer to its size (*kogamo* literally meaning 'small duck' in Japanese) or seasonal occurrence. For example, the French *sarcelle d'hiver* ('Winter Teal') and the Dutch *wintertaling* both refer to its main period of presence in these and other countries in western and south-western Europe. The Arabic name of the species also refers to winter. Interestingly, its likewise small-sized close relative Garganey *Anas querquedula* is called *sarcelle d'été* in French and *Zomertaling* in Dutch, both meaning 'Summer Teal'. Garganey winter in Africa and spend only a few summer months on the breeding grounds in Europe and northern Asia. The scientific name of Garganey, *Anas querquedula*, is derived from the Latin (where *querquedula*

FIG. 1.1 *Illustration of a Teal in Gesner (1555). Most of the plumage features are those of a Common Teal (general silhouette, pale spotted chest, white line on the greater coverts). Picture provided by Göttingen University (SUB Göttingen, 2 ZOOL I, 7115:3 RARA).*

means either Teal or 'small duck'). Before Linnaeus, it is not even certain that Common Teal and Garganey were considered to be distinct species. Belon (1555), for instance, only considered one such small duck species, whose description clearly corresponds to that of a Common Teal. However, he also mentions that the species is called *Garganei* in Milan (still used locally in the form *Garganèllo*).

Although Linnaeus considered Teal and Garganey to belong to the same genus as the Mallard, Teal and other smallish dabbling ducks have sometimes been distinguished from the other *Anas* species and placed in a subgenus of their own ('sous-genre sarcelle', as opposed to 'sous-genre canard' in Buffon (1853), though the distinction between species and plumage types was apparently not very clear at that time). Later, the genus *Querquedula* was often used for these small ducks; *Querquedula crecca* for the Common Teal and *Querquedula circia* for Garganey (for example in Stryan 1891, Rogeron 1903, Ternier 1922).

The origin of the name 'Green-winged Teal' for the American species/subspecies (see below) may seem puzzling to non-American readers, as *carolinensis* has no more green in its spread wings (nor males on their heads) than do European *crecca*. However, Green-winged Teal are sympatric with

TABLE 1.1. *Names for Common Teal in various languages. Alternative names exist in some cases (for example, Chinese and Russian), only one being given here.*

Albanian	Rosa kere	Italian	Alzavola
Arabic	بطة الشتاء	Japanese	Kogamo (コガモ)
Chinese	xiǎo shuǐ yā (小水鴨)	Latvian	Krīklis
Croatian	Patka kržulja	Lithuanian	Rudagalvė kryklė
Czech	Čírka obecná	Maltese	Sarsella
Danish	Krikand	Norwegian	Krikkand
Dutch	Wintertaling	Polish	Cyraneczka
English	Eurasian/Common Teal	Portuguese	Marrequinha-comum
Estonian	Piilpart	Russian	Chirok-svistunok (Чирок-свистунок)
Finnish	Tavi	Slovakian	Kačica chrapka
French	Sarcelle d'hiver	Slovenian	Kačica chrapkavá
German	Krickente	Spanish	Cerceta común
Hungarian	Csörgő réce	Swedish	Kricka
Icelandic	Urtönd	Turkish	Çamurcun
Irish	Praslacha	Welsh	Corhwyaden

other smallish duck species over large parts of North America, particularly the Blue-winged Teal *Anas discors*, which does sport a lot of blue in its plumage. The etymology of the scientific name of the Green-winged Teal is more obvious, though: *carolinensis* refers to Carolina, the states in the south-eastern United States where European settlers first observed these birds. Its French name has the same root (*Sarcelle de la Caroline*).

The latest taxonomy, based on genetic distance between species, clearly demonstrates that Common Teal belong to the same clade as other 'teal', such as Green-winged Teal *Anas carolinensis*, Speckled or Yellow-billed Teal *A. flavirostris flavirostris* and Sharp-winged Speckled Teal *A. flavirostris oxyptera* (Gonzales *et al*. 2009). Nevertheless, these species are also very close genetically to other dabbling ducks such as the different Pintail species (Figure 1.2).

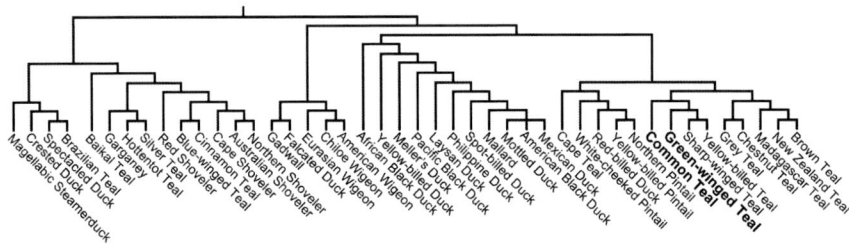

FIG. 1.2 *Phylogeny of the tribe Anatini (dabbling ducks) after Gonzales* et al. *(2009). Note that Garganey A.* querquedula, *which was long thought to be a close relative of Teal, is only distantly related to according to this phylogeny. Note also that the molecular data on which this tree (cladogram) is based suggest that the Speckled Teal A.* flavirostris *is the sister group of Green-winged Teal A.* carolinensis, *and that they together comprise the sister group of Common Teal A.* crecca. *See the original figure for analytical details.*

BOX 1.1. How many Teal species are there? The conventions used in this book

Whether or not the American Green-winged Teal and the Eurasian Common Teal should be considered as two separate species or rather as two subspecies of the same species has long been and is still debated. The official taxonomy in North America is symptomatic of this: American and Eurasian Teal were initially considered as two separate species, first *Anas crecca* and *Anas carolinensis* (American Ornithologists' Union 1886), later *Nettion crecca* and *Nettion carolinensis* (American Ornithologists' Union 1899). After being repatriated to the genus *Anas*, still as two distinct species (American Ornithologists' Union 1944), they were 30 years later grouped back into one single *Anas crecca* species with two subspecies (American Ornithologists' Union 1973). It was eventually proposed to re-split the two taxa into two separate species (Banks *et al.* 2002) but this has not yet been ratified, so that the official Check-list of North American Birds is still the 1998 version, which considers Common and Green-winged Teal as subspecies of nominate *Anas crecca* (American Ornithologists Union 1998); that is, *A. c. crecca* versus *A. c. carolinensis*.

In contrast, Common and Green-winged Teal have been considered as separate species for more than a decade by most European authorities (British Ornithologists' Union, Scott & Dickson 2001; European Records and Rarities Committees, Crochet & Joynt 2011) and by the International Ornithologists' Union (formerly International Ornithological Committee) alike (Gill & Donsker 2012).

At any rate, nowadays the two taxa are generally treated as separate species in reference texts (for example in Ogilvie 2002, Fox 2005b), whereas a single account was generally provided for the two subspecies only a few decades ago (Cramp & Simmons 1977, Johnson 1995). The scientific ecological journals and their contributors have been more consistent in treating the taxa separately over time.

Controversy over Teal taxonomy arose from the subtle but notable differences in male plumage, despite the general similarity in overall appearance and behaviour (Figure 1.3, see colour section). Conversely, Mallard look the same and behave in similar ways on both sides of the Atlantic, and no one seems to question the fact that Mallard belong to the same single Holarctic species, despite the fact that clear genetic differences have been demonstrated between a Eurasian and an American clade (Kraus *et al.* 2011).

Genetic analyses of mitochondrial DNA (mtDNA, the fraction of genetic information occurring only in the mitochondria of the cells, and which is transferred to the next generation from the mother only) have been carried out and used to determine if Common and Green-winged

Teal should be considered as two distinct species. Zink *et al.* (1995) found significant genetic differences between Teal on either side of the Bering Sea, but they did not deem them large enough to motivate a split into two species. Later analyses, however, demonstrated a level of genetic divergence (5.8%) that is similar to that between Mallard and Northern Pintail (5.7%), implying that *crecca* and *carolinensis* should be treated as proper species (Johnson & Sorenson 1999, see also Johnsen *et al.* 2010). The Aleutian Teal *Anas crecca nimia* was considered to be genetically indistinguishable from nominate *crecca* by these authors after molecular analysis, but only two specimens of the former were analysed.

Johnson & Sorenson's results, combined with plumage analyses and behavioural studies, played an important role in the British Ornithologists' Union deciding to split Teal into two species (see Sangster *et al.* 2001 about the pros and cons of lumping versus splitting taxa). A recent analysis, however, challenged the status quo: Peters *et al.* (2012) demonstrated that gene flow was limited between American and Eurasian Teal with respect to mtDNA, but that interchange was far more frequent for nuclear DNA (nuDNA, the fraction of genetic information in cell nuclei, transferred to the next generation by both parents); nuDNA types could be distinguished in this study but did not match geographic barriers, with some birds having the nuclear genetic signature of the other continent – that is, gene flow was detected. Such flow was limited or absent in the direction from America to Eurasia (*c.* 1 individual per generation), but much larger in the other direction, corresponding to *c.* 20 individuals emigrating from Eurasia to America per generation (Figure 1.4). This matches Palmer's (1976) statement of Eurasian Teal 'occurring as a straggler, sometimes evidently mating with local [American] birds'. From their genetic analyses, Peters *et al.* (2012) concluded that the two clades had probably started diverging 0.4 to 3 million years ago, and that the process of speciation has gone on since then. Nevertheless, the same authors argued that gene flow due to interbreeding between the two taxa is still too significant to consider them proper distinct species.

The latter analyses may cause the current splitting of Green-winged and Common Teal to be reconsidered yet again in the future. This notwithstanding, we decided to consider Common Teal *Anas crecca* and Green-winged Teal *A. carolinensis* as two separate species in this book, for practical purposes if nothing else. Although many aspects of the ecology of the two taxa have been studied on both sides of the Atlantic, and often yielded comparable results, there are also some important differences. For example, North American research has largely been carried out at the population level, while European studies tend to focus on the level of individuals. To provide a comprehensive view of the ecology of these

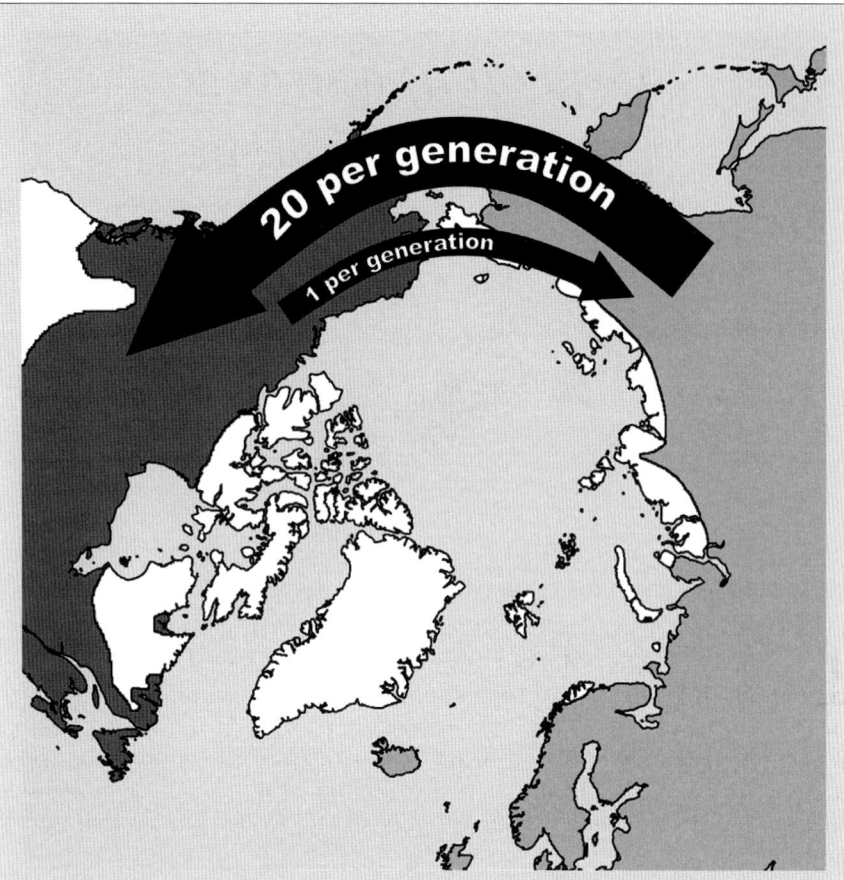

FIG. 1.4 *Polar view of the globe illustrating nuclear DNA gene flows between Common (pale grey) and Green-winged Teal (dark grey), expressed as the number of individuals switching from one population to the other per generation (based on results of Peters et al. 2012).*

birds, scientific sources from the two continents have been combined for the two species, and whenever possible it is stated whether the given information refers to Common or Green-winged Teal. 'Teal' alone is used in this book when the information explicitly refers to both taxa or when no distinction has been made between them by the source.

CHAPTER 2

General appearance and identification

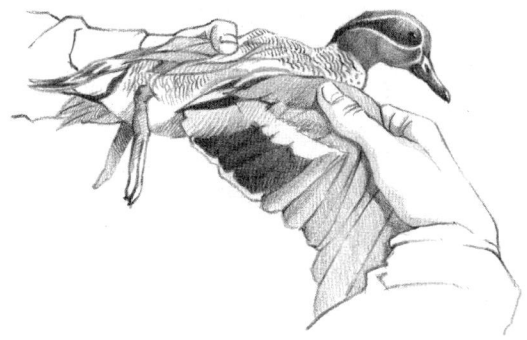

Like most dabbling ducks, Teal show a strong sexual plumage dimorphism. In other words, females and males look strikingly different most of the year. While males are brightly coloured when in breeding plumage, females wear mainly more uniformly brown colours. Juveniles of both sexes are similar to females during their first months of life, after which first-year males moult to a breeding plumage very similar from a distance to that of adult males.

DISTINGUISHING TEAL FROM OTHER DUCKS

In breeding plumage, male Common Teal and Green-winged Teal cannot be confused with any other species in western Europe or North America, respectively. The head is brownish-red with a metallic green patch around the eye and continuing down the neck. The breast is beige and densely spotted, the flanks are mottled grey and the undertail shows a yellow triangle bordered with black. The bill is black and the legs are dark grey. Male Common Teal can easily be distinguished from male Green-winged Teal: the former has a wide horizontal white line on each side of the body, formed by white scapulars, whereas these feathers are greyish in Green-winged Teal (Figures 2.1 (colour section) and 2.2). Male Green-winged Teal have no such white horizontal line but instead sport a conspicuous vertical white bar on each side of the breast (thus made up by body rather than wing feathers), which is also visible in flight. The green and brown parts of the head are generally

FIG. 2.2. *The white horizontal line on the side in Common Teal is formed by the scapulars (white circle on top photograph). Only Common Teal bear such white colour on the scapulars (bottom right), while scapulars of Green-winged Teal are vermiculated grey (bottom left). Photographs by Jean-Baptiste Mouronval/ONCFS.*

separated by a distinct cream-coloured border in Common Teal (bending down to the base of the bill), a feature generally less visible in Green-winged Teal (Pyle 2008; Figure 1.3, see colour section). The tips of the secondary coverts are cinnamon-brown in Green-winged Teal, as opposed to white or pale cinnamon in Common Teal, which may help species identification in the hand, especially in females, according to Jarvis (1966). Aleutian Teal (*A. c. nimia*) are similar to Common Teal, with males having a horizontal white line below the mantle rather than a vertical white bar on the side of the breast.

The only other duck with a somewhat similar plumage pattern is the Falcated Duck *Anas falcata* from Asia. Male Falcated Ducks have a generally grey body and a brown-red head with a metallic green patch around the eye, but can easily be distinguished from Common and Green-winged Teal by the tuft at the back of their head as well as the elongated, curved tertials which gave this species its name (*falx* = sickle in Latin). Falcated ducks are also bigger birds, being twice as heavy as Teal.

Female Teal have a dull brown plumage, like females of most dabbling duck species (Figure 2.3, see colour section). The difference between cryptic females and brightly coloured males can be explained by a process of sexual selection of the brightest males, and of natural selection of the least colourful females, since females alone incubate the eggs and raise the ducklings (see Chapter 6) and face high predation risks during this period. At close range, female Teal body feathers are a clear mixture of dark brown and light beige. The only bright colour is on the upperwing, which bears a patch of metallic green edged with creamy white on the secondaries (also called speculum or wing-flash marks; see Figure 2.5, colour section), but this is often not fully visible in non-flying birds. The plumage pattern of female Teal is broadly similar to that of female Mallard, Northern Pintail and Gadwall *Anas strepera*, to mention just a few species. However, female Teal have pale undertail feathers that form a clearly visible whitish horizontal line on each side of the rump (see Figure 2.3, colour section). More than plumage and colours, it is therefore the smallish body and the silhouette that can help in distinguishing between females of these species. If there are other ducks present to compare with, the much smaller size will clinch the identification of female Teal in Europe, except where female Garganey can be expected. Given reasonable views, however, the Garganey has a more contrasting pattern, with longitudinal striping on the head, including bright horizontal stripes above and below the eye, and a pale lore (that is, the spot at the base of the bill). Female Teal have a greyish-brown bill, often with some orange at its base. Adult females show more dark spots on the bill than do juveniles, which sometimes have no spots at all. Females of Common and Green-winged Teal are virtually impossible to distinguish from a distance, although the colour of the tip of the secondary coverts may be a criterion for birds in the hand (see above).

In North America, only Baikal Teal *Anas formosa*, Blue-winged Teal and Cinnamon Teal *Anas cyanoptera* are of comparable size to Green-winged

Teal. No confusion is generally possible for males in breeding plumage. Females and eclipse males (having 'the camouflage plumage' they temporarily acquire after breeding; see below) of the latter two species can generally be easily distinguished from Green-winged Teal by their bluish/greyish greater coverts (hence the generic term 'Blue-winged ducks', also comprising Northern Shoveler *Anas clypeata*). Baikal Teal, on the other hand, have metallic green wing-flash marks resembling those of Green-winged Teal, but females can easily be told from female Green-winged Teal by their white lores.

Albino Common Teal and Green-winged Teal have both been reported in the wild (Sotnikov 1999 and M. Weegman, pers. comm., respectively). First-generation Common Teal x Green-winged Teal male hybrids are readily identified as they show both vertical and horizontal white lines on the sides of the body, albeit somewhat blurred compared to pure-bred birds, and have therefore been reported several times (review in Palmer 1976, detailed description in Vinicombe 1994).

Age and sex determination

The wing-plumage criteria used to age and sex Green-winged Teal and Common Teal are generally similar in the two species. Here we only describe the most straightforward criteria, which should allow determining the sex and age of Teal in the hand in most cases. More detailed descriptions and other (sometimes hard-to-use) criteria are given in specific identification guides such as Rousselot & Trolliet (1991), Carney (1992), Baker (1993) and Pyle (2008), sources from which much of the information presented here was compiled. Common and Green-winged Teal are among the most difficult dabbling duck species to age (sexing is generally less problematic) with certainty, especially females, and most of the criteria presented here suffer from numerous exceptions.

After breeding, male Teal moult to an eclipse plumage very similar to that of females, though somewhat darker and with greater contrast. All wing feathers are moulted simultaneously, so the birds cannot fly for a relatively extended period averaging 21 days (Sjöberg 1988, Hohman *et al.* 1992). Just like the female plumage, the cryptic eclipse ('basic' in North American terminology) plumage of males is supposed to have evolved as an adaptation to predation risk during this flightless period. The primaries grow as much as 5mm per day (Sjöberg 1988), which shortens the flightless part of the moulting period. Males moult into a fresh breeding ('alternate' in North America) body plumage in autumn but then retain their wing feathers, which are only moulted once a year (Rousselot & Trolliet 1991). Individual condition plays a major role in male autumn moulting phenology (the actual timing of moult), although this has not been clearly investigated. What is clear in the field is that some birds still show no sign of breeding plumage when others have already completed their moult. Similarly, first-year birds generally moult later than adults. Detailed study of moult phenology in male Teal has demonstrated the paramount importance of thyroid and gonadal hormone peaks for moult initiation (review in Bluhm 1988). The sequence and phenology of moult is similar in females, although the general appearance of the plumage does not change during successive moult events.

From a distance it is very difficult to tell eclipse males from females. At this time sex determination is easiest with the bird in the hand, because the wing plumage, which differs somewhat between males and females, remains similar throughout the year in adult Teal, as in most dabbling ducks. The number of speculum feathers (secondaries) sporting metallic green for most of their visible length is a fairly reliable criterion in the sexing of Teal (Figure 2.5, see colour section): birds with more than four feathers of this type are males, while birds with fewer than four are usually females. However, sex cannot reliably be told by this criterion when there are exactly four metallic green secondaries. The most distal tertial should then be examined: in both Common and Green-winged Teal there is a distinct border between the shiny black of the outer part of the feather and the greyish-brown inner part in males, whereas the two colours change gradually between the two ends of the feather in females, with a more brownish aspect overall (Figure 2.6, see colour section). If still in doubt after wing examination, the rest of the plumage should always be examined with caution for other signs of male feathers (for example, the first vermiculated flank and scapular feathers).

Reliable distinction between adult and juvenile Teal is often impossible from a distance, and it is still difficult even in the hand. The only criterion providing absolute evidence of juvenile age is the presence of notched tail feathers (Figure 2.7). The central axis of the feathers (called the 'rachis') is the same for duckling down and for the first juvenile tail feather: when the feather has started growing, the end of the rachis gets broken when the down falls out, hence typically leaving a V-shaped feather. Conversely, completely new feathers are grown during subsequent moult events, with barbs all along the rachis, producing a pointed end. The problem is that juvenile Teal

FIG. 2.7 *Tails of Teal: a juvenile female with all feathers (rectrices) notched (left), a moulting juvenile male (note the four pointed fresh feathers in the middle of the tail, compared to some older lateral feathers still being notched; centre) and an adult female with only pointed feathers (right). Photographs by Jean-Baptiste Mouronval/ONCFS.*

gradually moult tail feathers during their first autumn, then acquiring pointed tail feathers. Whilst the presence of one or more notched feathers is always evidence of a juvenile bird, the presence of only pointed tail feathers therefore does not necessarily demonstrate adulthood.

Other criteria have to be examined for Teal that have pointed tail feathers. The tertials of adults are always longer than the fifth primary when the feathers are fully developed. The middle coverts are also large and of squarish shape in adults, and tend to be narrower in juveniles. This is, however, often difficult to assess on such small feathers. A problem, again, is that juveniles moult their tertials during their first autumn, so the criteria above are no longer reliable in winter (and age determination is therefore difficult during the transition period between these two seasons). In winter, a combination of criteria helps in distinguishing adults from juveniles: the greater tertial coverts are larger in adults than in juveniles before wing moult, and the clear white tip of the greater coverts gradually broadens from the body to the distal end of the wing in the former age class. In juveniles the white colour of such tips is more transparent, and forms a more dotted or irregular white line across the wing. Because juveniles gradually moult their tertial feathers but not the rest of the wing, the colour of such coverts may differ between adjacent feathers, and/or look fresher than on the rest of the wing coverts. The difference between greyish moulted tertials and more brownish greater coverts is remarkable in juveniles males, but is often negligible in juvenile females. Practice is necessary to reliably age Teal from their plumage in winter, and ageing should always rely on multiple rather than single criteria.

It is not possible to age adult Teal beyond the second calendar year. Individuals will therefore be considered as 'juveniles' or 'first-year birds' from

FIG. 2.8 *Standard terminology used to age Teal in Europe (above bars) and in the Americas (below bars). 1A: first calendar year; +1A: later than first calendar year; 2A: second calendar year; +2A: later than second calendar year; HY: hatching year; SY: second calendar year; AHY: after hatching year; ASY: after second calendar year.*

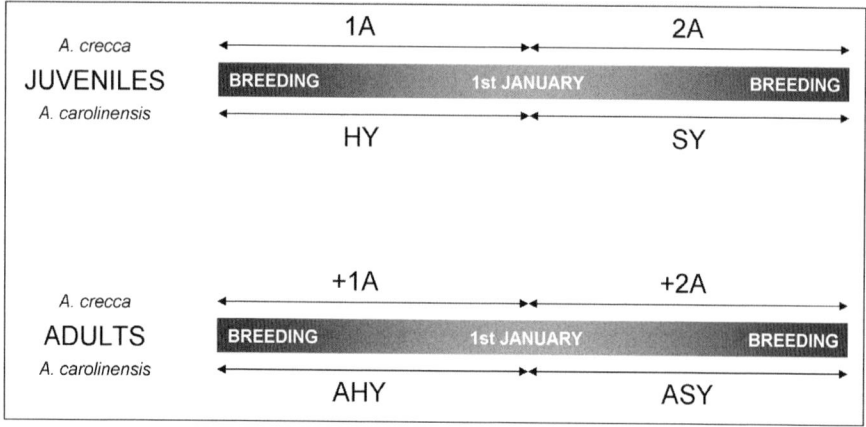

hatching to the next breeding season, and are then called 'adults' (Figure 2.8). Juvenile males acquire breeding plumage during their first winter.

In addition to using plumage, specialists are able to sex and age Teal (and most other Anatidae) by means of cloacal examination (Figure 2.9). Cautiously opening the cloaca exposes the genitalia in these birds, with males having a penis that is usually easy to identify. The penis is the size of a rice seed in juvenile males during their first autumn, and gradually grows to up to

> **BOX 2.1. Teal plumage: waterproof and in need of care**
>
> Bird plumage has many functions, from its obvious role in flight and in social interactions, to its role as camouflage (cryptic female versus bright male plumage; see above) and, of course, as insulation in the regulation of body temperature.
>
> To waterbirds plumage is also important for buoyancy (duck feathers are particularly well covered with oil and have exceptional floating ability). It has long been thought that the extraordinary waterproofing is due to oil secretion of the uropygial gland, close to the cloaca, oil that birds collect with the bill and spread over their feathers by preening (Jacob & Zisweiler 1982). However, recent work shows that the waterproofing is actually not associated with these oil secretions as such, but is due to the very tightly integrated structure of the feather barbs. The role of the oil secretions is rather to protect this feather structure, either by facilitating preening or by limiting bacterial feather degradation (Giraudeau *et al.* 2010a).
>
> Daily plumage care is naturally, therefore, a major activity performed by dabbling ducks (Figure 2.11). More than one hour per day is spent preening by Teal on average, increasing to up to two and a half hours in September, when birds start to moult into breeding plumage (Tamisier & Dehorter 1999).

FIG. 2.9 *Age determination by examination of the cloaca, from left to right: juvenile male during early autumn, juvenile male in January, adult male with much bigger penis, but more difficult to evaginate from the cloaca, and adult female with apparent oviduct opening.*

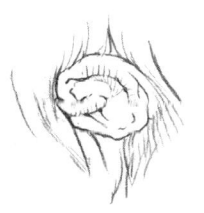 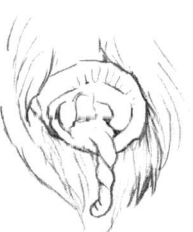 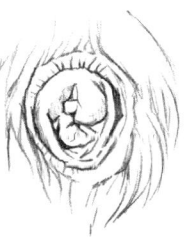 

FIG. 2.10 *Abdominal manipulation to sex Teal by the sound produced.*

10cm long and of spiralled shape in adults. It is nevertheless often more difficult to evaginate the penis out of the cloaca in adults than in juveniles. Obvious indications of a female are the lack of a penis and the observation of the oviduct's end, although detection of the latter requires some practise. Because most female Teal lay eggs at one year of age, adult birds will have an opened oviduct, whereas the oviduct will still be closed in juveniles until they breed. This permits fairly reliable distinction between females-of-the-year and older females throughout the year.

Age of dead birds can be easily determined by dissection, where the presence of a bursa of Fabricius (a lymphoid organ in the cloaca) will indicate a juvenile bird, while adults will have larger testes and ovaries (see for example photographs in Rousselot & Trolliet 1991).

Another way of sexing live birds is syrinx morphology, which leads to clearly different sounds uttered by males and females. Females generally produce a loud *kwack*, while males emit a lighter, flute-like sound. Briefly and distinctly pressing the abdomen from the back to the front is a harmless technique to push out some air, and the sound will clearly distinguish the bird's sex. Sounds thus produced will always be high-pitched in males and low-pitched in females

(Figure 2.10). This technique is effective as early as when birds are a few weeks old.

Measurements

Teal are notably smaller than all other dabbling ducks, with an average body mass of *c.* 320g in adult males, and 290g (*A. crecca*) or 310g (*A. carolinensis*) in adult females (Figure 2.12; see also Appendix 1). Teal body mass shows strong variation over the year, with a minimum after autumn migration and in late winter.

Wing length also differs between males and females: the mean flattened wing length is 191 and 182mm for adult males and females, respectively, with geographic differences. Juveniles have shorter wings than adults (Figure 2.13).

FIG. 2.11 *Preening female Green-winged Teal. Ducks always take the greatest care of their plumage, which plays a major role in thermal insulation and waterproofing. Photograph by Don Delaney.*

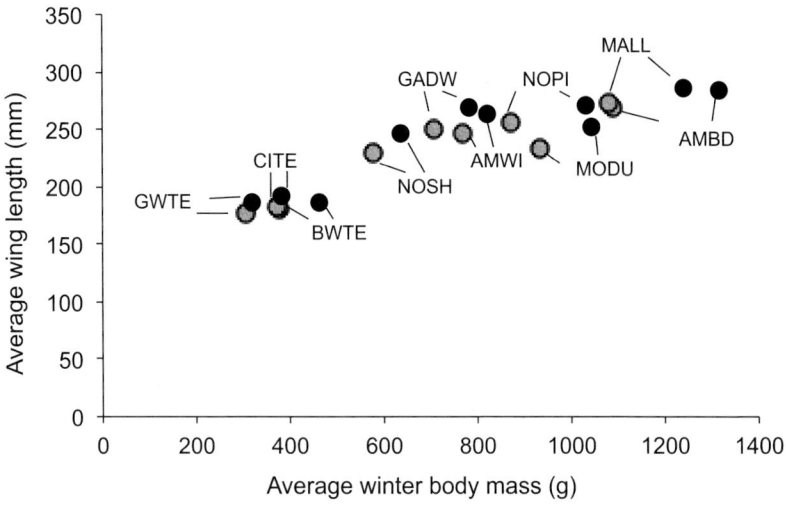

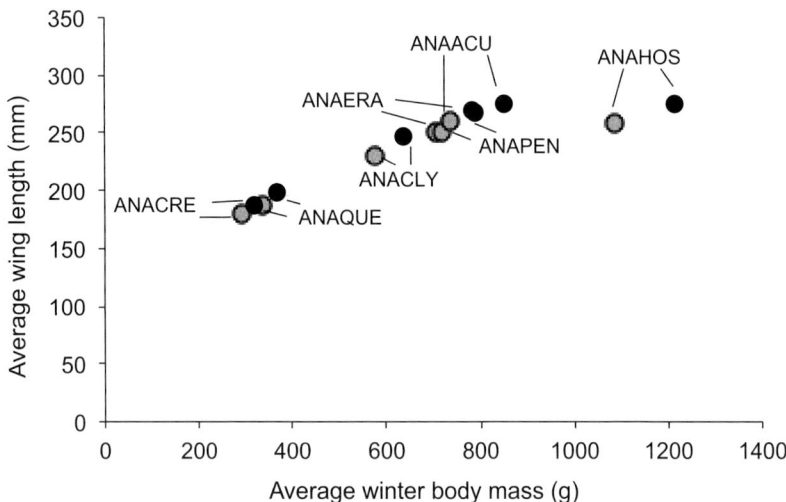

FIG. 2.12 *Mean winter body mass and wing length of males (black dots) and females (grey dots) of the species in the dabbling duck guilds of North America (top, 10 species) and Europe (bottom, 7 species). Species acronyms follow the usage on each continent. North America: GWTE: Green-winged Teal* A. carolinensis; *CITE: Cinnamon Teal* A. cyanoptera; *BWTE: Blue-winged Teal* A. discors; *NOSH: Northern Shoveler* A. clypeata; *GADW: Gadwall* A. strepera; *AMWI: American Wigeon* A. americana; *NOPI: Northern Pintail* A. acuta; *MODU: Mottled Duck* A. fulvigula; *MALL: Mallard* A. platyrhynchos; *AMBD: American Black Duck* A. rubripes. *Europe: ANACRE: Common Teal* A. crecca; *ANAQUE: Garganey* A. querquedula; *ANACLY: Northern Shoveler* A. clypeata; *ANAERA: Gadwall* A. strepera; *ANAPEN: Eurasian Wigeon* A. penelope; *ANAACU: Northern Pintail* A. acuta; *ANAHOS: Mallard* A. platyrhynchos. *Data after Kear 2005.*

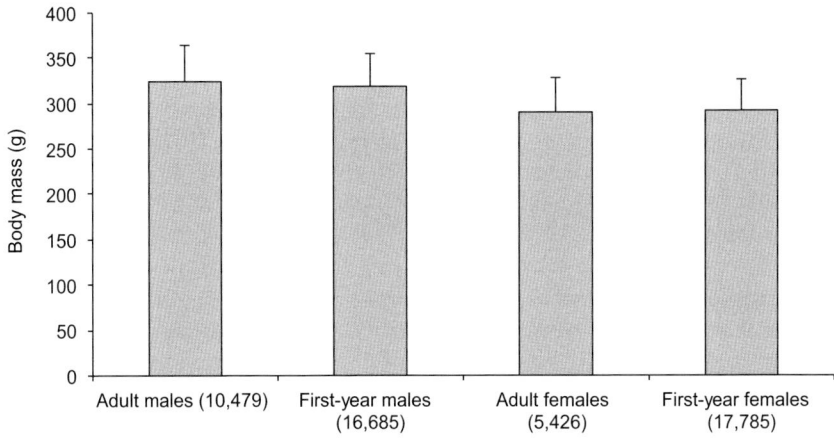
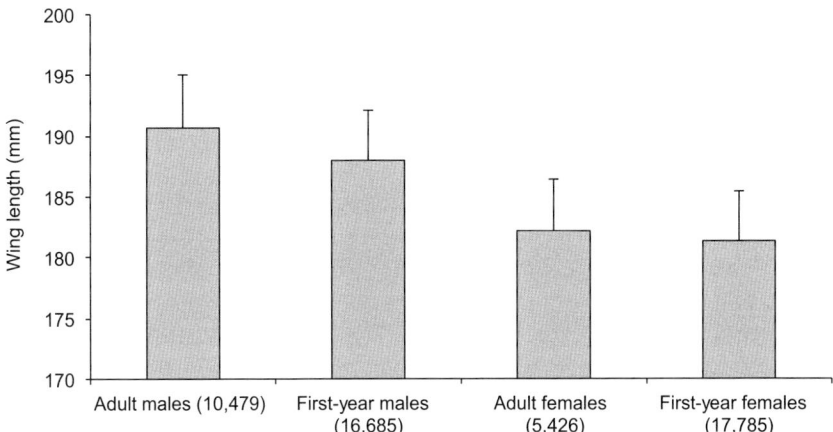

FIG. 2.13 *Mean body mass (g, top graph) and wing length (mm, bottom graph) for Common Teal by sex and age class. Vertical bars show standard deviation. Based on measurement of 50,376 wintering Teal by Tour du Valat staff, Camargue, southern France (Tour du Valat, unpublished data).*

The small size of Teal has important consequences for their relative energy requirements as well as for thermoregulatory needs in winter, presented in Chapter 5.

Friedmann (1948) considered the Aleutian Teal *A. c. nimia* to be larger than nominate *A. c. crecca*, especially adult males (mean wing length of 13 adult males: 193.1mm). This even led Delacour (1956) to consider Aleutian Teal as the eastern end of a clinal increase in body-size gradient among Common Teal populations. The lack of a genetic difference between Common and Aleutian Teal (see Chapter 1) may support this notion.

CHAPTER 3

Distribution and numbers

As for most waterfowl, winter surveys of Common Teal are carried out annually by thousands of volunteers in Europe, in addition to national networks of professionals in some countries (Box 3.1). In North America, however, numbers and trends of Green-winged Teal are generally based on highly standardised surveys carried out during the breeding season by federal government agencies and their partners. The distribution of the two Teal species is limited to the northern hemisphere, where they breed widely within a broad band of latitudes in North America and northern Eurasia. Apart from a few regions where they both breed and winter (for example the United Kingdom, or some areas in the north-western United States), Teal are strictly migratory and winter at lower latitudes, spread out over vast areas, except where habitats are completely inhospitable (such as extensive agricultural regions, deserts and mountain chains) (Figure 3.1).

SUBPOPULATIONS AND LONG-TERM TRENDS IN NUMBERS

The worldwide Teal (both species combined) population is estimated to be nearly 7,750,000 birds, of which *c.* 3.5 million are Green-winged Teal in North America (USFWS 2012), 10,000 are Aleutian Teal *Anas crecca nimia*, and more than 4 million are Common Teal in Europe, Asia and Africa (Wetlands International 2012).

Green-winged Teal breed throughout Alaska and most of Canada, but also in the north-western and north-central states of the contiguous USA (east of the prairies only in Minnesota, Wisconsin and Michigan) (Table 3.1). They winter throughout the western USA, whereas in the east the northern limit of regular winter occurrence runs from Oklahoma through the mid-Mississippi Valley to the northern parts of the eastern seaboard. Green-winged Teal

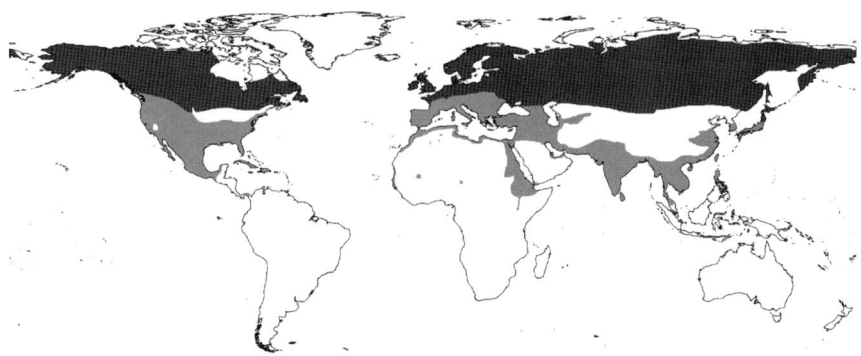

FIG. 3.1 *Common Teal (Eurasia) and Green-winged Teal (North America) geographic ranges. Dark grey areas show the breeding range, pale grey areas indicate the most important non-breeding range.*

TABLE 3.1. *Breeding Green-winged Teal population estimates for 2012 (from USFWS 2012, with permission). Minnesota, Wisconsin and Michigan are not covered by the annual breeding duck population surveys.*

Region	Breeding population size (pairs)
Alaska – Yukon Territory	352,500
Central & Northern Alberta – North-eastern British Columbia – North-western Territories	783,500
Northern Saskatchewan – Northern Manitoba – Western Ontario	68,000
Southern Alberta	137,000
Southern Saskatchewan	248,500
Southern Manitoba	78,500
Montana – Western Dakotas	9,500
Eastern Dakotas	58,500
TOTAL	1,735,500

winter in most Mexican states (Johnson 1995). No subpopulations are distinguished, but birds are considered to follow four main flyways, labelled from west to east are the Pacific, Central, Mississippi and Atlantic flyways (Bellrose 1976, Johnson 1995).

Two Common Teal subpopulations are generally distinguished in western Europe based on wintering area: one in the north-western part of the continent and one in the Mediterranean/Black Sea region (Scott & Rose 1996). Common Teal winter in all European countries, from the Baltic to the Iberian Peninsula, although in very small numbers in areas where ice forms regularly on lakes and wetlands. Several tens of thousands of individuals breeding in western Europe winter in North Africa annually (Delany *et al.* 2008). The species is rare in sub-Saharan western and central Africa (Brown *et al.* 1982). Common Teal broadly breed north of the 45th parallel, with much of the European breeding population occurring in Finland and Russia (Figure 3.1, Table 3.2). The species also occurs in Asia, breeding throughout the former USSR and wintering in distinct subpopulations in south-west Asia/north-east Africa (south to *c.* 6°N along the Nile Valley in Sudan, Brown *et al.* 1982), in southern Asia over India and Pakistan (also including Sri Lanka and the Maldives islands, Ali & Ripley 1968), and in eastern and south-eastern Asia, south to Indonesia (details in Dement'ev & Gladkov 1952, Figure 3.1 and Box 3.2). Although it was launched as recently as the late 1980s and still needs to be further developed in some areas, the Asian Waterbird Census already provides estimates for Common Teal wintering numbers in countries of these flyways (Li & Mundkur 2004, Table 3.3). A more comprehensive coverage of this part of the world is still needed, but incredibly high numbers of Teal have already been recorded at some Asiatic sites (77,500 individuals in Bundan Wari, Pakistan, in 1992; 27,500 at Chilika Lake, India, in 1996; Li *et al.* 2009). A list of the most important sites for the species in the AEWA and North American regions is provided in Appendices 3 and 4. No fewer than 72 sites have been recorded as being of international importance for Common Teal in Asia (Li *et al.* 2009).

The distinction made between subpopulations based on wintering areas (for example north-west European and Mediterranean/Black Sea) has a mostly operational goal: it aims to facilitate consistent management at the considered geographic scale. It is according to such 'population' sizes in particular that internationally important sites are identified, as for example those regularly hosting more than 1% of the total number of individuals. The problem with Teal is that they have a wide distribution and move a lot even within winters, so such immediate 1% thresholds may not be truly meaningful for these species. Furthermore, these subpopulations are clearly not isolated

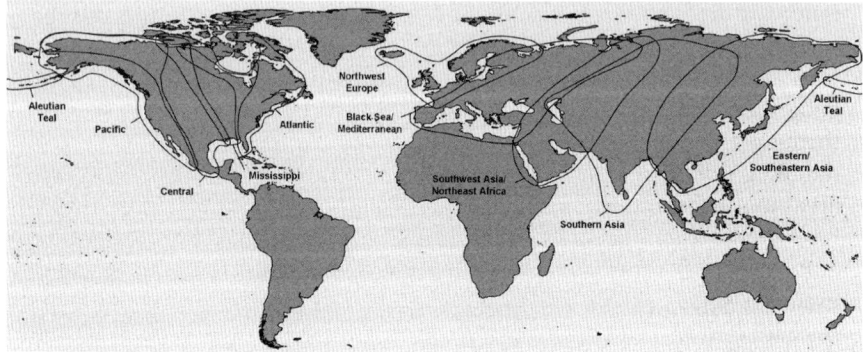

FIG. 3.2 *Extent of flyways in Common and Green-winged Teal. Based on information in Bellrose (1976), Scott & Rose (1996) and Miyabayashi & Mundkur (1999).*

from each other: an analysis of recoveries of rings fitted onto Common Teal in the Camargue, southern France, in the alleged 'Mediterranean/Black Sea' subpopulation, showed that almost 20% of the recoveries were actually in the 'north-west European' subpopulation (Guillemain *et al.* 2005a). Exchange rates between the other proposed subpopulations in Eurasia, or between flyways in North America, have never been quantified, but may be equally great. There are thus good reasons to believe that the future flyway concept will mainly be a convenient management tool, with hotspots of key sites or areas (NAWMP 2012), rather than an illustration of a real spatial genetic structure; for example, a recent worldwide study of Mallard based on a combination of genetic approaches showed that there is virtually no geographic structure in that species (Kraus *et al.* 2013).

> **BOX 3.1. History of waterfowl surveys**
>
> The first regular and large-scale winter waterfowl surveys were initiated in 1935 in the United States. Such surveys are made in December or January, but do not necessarily cover all wintering sites. A total number at the flyway scale is provided (Baldassarre & Bolen 2006). In 1955, coordinated breeding population surveys were initiated at the continental level and they have been running since. These are based on a statistically validated sample of transects (lines along which the birds are counted) in an area covering parts of Alaska, Canada and some northern states in the USA, notably the northern prairie states. Ducks are counted during the breeding season, mostly from aircraft. There is also extensive validation of these aerial data by ground-based surveys in representative areas. The surveys are coordinated by the United States Fish and Wildlife Service. In 2012, 3,471,000 individual Green-winged Teal were estimated during the breeding survey (USFWS 2012).
>
> These breeding population surveys are considered a more reliable source of data for estimating population sizes of North American waterfowl than winter duck counts (to the degree that the relevance of midwinter surveys has been questioned, Heusmann 1999).
>
> In Europe, coordinated midwinter waterfowl surveys at a large geographic scale were initiated in the 1960s. Again, this is only possible through the work of thousands of observers. A large proportion of the dabbling ducks (and most Common Teal) breed in sparsely populated areas in northern Europe, especially in Russia, so that surveying in winter has for a long time been the only feasible way to estimate population size and change. The aim of such winter counts is not to get an absolute and exact number of birds anyway, since this is an impossible goal to reach. Instead, by use of a standardised protocol, winter counts aim at obtaining a reliable index of population size that can be compared from year to year. Such values are used to assess population trends, as well as to

determine the sites of national and international importance. Estimating the densities and actual number of breeding Common Teal pairs, in Europe and further east, is an important goal for future conservation and management.

European winter surveys have existed since 1967. They were first conducted by the Bureau International de Recherche sur la Sauvagine (Hémery *et al.* 1979), founded in the 1950s and later transformed into the International Waterfowl Research Bureau, then in 1995 into Wetlands International, a non-governmental organisation now based in the Netherlands. More than one hundred countries are now taking part in the International Waterbird Census in Africa, Asia and Europe each 15 January.

In France, duck counts at some key wintering sites were initiated early, such as those carried out by the Tour du Valat in the Camargue from the early 1950s (Hoffmann 1955, Hoffmann & Penot 1955). Later, Camargue ducks were monitored by the same observer (A. Tamisier, CNRS) for 38 years in monthly aerial surveys covering the whole Rhône Delta and running from September to March each season (Tamisier & Dehorter 1999). Tamisier's protocol is used by the Tour du Valat in the present day, providing the Camargue with one of the longest European duck-monitoring datasets. However, this time series of population counts is still shorter than that from the Tipperne reserve in Denmark, where ducks have been counted several times per year continuously since 1929 (Meltofte & Clausen 2011).

FIG. 3.3 *Monitoring of waterbirds as part of the 1974 mid-January count of the Bureau International de Recherche sur la Sauvagine, south of Tan-Tan, Moroccan Atlantic coast. Photograph by Alan Johnson.*

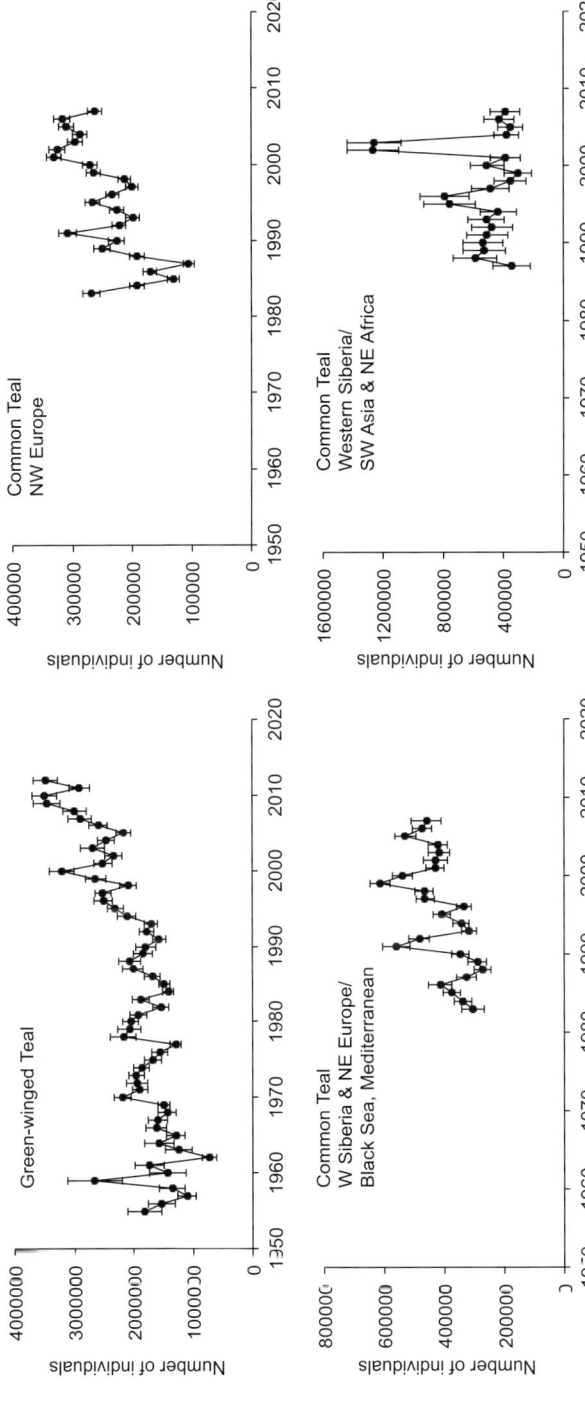

FIG. 3.4 *Trends in numbers of Green-winged Teal (based on spring pair surveys, estimate ±standard error) and of Common Teal (based on winter counts, estimate ±95% confidence interval). Apart from the W Siberian/SW Asia & NE Africa region, the other Teal populations are clearly increasing in numbers. Data courtesy of United States Fish and Wildlife Service and Wetlands International, respectively. See U.S. Fish and Wildlife Service (2012) and Delany et al. (2008) for details on analytical methods, especially the way missing data were imputed.*

TABLE 3.2. *Mean Common Teal numbers in European countries during winter and breeding seasons. Winter values are the mean of available data per country for the years 2000–2005 (after Delany et al. 2008). Breeding season data are from BirdLife International (2004a) except where mentioned. Empty cells indicate countries for which no data on breeding population size were given in BirdLife International 2004a.*

Country	Numbers in winter (individuals)	Breeding population size (pairs)
Albania	8,900	<10
Armenia	240	
Austria	2,160	70 – 120
Belgium	35,630	500 – 700
Bosnia	70	
Bulgaria	2,840	10 – 20
Croatia	1,450	50 – 100
Cyprus	1,650	
Czech Republic	450	60 – 100
Denmark	1,710	200 – 300
Estonia	20	2,500 – 3,000
Finland	<10	150,000 – 200,000
France	106,550	200 – 500
Germany	18,820	3,700 – 5,800
Greece	72,860	
Hungary	9,880	10 – 20
Ireland	22,290	250 – 1,000
Italy	92,410	20 – 50
Latvia	<10	2,000 – 5,000
Lithuania	–	2,000 – 3,000
Macedonia	70	<10
Montenegro	750	10 – 20 *
The Netherlands	29,240	2,000 – 2,500
Norway	330	30,000 – 50,000
Poland	110	1,300 – 1,700
Portugal	11,020	
Romania	4,490	10 – 20
Russia (European part)	1,830	665,000 – 740,000
Serbia	500	10 – 20 *
Slovakia	490	10 – 40
Slovenia	1,090	<10
Spain	84,880	10 – 40
Sweden	300	100,000 **
Switzerland	3,710	<10
Turkey	8,710	600 – 900
United Kingdom	159,540	1,600 – 2,800

* Data provided for Serbia and Montenegro together.
** Ottosson *et al.* 2012

TABLE 3.3. *Numbers of wintering Common Teal recorded during the Asian Waterbird Census, 1997–2001. After Li & Mundkur (2004). Numbers indicated are minimum and maximum national annual totals over this period. The very high variability between annual totals partly reflects variation in site coverage.*

Country	Minimum number recorded in winter (year)	Maximum number recorded in winter (year)
Bangladesh	4 (1997)	1,192 (2001)
Bhutan	97 (2001)	110 (2000)
China (mainland)	86 (2000)	162 (1999)
Hong Kong	2,557 (2001)	5,411 (1999)
India	2,248 (2000)	55,056 (2001)
Japan	9,115 (2000)	19,311 (1999)
Malaysia	–	300 (2001)
Myanmar	415 (2000)	1,478 (2001)
Nepal	379 (2001)	936 (2000)
Pakistan	16,563 (1998)	82,399 (2001)
Republic of Korea	3,881 (1997)	30,504 (1999)
Sri Lanka	14 (2001)	22 (1998)
Taiwan	20,629 (2001)	40,617 (2000)
Thailand	–	44 (2000)
Vietnam	1 (2001)	248 (1999)

BOX 3.2. Population sizes and trends

All estimates given below refer to the number of individuals in winter in each subpopulation/flyway for Common and Aleutian Teal, while numbers for Green-winged Teal refer to breeding individuals.

The **north-west European** subpopulation comprises *c.* 500,000 individuals in winter, while the size of the **Mediterranean/Black Sea** population is estimated at between 750,000 and 1,380,000 birds. Further east, the **south-west Asia/north-east Africa** subpopulation is composed of 1,500,000 birds (Delany *et al.* 2008). In Asia, the **south-asian** subpopulation (India and Pakistan) holds *c.* 400,000 Common Teal, and the **east and south-east Asia** subpopulation (from the Ural Mountains to Indonesia) is thought to comprise 600,000–1,000,000 birds (Li *et al.* 2009). Other surveys suggest the latter subpopulation may hold only *c.* 360,000 Common Teal, of which half winter in Japan, close to 150,000 in China and the remainder in Taiwan and South Korea (Cao *et al.* 2008). As a comparison, the size of the total Green-winged Teal population in **North America** is estimated at 3,500,000 individuals (USFWS 2012), and the population of Aleutian Teal at 10,000 birds (Wetlands International 2012).

The Aleutian Teal population is considered to be stable (Wetlands International 2012). The same authority considers the continental North

American population to be declining, though USFWS (2012) instead considers Green-winged Teal to be increasing, with 2012 numbers even being 20% greater than those estimated in 2011. The eastern and southern Asian Teal subpopulations are thought to be declining, while the trend is unknown for the south-west Asia/north-east Africa subpopulation. The Mediterranean/Black Sea subpopulation is also assumed to be declining in its eastern part, while numbers are increasing both in the west of the Mediterranean and in the north-west European subpopulation (Delany *et al.* 2008, Wetlands International 2012).

BOX 3.3. Teal conservation status by listing authorities

CITES (*Convention on International Trade in Endangered Species of Wild Fauna and Flora*)

Given their abundance and favourable trends in numbers, Teal (either Common or Green-winged) are not listed by CITES among the species with a particular protection status regarding their potential overexploitation, in particular through trade (CITES 2011). Teal is therefore not in any of the three CITES Appendix lists.

IUCN (*International Union for Conservation of Nature*)

For the same reasons as above, IUCN attributed a 'Least Concern' status to Common and Green-winged Teal. This is the lowest level of extinction risk according to this organisation.

BirdLife International

BirdLife International considers Common Teal as 'Secure' and 'Non-SPEC', which corresponds to species with a favourable conservation status in Europe, but whose numbers (during the breeding season in the present case) are not concentrated within Europe. According to BirdLife International, the species has a 'Favourable' status at the scale of both the European Union and the European continent (BirdLife International 2004b).

CMS (*Convention on Migratory Species, or Bonn Convention*)

This convention lists most *Anas* species in its Appendix II, which comprises species not currently benefitting from immediate protection by the parties (as opposed to species in Appendix I, in which neither Common Teal nor Green-winged Teal is listed), but for which the parties should seek agreements concerning their conservation and management (CMS 2009).

Bern Convention on the Conservation of European Wildlife and Natural Habitats

The Common Teal is in Appendix III of protected wildlife species in the Bern Convention, according to which hunting is possible in all contracting states, but hunting activity has to be regulated so as to ensure sustainability.

AEWA (*African-Eurasian Waterbird Agreement*)

According to AEWA, Common Teal populations from 'North-Western Europe' as well as from 'West Siberia, North-East Europe, Black Sea & Mediterranean' are in 'Column C, category 1', which signifies the lowest level of risk despite the fact that these populations 'could significantly benefit from international cooperation'. Conversely, the Common Teal population in 'Western Siberia/SW Asia & NE Africa' is in 'Column B, category 2c', which is for those populations 'showing significant long-term decline' (AEWA 2008).

European Directive 2009/147/CE, formerly 79/409/CEE, called 'Birds Directive'

The Common Teal is listed in the 'Birds Directive', which 'relates to the conservation of all species of naturally occurring birds in the wild state in the European territory of the Member States to which the Treaty applies', and 'covers the protection, management and control of these species and lays down rules for their exploitation'. In practice, exploitation of these species is generally non-commercial and related to hunting.

Listing of Green-winged Teal in North America

In a manner similar to that for Common Teal in Europe, Green-winged Teal is not listed in the *Federal List of Endangered and Threatened Wildlife and Plants* in the United States, which is the first step towards inclusion in the list of endangered species that can get a special protection status. Regardless of this, the Green-winged Teal is protected by the *Migratory Bird Treaty Act* which, for example, prohibits the collection of its eggs (as for Common Teal in Europe). Green-winged Teal is listed as a huntable species under some conditions.

In Canada, the Green-winged Teal is not being considered under the COSEWIC 'species at risk' listing process, and is listed among the huntable species under the *Migratory Birds Convention Act* of Canada. For similar reasons, the Green-winged Teal is not listed among the species at risk in Mexico under *NORMA Oficial Mexicana* NOM-059-SEMARNAT-2010.

FIG. 3.5 *Ratification of the Ramsar Convention in Iran, 1971. The Ramsar Convention has since then been the most important tool for wetland conservation, especially in Eurasia. More than 40 years after its inception it is still frequently referred to in conservation work in many European countries. From left to right: M. Eskander Firouz, Director of the Game and Fish Department of Iran, organiser of the conference; Dr Shri A. Hejmadi, from the Ministry of Food, Agriculture, Community Development and Cooperation of India, signing for the Republic of India, and Prof. G.V.T. Matthews, Director of the International Waterfowl Research Bureau (Tour du Valat photographic archives).*

CONSERVATION STATUS

The wide geographical distribution of Common Teal and its overall large population size, including some increasing subpopulations, provide the species with a favourable conservation status according to the international conventions and agreements (Box 3.3). Similarly, the Green-winged Teal breeds mainly in wetlands of the North American boreal forest, and as a consequence it has not been affected by breeding habitat loss linked with agricultural development to the same extent as is the case in prairie-nesting ducks (Tesky 1993). It is therefore not listed in lists of species at risk in North America.

CHAPTER 4

Movements and habitat use

THE ANNUAL CYCLE

The Teal is a strictly migratory bird, with most populations undertaking journeys of 2,000–4,000km from breeding areas to reach their winter quarters. In restricted areas, Teal both breed and winter in significant numbers (in the United Kingdom, for example). However, summer and winter visitors are then considered to be different individuals, so that there are no resident Teal populations, with the notable exception of the Aleutian Teal (Kenyon 1961). Teal therefore travel twice a year over extremely different habitats from the northern to the southern parts of their geographic range. As a consequence, over the annual cycle Teal must make themselves equally at home in forest ponds, lakes, rivers, marshes, estuaries, coastal lagoons and even the open sea. Since most Teal do not live very long, there is rarely any individual experience to build on. Birds may benefit from flock cohesion and learn from others, but in any case require a tremendous adaptability on a daily basis to find food and avoid becoming food for others.

The annual cycle of Teal is thus centred around two energy-costly migrations, the success of which depends on major physiological adaptations in the short term, which in turn are made possible by evolutionary (long-term) adaptations. Accordingly, Teal body mass is low after the two migration episodes, especially so after autumn migration. Both migrations are followed by intense foraging activity to replenish body reserves. It is thus of crucial importance to the birds to match their migration periods to the general environmental conditions (so as to benefit from tailwinds during flight, and to avoid arriving too early on still inhospitable breeding grounds in spring; Arzel *et al.* 2009). At the same time, 'the early bird' has advantages in each phase of the annual cycle, for example in getting the best breeding sites

(Elmberg *et al.* 2005) or access to food before intraspecific competition may start. Social relationships change very much between seasons: in winter Teal are gregarious, sometimes gathering in flocks of several thousands or tens of thousands in areas where pairs form. This is also true for certain staging sites, in autumn as well as in spring. Once back at their breeding sites, Teal become territorial and they generally occur at very low density within the breeding range, especially in regions dominated by oligotrophic (that is, nutrient-poor) wetlands, as is the case in much of the boreal (northern) zone (Ottosson *et al.* 2012).

THE MIGRATIONS

Migration routes

Ducks and geese are textbook examples of migratory birds. In many cases they do not spend more than a few months within any one part of the annual geographic range. As a result of massive ringing operations in western Europe, especially since the 1950s, the migratory journeys of the Common Teal are well documented. Birds wintering along the Mediterranean coast of western Europe mostly travel south of the Alps in spring (though some travel along the Rhône Valley directly north to Belgium or Germany), spend some days in northern Italy, probably to replenish their reserves or to take some rest (Spina & Volponi 2008), then move across central and eastern Europe towards their Russian breeding grounds. These birds come back via either the same or more northerly routes, along the Baltic Sea then following the Danish and Dutch coasts (Callenge *et al.* 2010, Box 4.1). That Teal wintering in France mainly come from NW Russia and adjacent parts of Finland has recently been confirmed by analysis of radioactive isotopes, in this case chemical signatures in feathers showing the general geographical area where they were grown, and hence where the sampled bird hatched or moulted (Legagneux *et al.* 2012, Guillemain *et al.* 2014). Teal wintering along the Mediterranean coast and on the Iberian Peninsula thus show loop migration behaviour to some extent (Klaassen *et al.* 2010). Similarly, some of the birds wintering in the Netherlands or in Britain may also migrate across northern Italy and the Balkans in spring (Wolff 1966, Ogilvie 2002, Figure 4.2), even if a more natural spring migration route would be along the North Sea and Baltic coasts. Teal wintering in north-western Europe tend to breed in Fennoscandia, although their breeding range overlaps with (and cannot be statistically distinguished from) that of the birds wintering along the Mediterranean coast and more likely breeding in Russia (Guillemain *et al.* 2009a). Further east, massive ringing operations from the mid-1960s to the mid-1970s in India revealed that Common Teal wintering on the Indian subcontinent mostly originated in central Russia, especially from longitudes between 60 and 120°E and from latitudes above 50°N, though some rings were reported from Iran and Afghanistan (Ambedkar & Daniel 1990). Birds wintering in Japan may chiefly originate from breeding areas in Kamchatka (Dement'ev & Gladkov 1952).

In North America, Green-winged Teal use four main flyways (Figure 3.1; detailed list of states in which the species breeds or winters provided by AOU 1998). Birds breeding in Alaska (excluding Aleutian Teal) migrate south to British Columbia or California through the Pacific flyway (Munro 1949), while those nesting in central Canada head south through the Central or the Mississippi flyways through central USA, down to major wintering grounds along the coasts of Texas and Louisiana. Finally, birds breeding in Québec and other parts of eastern Canada (from the eastern half of Ontario to the Atlantic Ocean) move south along the Atlantic flyway to the eastern United States (Moisan 1966, Moisan *et al.* 1967, Johnson 1995). As for the European flyways (Guillemain *et al.* 2005a), some Common Teal completely switch from one flyway to another, a phenomenon called 'abmigration' (Thompson 1931). This was detected very early in Green-winged Teal (Low 1949). For example, a Green-winged Teal ringed in California was recovered two years later in Labrador (Moisan *et al.* 1967). Similarly, two Common Teal ringed in Iceland during the breeding season were recovered the next year in Fennoscandia (Lebret 1947). In North America as in Europe, flyway delineations therefore do not necessarily represent discrete Green-winged Teal populations, although a large proportion of the populations are likely to be faithful and remain within a given flyway. Such flyways rather have a role in providing convenient units for management and conservation. In Mallard, the only dabbling duck so far for which genetic evidence is available to evaluate the flyway concept at the Holarctic scale, two genetic clades could be distinguished, one in Eurasia and another in North America, but within these clades no genetic support was found for further distinctions into flyways (Kraus *et al.* 2011).

Transatlantic flights

Observations of Green-winged Teal have been increasing for a long time in the western Palearctic, reaching *c.* 80 records per year in the 2000s (De Juana & Garcia 2010). Although some of these individuals may be escapees from public parks or private duck collections, geographic and temporal distributions of records as well as the occurrence of American ring recoveries in Europe (Dennis 1994) support the hypothesis of a genuine American origin for several of these birds. It is not well known, though, whether the increasing trend is mainly due to better ornithological coverage or to increased vagrancy. Early records also mention Green-winged Teal in north-eastern Siberia (Dement'ev & Gladkov 1952). It is uncertain if these represent trans-Pacific flights proper, or are instead the result of individuals simply following the north-east Pacific coast to the western end of Alaska and eventually crossing the Bering Strait.

Similarly, the Common Teal has often been recorded in North America (for example Brock 1907, Norton 1911, Cruickshank 1936, detailed review in Phillips 1923), and it is now considered a regular vagrant in many states and provinces in eastern North America. As early as 1979, Toweill also listed 40 records of Common Teal in western states, from British Columbia to California. Based on the knowledge that Aleutian Teal is mostly resident, he suggested (as also Jarvis 1966) that these records were of Common (*crecca*) rather than Aleutian Teal, but did not speculate whether the birds had a European or Asian origin. Ringing recoveries also document the occurrence of east–west transatlantic vagrancy (Atkinson *et al.* 2006). However, it has to be kept in mind that ring recoveries over the Atlantic are extremely rare: close to 400,000 Green-winged Teal have been ringed since 1960 in Canada and the United States, yielding more than 30,000 recoveries (USGS 2011), of which only 4 were in Europe (1 in Iceland and 3 in the British Isles, Dennis 1994). Similarly, of several hundred thousand Common Teal ringed in Europe, only two birds ringed in the British Isles (Tuck 1968, Atkinson *et al.* 2007, du Feu *et al.* 2009) and one from Denmark (Bønløkk *et al.* 2006) were recovered in North America, all in Newfoundland. Hence, transatlantic journeys should be seen as exceptional movements rather than regular migration (see De Juana & Garcia 2010).

BOX 4.1. Annual movements of Common Teal wintering in the Camargue, southern France

Bird-ring recoveries are often represented as dots on a map, which allows flyways (or principal 'migration corridors') to be determined if birds follow well-defined routes. However, such mapping becomes less useful when there are so many recoveries that the whole map is covered with dots. This is the case for Teal wintering in southern France, due to very large numbers of birds having been ringed and high harvest rates providing abundant ring recoveries (Figure 4.1).

In addition, it is often impossible to determine if the lack of recoveries in an area or during a period is due to a genuine lack of use of that area by birds, or if there was simply no one there to recover and report rings. To somewhat overcome these problems, kriging techniques (a method of extrapolation of the data) are often used to smooth the distribution of recoveries over maps. Accordingly, instead of showing the exact location of each recovery, kriged maps derived from such extrapolations show the likelihood of recovering a ring in different areas, for each coordinate on the latitude and longitude axes. If a village lacks recoveries while birds were recovered in all neighbouring areas, a krigeage map will show that the probability of recovering rings in that village is actually high, even if recoveries have never occurred. To further improve the analysis of ring recoveries, kriging can now be done to simultaneously compute the probability of recoveries over time and over geographic coordinates; the likelihood of recovering a bird in an area for a given period is calculated from the recoveries made during the previous and following time intervals, both there and in the surroundings. Maps for each time period can then be portrayed as films, providing a continuous view of bird movements as inferred from changes in the location of recoveries over time.

Such a promising approach has been applied to the ring recoveries of Teal initially captured in the Camargue. Beyond the hard-to-interpret basic map presented below (Figure 4.1), this provides a dynamic view of where birds are likely to be (recovered) during each time period, and how quickly they move within the flyway throughout the year (Figure 4.2). This illustrates the loop migration described in the previous pages (see also Lebreton 1973), and also shows that Common Teal migrate very quickly to their breeding grounds in spring, but migrate back at a much slower pace over an extended period of time during autumn, as also documented for Green-winged Teal in North America (Johnson 1995, Callenge et al. 2010).

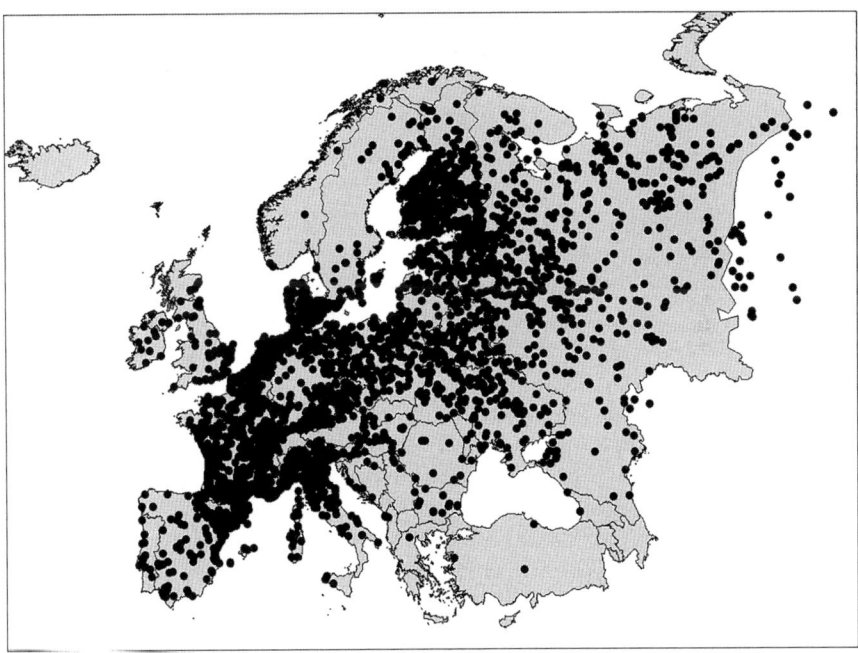

FIG. 4.1 *Recoveries in Europe and adjacent areas of Common Teal initially captured in Tour du Valat, southern France. After Guillemain et al. (2005a), with kind permission from John Wiley & Sons, Inc.*

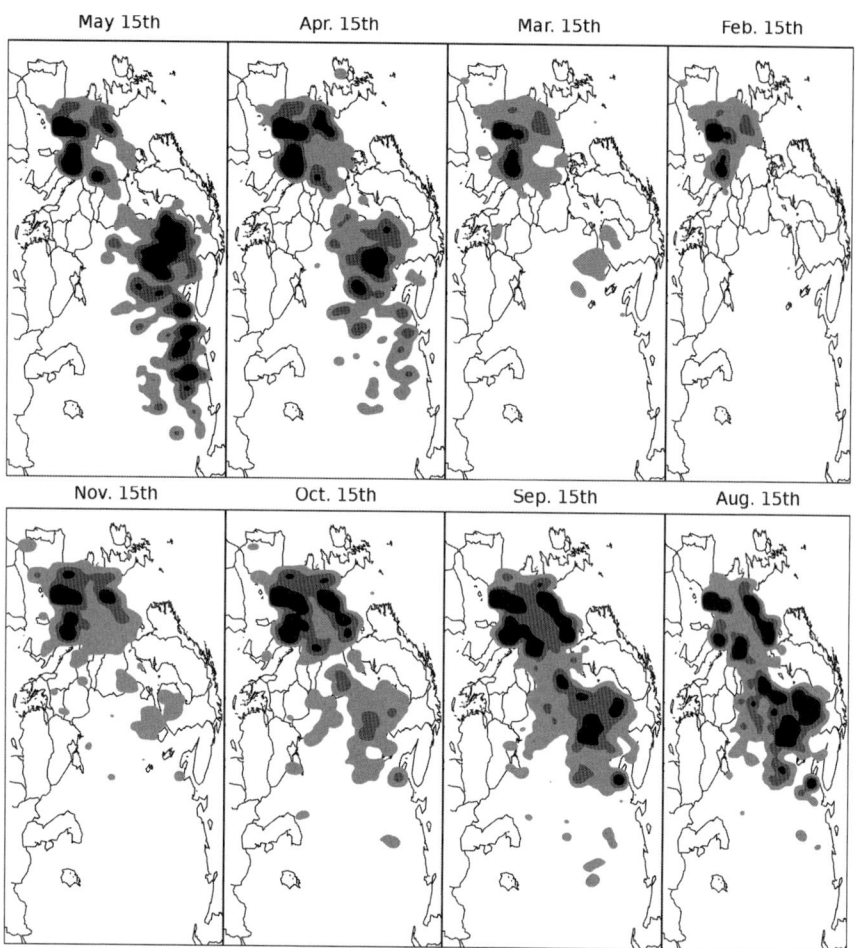

FIG. 4.2 *Footage from a film showing ring recoveries in western Europe of Common Teal initially ringed in Tour du Valat, southern France. Based on kriging techniques, darker areas show greater probability of ring recovery. With kind permission from Springer Science+Business Media:* Journal of Ornithology, *A new exploratory approach to the study of the spatio-temporal distribution of ring recoveries: the example of Teal (*Anas crecca*) ringed in Camargue, southern France, volume 151, 2010, pages 945–950, C. Calenge, M. Guillemain, M. Gauthier-Clerc & G. Simon, Figures 1–2. The data used in this figure are the same as in Figure 4.1.*

Migration timing

Male Teal abandon their mate during incubation to moult locally or after migrating to specific moulting sites (detailed description of duck moult migration in Salomonsen 1968). Major moulting areas for Common Teal in Europe include the Volga Delta (Scott & Rose 1996) as well as some sites in the Baltic States and western Russia (Viksne *et al.* 2010). In North America, the Delta Marsh (Manitoba) has long been recognised as a major moulting area for Green-winged Teal (Hochbaum 1955, Sowls 1955). Departure date of male Teal from the breeding grounds therefore depends on breeding phenology, which itself depends on the geographic position within the flyway (the further away from the wintering grounds, the later the laying date). Male departure to moulting sites has, for example, been recorded during the first part of June in the Kirov area, 800km north-east of Moscow (Sotnikov 1999), while it occurs mostly during the second half of this month along the Tobol River, further east (Blinova & Blinov 1999). In the Volga Delta, one of the main moulting areas for males, the number of moulting birds peaks during the first 10 days of August (Dement'ev & Gladkov 1952). Autumn migration from the north-eastern European breeding sites (females and juveniles) and moulting sites (males) occurs between August and early October (Sotnikov 1999, Zarudnyy 2003), and most birds reach their wintering grounds in October or November (Lebret 1947).

In North America, birds similarly depart from the breeding grounds in late August, and reach their southern wintering areas in October–November (Munro 1949, Moisan *et al.* 1967, Bellrose 1976, Johnson 1995).

Besides this general pattern, Teal from both continents show marked inter-individual and inter-annual variation in migration phenology; the overall distribution of birds within their flyway is skewed towards lower latitudes during colder winters (Ridgill & Fox 1990). During seasons with more normal weather, some birds (likely those that will become the earlier breeders) move down the flyway and can reach their wintering quarters as early as late August or early September. Other individuals migrate considerably later, initially making use of food available in more northerly areas, then gradually moving south in a series of short hops (Dalby 2013). Consequently, the autumn migration of Teal spans a very long period, with some birds not reaching their wintering grounds until December (Bellrose 1976, Guillemain *et al.* 2006). This is only a couple of weeks before the first individuals depart for their breeding grounds, since spring migration starts during the first days of February (Figure 4.3). The population of Teal in a winter quarter is thus permanently fuelled by the arrival of new individuals. This has major consequences for population dynamics, as explained in Chapter 8. Many birds harvested by hunters are replaced by new migrants, so that the apparent stability of bird numbers within a season may actually mask a process involving an ongoing turnover of individuals within the population.

Just as in Europe, early spring departure dates have been recorded in North America, where the southern breeding sites (for example Manitoba) are reached by March or early April, while more northerly breeding sites are reached in May (Bellrose 1976). In western Europe, breeding sites of Common Teal in Fennoscandia are reached by late March in the southern parts to early May in northern boreal forest. In the latter areas Common Teal often spend one to three weeks on flooded fields and wetlands, or on open water in estuaries and rivers, until the nearby breeding ponds become ice-free.

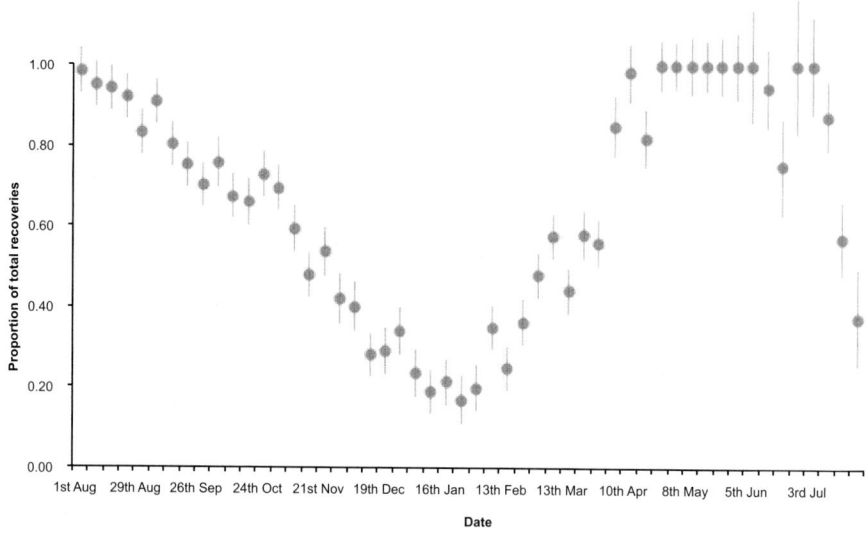

FIG. 4.3 *Mean proportion of Teal recoveries from areas outside the Camargue to the north and east (by week; vertical bars show standard errors). Spring and autumn migrations are clearly observable by steep changes in that proportion. After Guillemain et al. (2006), with kind permission from Wildlife Biology.*

Common Teal nesting in Russia close to Kazan, *c.* 700km east of Moscow, settle as breeding pairs in the second half of April (Artemyev & Popov 1977), while birds moving further to the Ural Mountains region reach their breeding grounds no earlier than the second half of May (Blinova & Blinov 1999).

It is clear from the recoveries of Camargue-ringed birds that spring migration is both quicker (Figure 4.3) and made along a straighter route (Figure 4.2) than autumn migration. These patterns are consistent with the finding that early spring-arriving Teal have higher breeding success (Elmberg *et al.* 2005), but there may also be some value in arriving early at the wintering grounds to start preparing for cold periods early (Tamisier *et al.* 1995). However, food abundance (especially seeds) is higher in autumn (Madsen 1988, Guillemain *et al.* 2000a), so that feeding conditions are probably good throughout the flyway and Teal do not have to hurry as much as in spring.

The migration flight

Like most dabbling ducks, Teal migrate mostly at night (Bellrose 1976, Svazas 1994, Blinova & Blinov 1999). Instantaneous flight speed tracked by radar was recorded at 19.7m/s (71km/h) by Alerstam *et al.* (2007), which is comparable to the 14m/s obtained from radar records for Garganey by Bruderer & Boldt (2001), or the range speed from Kvist *et al.* (1998) and Engel *et al.* (2006) for Common Teal flying in a wind tunnel (Figure 4.5). Based on ring recoveries, Hildén & Saurola (1982) provide a mean overall autumn migratory speed of

FIG. 4.4 *Teal may have to wait for their breeding lakes to become ice-free when they reach the breeding grounds in spring. Sometimes they have to endure icy and snowy conditions for a couple of weeks before they can settle on their breeding ponds. Photograph by Marcus Wikman.*

c. 60km/day, with maximum migration speeds of over 100km/day. Fransson & Pettersson (2001) provide maximum migration speeds of over 200km/day, and a Common Teal ringed in Portugal travelled 864km in two days (Rodrigues *et al.* 2006). The absolute record, however, seems to come from a radio-tagged Common Teal recovered in western France that was in Denmark the previous day (1,285km in 24 hours, a mean speed of more than 50km/h under the conservative assumption of a straight flight path and no stopover rest; Clausen *et al.* 2002). Some maximum recorded migration distances for Common Teal in western Europe are indicated in Table 4.1. It should be kept in mind that many of these ring recoveries occurred more than one year after ringing, so that the individual may not have covered that distance during a single migration journey. These distances are therefore presented with the purpose of showing how far within their geographic range birds can move. Interestingly, maximum ring-recovery distances are far greater for Green-winged Teal: despite similar numbers of ringed birds on the two continents (see above), there are 380 recoveries more than 4,500km from the ringing site in North America (Figure 4.6), as opposed to just a handful in Europe (Table 4.1). More than 20 American recoveries were made over 6,000km from the ringing place, generally in Texas, Louisiana or Mexico for birds ringed in Alaska. Such distances are of course strongly dependent upon the size of the breeding range, and on the spatial distribution of ringing effort and hunting pressure,

FIG. 4.5 *Marcel Klaassen measuring Common Teal flight characteristics of a live bird in a wind tunnel at the Ecology Building of Lund University in Sweden. Photograph by Magnus Elander.*

TABLE 4.1. *Some record distances between Common Teal ringing and recovery locations.*

Place of ringing	Place of recovery	Distance (km)	Reference
Kirov area, Russia	Morocco	5,631	Sotnikov 1999
Puydarrieux, France	Omsk area, Russia	5,116	Guillemain *et al.*, unpub.
San Jacinto, Portugal	Severouralsk, Russia	5,049	D. Rodrigues unpub.
Crowland, England	Sargatskoye, Russia	4,804	BTO unpub. *
Puydarrieux, France	Priuralskiy distr., Russia	4,684	Guillemain *et al.*, unpub.
Arles, France	Yarato Lake, Russia	4,630	Tour du Valat **
Abberton Reservoir, England	Surgut, Russia	4,541	BTO unpub. *
Arles, France	Mokrousova area, Russia	4,505	Tour du Valat **
Arles, France	Aksarka area, Russia	4,497	Tour du Valat **
Denmark	Newfoundland, Canada	4,447	Bønløkk *et al.* 2006

* The BTO Ringing Scheme is funded by a partnership of the British Trust for Ornithology, the Joint Nature Conservation Committee (on behalf of Council for Nature Conservation and the Countryside, the Countryside Council for Wales, Natural England and Scottish Natural Heritage), The National Parks and Wildlife Service (Ireland), and the ringers themselves.
** Data courtesy of the Tour du Valat ringing database.

so that the number of individuals recovered does not gradually decrease with increasing recovery distance, but rather has an irregular shape (Figure 4.6). It is unlikely that the greater maximum recovery distances for Green-winged Teal reflect a better adaptation to long-distance migration than in Common Teal, since the small difference in body weight between them should not be sufficient to affect migration distances to such an extent. Rather, these differences may simply reflect the geographic features of the two continents, and the position of the most attractive wintering grounds compared to the breeding range: if Common Teal were to migrate similar distances to Green-winged Teal they would end up in desert-like regions (for example the Sahara), obviously not an optimal decision. This sets natural limits to the likelihood of obtaining long-range recoveries in Europe, whereas they are more likely in North America and in east Asia. In addition to maximum ring-recovery distances as considered here, Viana *et al.* (2013) also recorded shorter travel distances during the autumn migration journey of Common Teal than of Green-winged Teal, which were also partly attributed to differences in the spatial configuration of the landscape.

It should be noted that massive ringing operations in India revealed that Common Teal do cross the Himalayan mountain chain, although it is suspected that they use the lower altitude offered by valleys and mountain passes rather than fly over the mountain tops (Ambedkar & Daniel 1990). When this hurdle is passed, Common Teal could migrate considerable distances over more hospitable habitats in this flyway, and these authors mention a ring recovery more than 6,400km away from the ringing site in

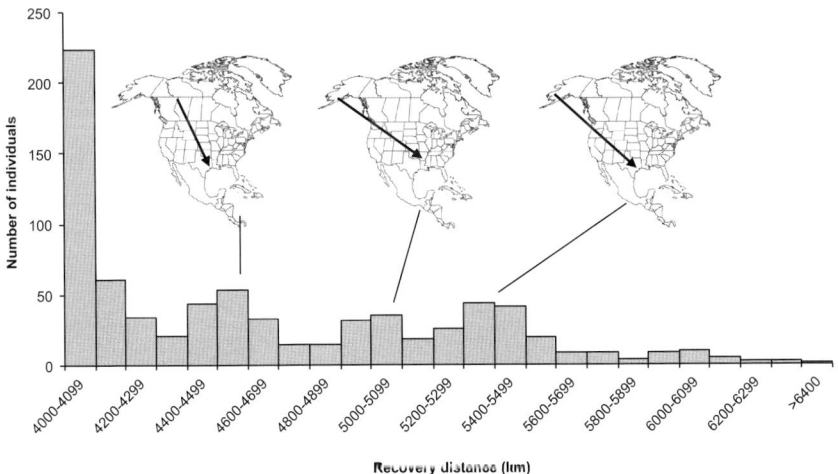

FIG. 4.6 *Distribution of ring-recovery distances for Green-winged Teal. The arrow on each map joins the mean position of ringing with the mean position of recovery for each class of (recovery) distance. Data courtesy of USGS Bird Banding Laboratory (2011).*

northern India. The shorter migration distances observed in Europe than those noted in North America are therefore unlikely to simply result from high mountains limiting Teal migration (for example the Ural Mountains).

Flight altitude has not been measured specifically for free-ranging Teal, but was *c.* 50 m in ducks (species not distinguished) according to radar tracking by Cooper & Ritchie (1995). Teal are considered to migrate in larger flocks than many other duck species (Bellrose 1976). Such large migratory flocks do often, however, gradually shrink during spring as birds disperse into the breeding range (Artemyev & Popov 1977). In spring most Teal migrate in pairs, as judged by the rather even sex ratio in arriving flocks (Berezovikov 2001), while the sex ratio is male-biased (though to varying degrees at different latitudes) on the wintering grounds (Moisan 1966, Bellrose *et al.* 1961, Guillemain *et al.* 2009b).

Differential migration

In addition to inter-individual differences in migration date for a given area, many bird species show 'differential migration', according to which a certain sex or age class migrates further away from the breeding grounds or at different times than others (Terrill & Able 1988, Alerstam & Hedenström 1998). In the case of northern hemisphere ducks, females often migrate further (that is, south) than males, so that the proportion of females in the wintering population decreases with increasing latitude (see for example Campredon 1983 for Eurasian Wigeon *Anas penelope*, or Carbone & Owen 1995 for Common Pochard). A comparison between populations of wintering Common Teal in Abberton Reservoir (England, 51°49'N) and the Camargue (southern France, 43°30'N) confirmed the existence of differential migration between the sexes. The proportion of males was greater in the more northerly area than in the southern one (64% and 58%, respectively, with age classes combined; Guillemain *et al.* 2009b), thus supporting earlier studies (Wolff 1966, Ogilvie 1983, Perdeck & Clason 1983 for Common Teal, Palmer 1976 for Green-winged Teal).

The differential pattern of distribution of the two sexes in Teal during winter therefore matches that of many other species, although the drivers behind this may be very different. Differential migration has so far mostly been studied in birds (especially passerines) whose main constraint in winter is to find enough food resources in order to survive an energetically demanding season, while not wintering too far away from the breeding grounds, so as to limit migration costs. Accordingly, three hypotheses are generally advanced to explain differential migration in birds (Cristol *et al.* 1999): 1) dominant individuals, often males, may monopolise the northern wintering grounds due to their competitive superiority, thus having the benefit of a shorter migration journey, 2) the larger individuals, again often males, owing to a lower surface area:mass ratio, are energetically better able to winter in colder northern areas, and 3) the individuals getting the greater benefits from an early arrival at the breeding sites are the only ones to accept the energy costs of wintering in colder northern areas. In many birds, males are typically dominant over females (Gauthreaux 1978), are larger than females (Ketterson & Nolan 1976) and have to migrate ahead of females to establish territories in spring (this obviously applies to species where pair bonds are established on breeding grounds after the birds' spring arrival) (Kokko *et al.* 2006). The situation is very different in Teal, especially in males, where a major additional requirement in winter is to secure a mate, since pairs form during this period (see Chapter 6).

The hypothesis that it is a greater advantage for males than for females to winter close to the breeding grounds hence does not hold in Teal, since the two mates first have to meet each other during winter, then migrate together in spring. Similarly, the body size hypothesis has been tested by comparing body mass by sex and age class between two wintering areas (England and southern France): as expected, male Common Teal were heavier than females, but British-ringed males were not heavier than those wintering in the Camargue (Guillemain *et al.* 2009b). Conversely, birds wintering in the Camargue were slightly heavier, suggesting that the body mass hypothesis does not explain the different proportions of males in Britain and France. Neither does the dominance hypothesis seem to make most sense in Teal: although male ducks are generally dominant over females within a given area (Hepp & Hair 1984), the only study directly addressing dominance relationships among unpaired Green-winged Teal found no difference between the sexes (Johnson & Rohwer 1998). Furthermore, it makes no sense at all that the most dominant males should tend to winter as far away from potential mates as possible.

While the traditional view is that wintering latitude is mostly driven by energetic considerations, with the 'best' males (larger, more dominant, and so on) wintering further north because travelling to more distant winter quarters entails significant migration costs, an alternative hypothesis in the case of Teal, and possibly other ducks, could be just the opposite. Migrating 500 to 1,000km further may not be that expensive for a Teal, nor would it involve a major time delay during spring migration (cf. the Danish bird that travelled 1,285km in 24 hours; Clausen *et al.* 2002). Furthermore, migrating towards more southerly wintering grounds may bring major benefits in terms of winter mating opportunities. Rather than wondering why only some individuals are able to winter in northern areas, as is often the case in passerines, for example, the question may be considered the other way round: why don't all males join southerly winter congregations of females? This opens interesting perspectives for future behavioural and physiological research to better understand how these birds trade individual survival against expected reproduction, and how some individuals may be constrained while making such strategic decisions.

Winter site fidelity

One popularly held view of the migratory bird is that there is a relatively fixed migratory timetable, for autumn departure as well as for spring return (Newton 2010). It has been shown above that such predictability of movements only exists to a very limited extent in Teal. Similarly, and probably because this is how migration was first studied (that is, J. J. Audubon observed ringed Eastern Phoebe *Sayornis phoebe* returning to the same place year after year, Box 4.2), migratory birds are expected to be faithful to sites where they can repeatedly be observed from one year to the next. Such fidelity to wintering sites (incorrectly termed 'winter philopatry', as philopatry should only refer to the place of birth, cf. Pearce 2007) has been demonstrated frequently in swans and geese. Conversely, ducks have generally been considered to have less faithful migratory traditions, leading to far lower fidelity to given winter quarters (Wolff 1966, review in Robertson & Cooke 1999). However, there could be a strong methodological bias behind this statement: while most winter site fidelity studies in geese and swans were based on colour marking (either colour rings or neck collars), all studies on winter site fidelity in ducks were based on metal ring recoveries, which are less effective in providing such evidence. The use of colour markers for ducks (nasal saddles that can be read from a distance without the need to physically recapture or recover the bird to examine the ring number) demonstrated that 57% of surviving Teal returned to the same site within the Camargue from one winter to the next (Guillemain *et al.* 2009c, Box 4.2). This is actually similar to recorded homing rates for geese in other areas (49 to 98% according to the review by Robertson & Cooke 1999).

As explained above, Common and Green-winged Teal can also completely change their winter quarters from one year to the next; that is, abmigrate. For example, Common Teal wintering along the Mediterranean coast one winter may well be recovered along the Atlantic coast the following year, and similar exchanges have been documented between the Pacific and Atlantic coasts in North America. Abmigration of this type is generally more common in male than in female ducks, and it is thought to be due to males following their mate to her breeding grounds if partners from two different flyways happen to winter and pair in the same area, for example because cold spells forced them both to the same southern location (Anderson *et al.* 1992).

Cold weather (winter) movements

In addition to having relatively predictable migration journeys, Teal are known to be among the ducks that are the most affected by winter cold spells (Lebret 1947, Wolff 1966, Lebreton 1973, Ogilvie 1983, Ridgill & Fox 1990). Cold spells in northern areas induce quick and massive shifts towards southern climatic refuges, areas where Teal can suddenly appear in large numbers. Females may respond more quickly than males under such scenarios (Bennett & Bolen 1978, Fox *et al.* 1992). At the scale of a continent, Teal will therefore winter at lower latitudes during a cold winter than in a mild one (Figure 4.7).

The ability of Teal to quickly respond to adverse weather is partly due to their ability to fly long distances fast as described above. That Teal must respond to severe cold spells is explained by their small size, which leads to relatively high relative energy requirements for thermoregulation compared to larger duck species (see Chapter 5). Energy requirements increase less rapidly than body mass does across species, smaller species having greater energy requirements per unit of body mass. As a case in point, Bennett and Bolen (1978) recorded a greater impact of harsh weather on survival in Green-winged Teal than in other duck species.

Moreover, like other dabbling ducks Teal do not dive to forage, which limits their foraging depth to *c.* 24cm when upending; that is, foraging by submersing the anterior part of the body (Thomas 1982). In other words, Teal are more restricted to shallow wetland habitats, which in turn are more prone to freeze, quickly limiting access to food. Cold weather thus increases Teal daily energy requirements dramatically, while at the same time their foraging habitats become inaccessible. This translates into a very clear and

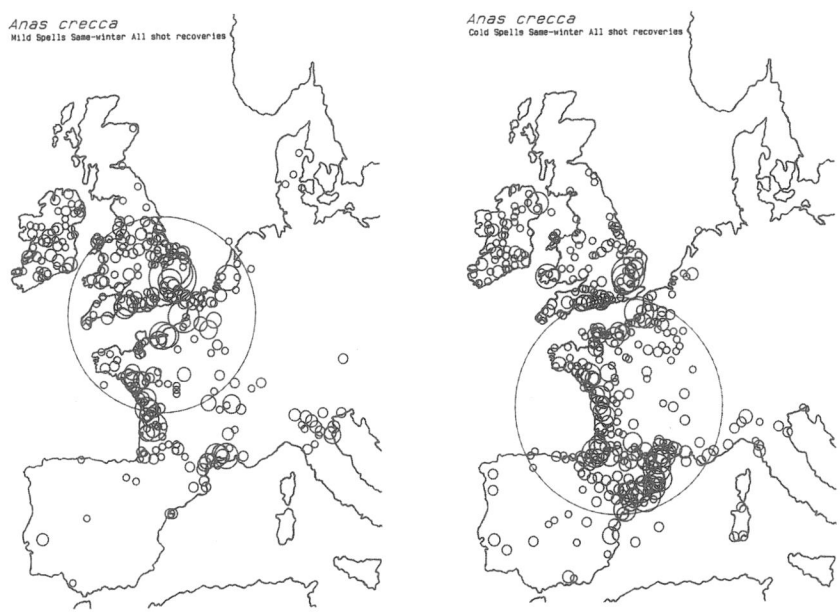

FIG. 4.7 *Common Teal ring recoveries in western Europe during mild (left) and cold (right) winters. The large circle represents the adjusted mean recovery position, circumference proportional to sample size (see initial reference for details). From Ridgill & Fox (1990), with kind permission from Wetlands International.*

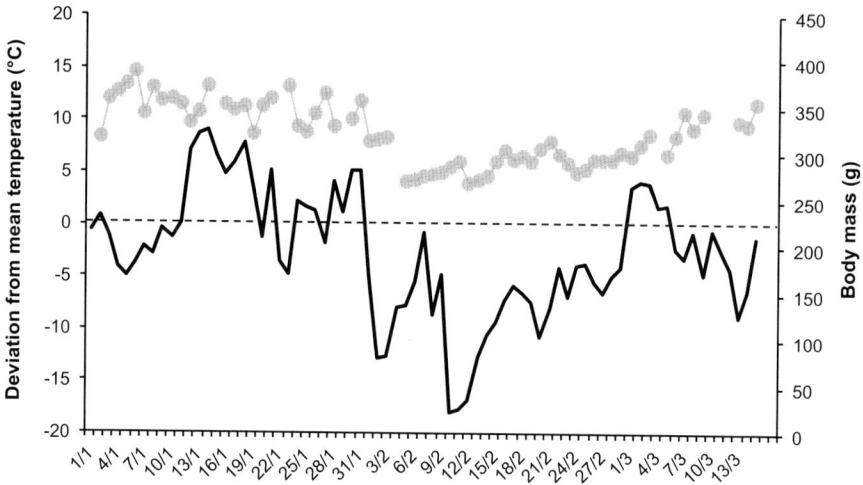

FIG. 4.8 *Changes in adult male Common Teal body mass over time during the February 1956 cold spell in the Camargue, inferred from live birds weighed during ringing operations. The black graph shows the difference between the daily minimum temperature and the 1954–1963 average minimum. Grey dots show daily mean body mass. Mean Teal body mass does not vary much during the first few days of cold weather, then shows a massive 15% drop in just 2 days due to local birds losing mass and lean birds arriving from northern areas to find shelter in the Camargue. Mean body mass remains low during the whole cold spell, but recovers afterwards (Tour du Valat, unpublished data).*

quick drop in body mass (Figure 4.8). Escape to warmer regions to the south is therefore the only option for Teal when freezing conditions last more than a few days. Baldassarre *et al.* (1986) suggest a fasting ability of *c.* 6 days for Teal in January, based on the assumption that they have stored 70g of lipid fat reserves.

BOX 4.2. Individual histories: marking techniques for Teal and other ducks

People have long marked migratory birds, often with the aim of finding out whether they would return to the same place from one year to the next; John James Audubon is considered a pioneer in this area, fitting a silver ring to Eastern Phoebe for this purpose in the beginning of the 19th century (Chansigaud 2009). Hans Christian Cornelius Mortensen, a Danish teacher, went further by introducing numbered rings with an indication of ring origin in 1890 (Preuss 2001), opening the way towards individual bird recognition and individual-based studies. While Audubon could only determine if the bird he ringed came back to the same place,

Mortensen's technique allowed foreign observers to send back the information when rings were recovered. In addition, individual marking allowed relating of individual recovery information to ringing data. Mortensen ringed a few ducks himself, but duck-ringing operations took off from the 1930s onwards (Lebret 1947, Munro 1949; Figures 4.9 and 4.10).

Ringing, however, has a major limitation: because metal rings are so small, it is generally impossible to read the individual code without having the bird in the hand, either dead or alive. Obtaining a ring recovery therefore depends not only on the presence of a ringed bird, but also on its likelihood of recapture or recovery and reporting. To overcome this problem, in addition to standard metal tarsal rings, birds are frequently marked in a way that makes it possible to identify them from a distance. Neck collars have been developed for geese and swans, coloured rings for waders, patagial tags for some raptors, and so ons (review in Calvo & Furness 1992). Patagial tags made of plastic flags pinned through the anterior part of the wing have been used in ducks, too (Anderson 1963, Baldassarre *et al.* 1988a), but with limited success as they apparently affected birds' behaviour (Szymczak & Ringelman 1986, Bustnes & Erikstad 1990, Brua 1998). Similarly, the bills of ducks can get caught in neck collars, preventing the use of this technique, while this problem is virtually absent in long-necked species like most geese (Marion & Shamis 1977).

Colour rings have been used for species of duck that spend a significant part of their time on land (for example Eurasian Wigeon, Mitchell 1997). This is, however, not the case for Teal and most other dabbling ducks, whose legs are rarely visible. For such species, visual markings on the bill have been developed. Nasal discs have been used since the 1960s, mostly in North America: small plastic discs are fitted to both sides of the bill with a nylon pin through the nostrils (Bartonek & Dane 1964). The observation of nasal-disced ducks getting caught in vegetation by their marks has led to the development of new markers covering the bill, named 'nasal saddles' (Sugden & Poston 1968). This method was introduced to Europe by David Rodrigues in Portugal in 1993 (Rodrigues *et al.* 2001), and later applied to ducks in a growing number of countries during the 2000s. Close to 8,000 Common Teal have been saddled so far in Europe, after a range of tests carried out both in aviaries and in the wild had shown that the method did not negatively affect dabbling ducks in terms of physiology and behaviour (Guillemain *et al.* 2007a). An average of 1,000 Common Teal are ringed in France annually, of which two-thirds are also fitted with a nasal saddle (Figure 4.11). Observations of these saddled birds ('resightings') provide valuable information for the calculation of their demographic rates (survival, emigration) and for studying habitat use (see main text).

Nasal-saddling programmes have now been set up for Common Teal in

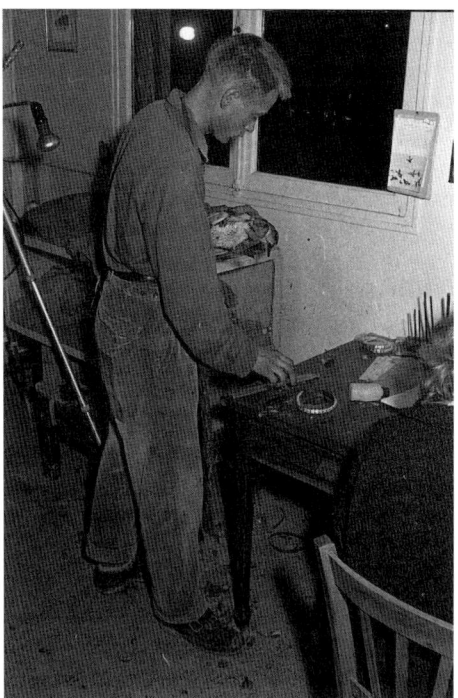

FIG. 4.9 *Ringing of a Common Teal in Tour du Valat, the Camargue, in the 1950s (Tour du Valat private collection).*

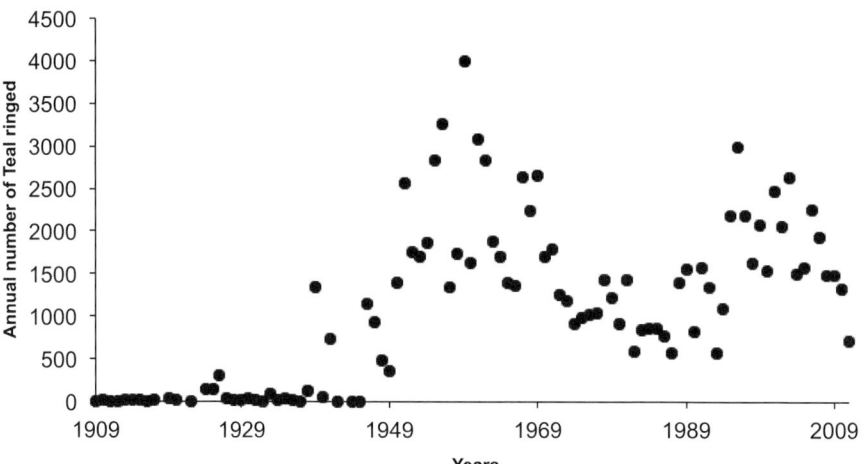

FIG. 4.10 *Annual number of Common Teal ringed in the United Kingdom and Ireland from 1909 to 2011. The first period with high annual totals occurred during a period of increased catching effort in the 1950s–1960s to assess duck flyways and estimate demographic parameters. Fluctuating numbers between the mid-1990s and mid-2000s illustrate fluctuating numbers of Common Teal at key trapping stations. Data courtesy of the British Trust for Ornithology, accessible at: http://blx1.bto.org/ringta/ringing-totals.jsp?archive_euringNo=1840&archive_year= ALL.*

Germany, Spain, Portugal, Sweden and France. Distinction of the country of origin is possible by means of saddle colour and alphanumeric coding. The various duck nasal-saddling schemes are coordinated at the European scale by Coimbra University in Portugal to avoid double use of individual codes (see http://www.pt-ducks.com/ for regularly updated information).

In addition to visual marking, the technical development and miniaturisation of electronic devices allow individual recognition of ducks (Figure 4.12). In the late 1970s, Tamisier & Tamisier (1981) used small radio transmitters (or 'tags') to study Common Teal habitat use in the Camargue. Each individual bird bearing a tag sends signals at its own radio frequency, and the observer looks for the presence of individual birds by scanning the range of transmitter frequencies. A directional antenna limits signal reception to when the antenna points to the bird. Each bird's precise position can then be determined by using the directional antenna from different places (by triangulation or paired determination of bearings).

VHF tags (radio-tracking) are still used (see for example Guillemain *et al.* 2002a), but increasingly tend to become replaced by satellite geologgers (based on the Argos or GPS system). Such loggers have now come down to a mere 10g, making it acceptable to use them on Teal (Figure 4.13). The logger regularly stores the geographic position of the

FIG. 4.11 *Common Teal with a nasal saddle. Photograph by Serge Lardos.*

bird, recorded via satellite, for up to several years in solar-powered tags. As opposed to a VHF tag, it is therefore not necessary to search for the individuals. Furthermore, the constraint of maximum distance of transmission (generally not more than a few kilometres from the receiver

FIG. 4.12 *Common Teal with a radio tag. The smallest tags (3.5g) are attached to the rectrices so that only the antenna is visible once the bird is released. Photograph by Pär Söderquist.*

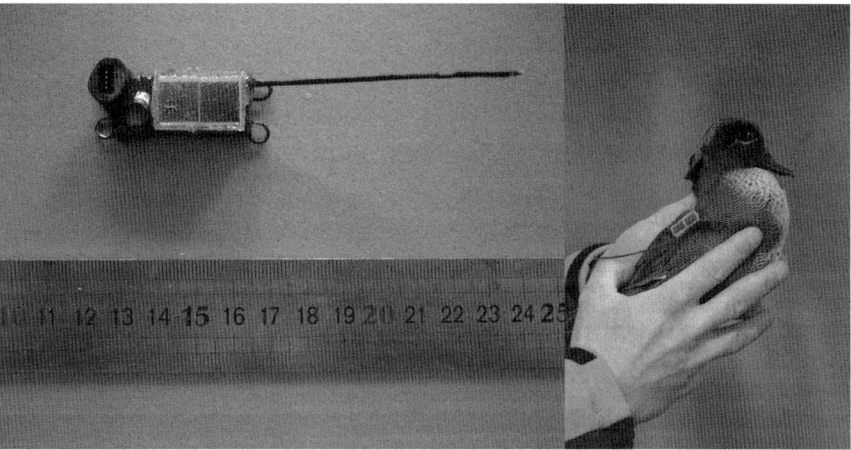

FIG. 4.13 *GPS tag of a type that can be fitted to a Teal (left, photograph by Matthieu Guillemain). The solar panels (under the transparent coating) increase tag life expectancy to several years. The tag is fitted onto the duck's back using a lightweight Teflon harness (right, photograph by Pär Söderquist). Birds with this type of GPS tag have to be recaptured to download the data.*

> for VHF tags) does not apply to satellite loggers, which can track birds throughout their annual range. Some satellite loggers regularly transmit their data to a satellite or terrestrial receivers (via Bluetooth or through the mobile phone network, for example). The smallest tags only record the geographic position and store the information, so that it is necessary to recapture the bird to download the data onto a computer.

Habitat use during the annual cycle

The breeding season

As the Common Teal and Green-winged Teal are such widespread and abundant ducks, their breeding pairs and broods can be found on a wide array of wetlands. Over most of the range smaller wetlands are preferred, making Teal a typical inhabitant of marshes, Beaver (*Castor canadensis* or *C. fiber*) ponds, tarns, potholes, tundra pools and slow-moving streams. Nevertheless, they also use larger lakes and wetland complexes, often where there are protective grassy or shrubby shoreline habitats. Wetland selection in ducks breeding in boreal regions is still a somewhat mysterious process, though, as many wetlands are without breeding birds, including Teal, in any given year. Moreover, a lake that is empty one year may very well have breeding Teal the next. As a general rule, lakes with steep littoral zones and little emergent vegetation are poor breeding habitats for Teal, whereas shallow lakes with protective emergent vegetation are preferred. In recent years there has been a growing interest in the role of predatory fish to explain absence–presence patterns of breeding Teal (Elmberg *et al.* 2010, Dessborn *et al.* 2011a; Figure 4.14), as large fish such as Pike *Esox lucius* frequently prey on ducklings and even small adult ducks (Solman 1945). By and large, breeding Teal are often unevenly distributed over the landscape, including in the core of their vast breeding range throughout the boreal regions of the northern hemisphere.

Compared to the large concentrations often observed at winter and staging sites, Teal disperse greatly on the breeding grounds, and the density of breeding pairs is usually low, although this largely depends on the size distribution of wetlands. As a case in point, Finnish data for Common Teal show that breeding density ranges from 17 to 55 pairs per km^2 wetland area in the smallest wetland class (1 ha, or $0.01km^2$), whilst in the same landscape, density in the class of largest lakes (>1000 ha, or >$10km^2$) is only 0.4–2.6 pairs/km^2 (Väisänen *et al.* 1998). Russian field studies also document this preference for small wetlands over large waterbodies (Sotnikov 1999). Another way to describe breeding density is as pairs per 1,000m of shoreline (Haapanen & Nilsson 1979, Elmberg *et al.* 2000). This method corrects for some of the effects related to perimeter length that arise when deep

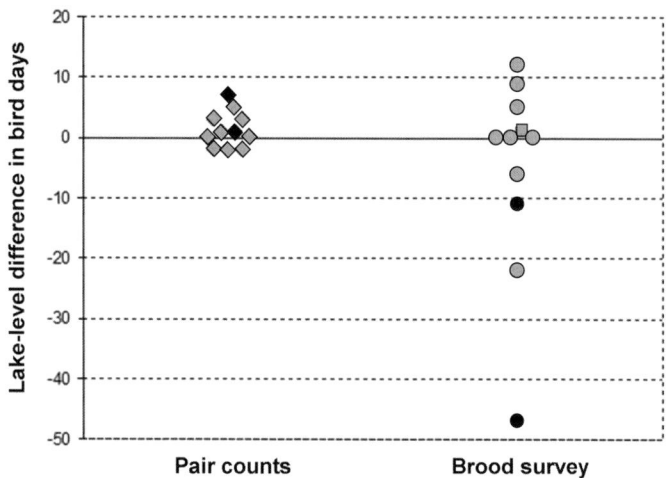

FIG. 4.14 *Eleven lakes in northern Sweden were used in an experiment to study the effect of Pike presence on breeding performance in Teal. All lakes were naturally fishless in the first study year. In the second study year large Pike were introduced to two of the lakes (black symbols), while the remaining nine served as controls (grey symbols). In each lake and study year, there were three pair counts to assess the use of lakes by Teal during the settling and nesting phase, and later three brood surveys to assess lake use by females and broods during the duckling stage. The y axis shows the change in lake use expressed as the number of bird days spent by pairs of ducks (left) and duck broods (right) on the lake, respectively. As can be seen, Pike introduction did not reduce lake use by Teal during pair counts, whereas a dramatic reduction was observed in the brood surveys. The experiment suggests that Teal do not detect or do not avoid Pike presence when deciding where to breed, and that Teal broods either suffer large predation loss or simply leave the lakes containing Pike. Redrawn after Dessborn* et al. *(2011a).*

oligotrophic lakes with bowl-shaped basins are compared to shallow, more productive lakes with saucer-shaped basins.

Variation in the estimates of breeding density decreases, though, as the scale of study increases. Recent national estimates from Sweden and Finland (Table 3.2; Väisänen et al. 2011, Ottosson et al. 2012) suggest that there are 0.25 breeding pairs/km² overall surface (land and water combined) in Sweden and 0.66 in Finland. The equivalent estimates from aerial transects over Fennoscandia in the 1970s were from 0.01 to 0.19 breeding pairs/km² total area, but with aerial transects regularly spread over the countries, not just over wetlands (Haapanen & Nilsson 1979). It is also not known how representative these figures are for very large wetland areas. Most of the breeding range of Common Teal as well as Green-winged Teal sits within remote and poorly studied areas in Russia and Canada. The fact that habitat type strongly affects Teal breeding density is evident also from the breeding duck surveys in Canada (Bellrose 1976): pair density ranges from 0.44 to 1.25 per km² total surface area (not only water area) in the Arctic and sub-Arctic deltas (which are probably better than average Teal habitats), and similarly from 0.37 to 1.08 pairs/km² further south in the aspen parklands, while the wetlands of the prairies only host 0.11–0.71 pairs/km² (0.12 pairs/km² total surface area in Evans & Black 1956, 0.11–0.29 per ha of water area in Duebbert & Lokemoen 1980), and those of the boreal forest only 0.11–0.41 pairs/km². Ross *et al.* (2002) even recorded as few as 0.08 pairs per km² general surface area in the boreal forest of northern Ontario. At the other extreme, densities as high as 1.4 to 2.1 pairs/km² water area have been documented in Kivach Nature Reserve, Karelia, Russia, during the 1950s and 1960s (Zimin *et al.* 1993), which is consistent with the mean 1.5 pairs/km² water area recorded by Pöysä (1985a) in Lake Vetsijärvi, Finland. Values as high as 4.4 pairs/km² water area were recorded in some lakes of the isolated and highly productive Mývatn area in Iceland (Gardarsson 1979). Fluctuations in habitat type or quality can also dramatically affect the density of nesting Teal: within the same 60 square miles (155km² total surface area) in British Columbia, Munro recorded 45 adults and 17 broods in 1938, but only 12 adults and 2 broods 10 years later (Munro 1949). In many areas Teal breed in reedbeds or along ditches (see for example Fox 1986), where relating breeding pairs to area of open water is obviously a less meaningful measure.

Habitat choices of nesting and brood-rearing Teal are naturally studied at the local level, where Common Teal in Finland stand out as the best-studied example. They can breed and rear broods until fledging on most types of wetland, although showing a marked preference for small to medium-sized waterbodies, and for mesotrophic–oligotrophic wetlands (those with medium to poor nutrient status/productivity). Breeding Common Teal can hence be found from sheltered parts of large oligotrophic lakes, to ditches and small tarns deep in the forest (Cramp & Simmons 1977). However, at the catchment level a few small wetlands account for a disproportionately large share of the annual production of young (Elmberg *et al.* 2005). Research from Finland also highlights Common Teal's marked preference for Beaver-created ponds (Nummi & Hahtola 2008, Table 4.2, Box 4.3), as is also observed for Green-winged Teal in North America (Brown *et al.* 1996, Baldassarre & Bolen 2006).

As a general rule, breeding density of Teal is highest where aquatic plants and shoreline vegetation are most abundant, as well as in invertebrate-rich wetlands, especially those with the most abundant emergence of Diptera (midges, mosquitoes, flies) (Haapanen & Nilsson 1979, Elmberg *et al.* 1993, Vartapetov 1998, Nummi *et al.* 2013; Figure 4.15, see colour section). Similarly, breeding Green-winged Teal preferentially select wetlands whose shallow areas are more extensive, with more phosphorus circulating in the system and where chlorophyll density is higher (Paquette & Ankney 1996). In sub-Arctic Québec, Canada, Décarie *et al.* (1995) also found Green-winged Teal brood density (*c.* 2.0 broods per 100km² of habitat on average) to be highest in vegetated lakes and small wetlands such as ponds, fens or bogs, which were used to a greater extent than their relative availability in the landscape.

The nest is usually built on the ground, within 100m of the nearest shoreline (Munro 1949, Fox 1986), sometimes right by the waterline (A. D.

TABLE 4.2. *Teal brood density (broods per km pond shoreline) and invertebrate abundance index (mean ± SE) on Beaver and non-Beaver ponds of the Finnish boreal forest. The differences between the two habitats are statistically highly significant: Beaver ponds create favourable conditions for crustaceans and insect larvae, themselves very attractive to breeding Teal. After Nummi & Hahtola (2008).*

Dependent variable	Beaver ponds (n = 11)	Non-Beaver ponds (n = 26)
Teal brood density	0.62 ± 0.10	0.05 ± 0.04
Invertebrate abundance index	300.1 ± 81.6	48.7 ± 5.4

BOX 4.3. An unexpected ecological relationship: facilitation of Teal by Beaver

Negative relationships between species are well known and often described, for example competition, parasitism and predation. Positive interactions, however, also exist when one or both interacting species benefits from the presence or behaviour of the other. The term 'facilitation' is used, for instance, when one species improves the habitat for another, without this being costly to itself (see Bruno *et al.* 2003).

Because the dams they build profoundly affect their environment, Beavers (*Castor canadensis* in North America and in European areas where it has been introduced, native *Castor fiber* in some other parts of Europe) are considered to be major 'ecosystem engineers' where they occur (Nummi & Hahtola 2008). Inundation caused by beaver dams kills most trees in the flooded area, and the new wetland is very productive for several years as the organic nutrients from the old forest floor and newly decomposing plants promote aquatic primary production (Figure 4.16). This, in turn, leads to a burst of food production (invertebrates, and seeds from aquatic plants) for Teal and their ducklings to exploit (Nummi 1992).

Nummi & Pöysä (1997) showed that Common Teal respond markedly (more so than Mallard) to the creation of Beaver dams, and use the resulting ponds for foraging and breeding. Similarly, Rempel *et al.* (1997) showed a strong selection of riverine Beaver pond marshes by Green-winged Teal. Recent work has explained why these habitats are so attractive for breeding: Teal pairs are more productive if breeding on Beaver ponds, both because they provide more abundant food for ducklings and because predation on ducklings is lower. Indeed, terrestrial predators are less efficient at catching young ducks in the shallow areas bordering Beaver ponds, and fish predators like Pike should not be present or should be of very small size in such habitats, which represent the early stages of trophic succession and maturation of wetlands. Daily mortality of Teal ducklings is therefore three times lower in Beaver ponds

FIG. 4.16 *Typical Beaver pond selected by breeding Common Teal in the Finnish boreal forest. Photograph by Johan Elmberg.*

> than in other forest ponds in Finland, so that brood density is 0.6 and 0.05 per shoreline kilometre for Beaver and non-Beaver ponds, respectively (Nummi & Hahtola 2008). A long-term monitoring survey conducted in 12 wetlands used by breeding Common Teal over 15 years in Finland showed that the best wetland in the study area was a pond created by a beaver dam (Nummi *et al.* 2010).
>
> Positive effects of Beaver damming to ducks are not limited to Fennoscandia or Europe: Green-winged Teal and other American species (for instance Wood Duck *Aix sponsa* and American Black Duck *Anas rubripes*) get the same benefit from Beaver ponds in North America (Brown *et al.* 1996, Baldassarre & Bolen 2006).

Fox, pers. comm.), and often under dense plant cover or in tussocks. Nevertheless, Teal nests are sometimes hundreds of metres away from open water (Cramp & Simmons 1977), and distances of up to 1.5km from the nearest waterbody have been recorded (Gerasimov *et al.* 2000). While Teal are among the shiest of all dabbling ducks in winter, human activity does not necessarily restrict them from breeding on otherwise favourable sites (see more about human disturbance in Chapter 7).

Adult Teal moult during the latter part of the brood-rearing period and into August. This part of the annual cycle has been very little studied and much remains to be learned. It is conventional wisdom that most successfully reproducing females moult where their brood is, or very close to this location (Hohman *et al.* 1992), though they would generally delay moult until their ducklings have become independent (Rohwer & Anderson 1988). Males, however, often leave breeding ponds for more distant locations, and it is probably more common for male Teal to choose moult lakes at a landscape or regional level. One such locality, Lake Brånsjön in northern Sweden, was studied by Sjöberg (1988). This lake offers an abundance of invertebrate food, superior to virtually all breeding lakes in the surrounding landscape, and hundreds of Common Teal – mainly males – gather there to moult. Flocks of moulting Teal typically use large wetland complexes, such as deltas of major rivers (for example the Volga River in Russia, Scott & Rose 1996) and more eutrophic (nutrient-rich) lakes with abundant food and sheltering vegetation, typically larger than most breeding ponds (Sjöberg 1988). In a description of a major moulting area in Denmark, Kortegaard (1974) suggested that food abundance and safety from predators were the main habitat features determining use of the area by Common Teal.

Migrations

As in dabbling ducks in general, the ecology of migrating Teal is more poorly known than that of wintering and breeding birds. This is equally true for autumn and spring migrations, including the whereabouts of Teal in the habitats used the last few weeks before settling on the breeding wetlands proper (Arzel *et al.* 2006). In terms of habitat, all wetland types, including intertidal areas and rivers (Figure 4.17), can be used by migrating Teal during these energy-demanding periods, as long as the depth is sufficiently shallow to permit foraging. In one of the few detailed studies of autumn habitat use by Common Teal, Bregnballe *et al.* (2009a) observed that birds remained in the same shallow areas throughout the 24-hour period, but switched to saltmarshes when the water level rose (this is associated with greater predation risk).

Although habitat selection processes of migrating Teal are poorly quantified, some major autumn migration stopover areas have nevertheless been identified. Outstanding among such known places are the mouth of the River Ob and the Kanin Peninsula in Russia, the Wadden Sea (The Netherlands, Germany and Denmark) and Sultansazligi in Turkey (Scott & Rose 1996, Viksne *et al.* 2010), where tens of thousands of Common Teal can be simultaneously observed (see also Appendix 4). Teal do not tend to gather in large groups to the same extent in spring as they sometimes do in autumn, but astonishing concentrations of up to 50,000 Common Teal have been reported in spring in European Russia and the Baltic States (review in Viksne *et al.* 2010). What little is known about spring-staging Teal (cf. Arzel &

FIG. 4.17 *Teal do not only rely on standing waters but can also be observed in river habitats, as here in Kamo River, Japan ('Duck river' in Japanese). Photograph by Didier Buysse.*

Elmberg 2004 and references therein) almost exclusively refers to areas representing the first half of the northbound migration. In their study of a temporary guild of six species of spring-staging dabbling ducks halfway up the flyway, Arzel & Elmberg (2004) found that Common Teal were generalists in habitat choice, with a very high spatial overlap with Mallard in microhabitat use. They also found that Common Teal preferred the shallowest parts of the eutrophic wetland being studied, and that they foraged by both day and night. Similarly, DeRoia (1989) observed Green-winged Teal to preferentially select shallow ponds around Lake Erie during both spring and autumn migrations, especially ponds where invertebrate density was higher than in unused control ponds.

With respect to reproduction, animals can be placed along a gradient from purely 'capital breeders' *sensu* Drent & Daan (1980) – that is, those that fuel their reproduction solely by body reserves, sometimes stored long before actual reproduction – and 'income breeders' at the other extreme, which use only the energy they can acquire on the breeding grounds. Teal are considered to be on the 'income breeding' side of this gradient, although they are not pure income breeders, as they start to acquire energy for reproduction during the last legs of their spring migration (see Chapters 5 and 6). It therefore remains a serious knowledge gap that so little is known about the ecology of staging Teal approaching their breeding grounds (Figure 4.18), as breeding success and hence recruitment are likely to depend

FIG. 4.18 *Very little is still known about the ecology of spring-staging Teal, especially their whereabouts and habits during the last few weeks before settling on their breeding wetlands. River deltas and estuaries in the boreal biome may have a crucial role in the annual cycle by providing foraging opportunities at this time. The picture shows the Umeälven River Delta in northern Sweden, a wetland site of international importance used by thousands of Common Teal in spring before ice-melt on breeding ponds further north. Photograph by Staffan Müller.*

to a large extent on the availability of foraging habitat and food, and thus foraging success, in the few crucial weeks before egg-laying starts. This in turn is not only an issue of habitat, but is also related to between-year variation in spring weather and to disturbance from predators (Arzel *et al.* 2014). The largest congregations of late-staging Teal have been reported from river deltas and estuaries in the boreal biome, on shallow wetlands that become ice-free earlier than surrounding ponds and lakes.

At these preferred spring late-staging areas Common Teal typically perform local daily movements between foraging and roost sites, probably broadly similar to what is observed on winter sites (see next section). They readily switch between foraging in natural and agricultural habitats, but the relative importance of these remains unknown. Pilot studies suggest that little invertebrate food is accessible here at this time of year, and that the previous year's seeds from aquatic plants may be an important source of energy, even though pre-breeding Teal would also require high-protein animal food to prepare for egg-laying. At these late-spring migration sites Common Teal typically associate with dense flocks of Mallard and Eurasian Wigeon when foraging and roosting. Virtually nothing is known about the duration of staging and time use in Teal as they approach their final breeding sites. Much of what we can say remains inferential and guesswork until spring-staging Teal are studied by telemetry and other methods that facilitate evaluation of nutritional, energetic and behavioural strategies of pre-breeding individuals.

Winter areas

Wintering Teal, like most dabbling ducks, show very specific habitat-use strategies: birds often commute twice daily between sites of very different types used for either diurnal or nocturnal activities. During daylight hours, they gather in large flocks at day-roosts for comfort and social activities (resting, preening, pair formation; Figure 4.19). From these day-roosts, Teal then spread out in small groups at dusk to nocturnal foraging grounds (Tamisier 1976, 1978; Box 4.4). Teal day-roosts are typically large wetlands (lakes, open waterbodies, reservoirs, and so on), while nocturnal foraging sites are more frequently shallow marshes with denser vegetation cover, or flooded pastures. Such diurnally and nocturnally used habitats may not always be very distant from each other: during a radio-tracking study in western France virtually all granivorous ducks such as Teal were observed to leave their day-roost at night for distinct foraging habitats, but the latter could be only a few

FIG. 4.19 *Thousands of Common Teal at a winter day-roost at Tour du Valat, Camargue, southern France. Bird numbers and density sometimes reach very high values at such sites. Photograph by Alan Johnson.*

BOX 4.4. Daily habitat-use strategies by wintering Teal: the 'Functional Units'

Dabbling ducks face numerous and sometimes conflicting demands in winter, forcing them to develop complex habitat-use strategies.

Teal have a seed diet in winter, and their small size translates into high relative energy needs (energy required for maintenance per unit of body mass; see Chapter 5), requiring long foraging and digestion times. Teal use seeds shed by plants (terrestrial and aquatic) found in the upper sediment of waterbodies, favouring selective foraging in shallow wetlands with abundant seed-producing vegetation. However, reduced peripheral visibility in vegetated marshes decreases detection of aerial predators such as harriers/marsh hawks *Circus* spp. Consequently, these foraging habitats are preferentially used at night, when predation risk from raptors is much lower.

To overcome the problem of diurnal predation risk, Teal switch to more open habitats during the day, where they can detect approaching raptors from a greater distance and adopt appropriate escape-flight behaviour if necessary. The concentration of Teal in such sites also enhances predator detection, providing greater safety due to flock benefits. On the other hand, large concentrations of densely packed prey are also particularly attractive to predators (Guillemain *et al.* 2007d), so the frequency of fly-overs by, for example, Peregrine *Falco peregrinus*, Yellow-legged Gull *Larus michahellis* or Marsh Harrier *Circus aeruginosus* can reach up to 130 per day (recorded in western France by Fritz *et al.* 2000, see also Johnson & Rohwer 1996; Figure 4.20). Each raptor fly-over normally triggers escape flight in the Teal, otherwise these ducks would certainly be individually attacked. It is therefore unwise for Teal to indulge in risky foraging activities, such as feeding with their eyes underwater, even when they gather in large flocks during the day. Not only does such foraging increase predation risk, but individuals cannot monitor the behaviour of neighbours to assess the general level of predation risk either. Food availability in such diurnally used sites is anyway often limited, owing to deeper water limiting food access and to rapid food depletion due to the large concentrations of ducks. During the day, Teal therefore use such sites to rest, preen or engage in social activities such as pair formation, hence the term 'day-roost'.

Teal are often forced to use quite different habitats on a daily basis. The combination of a locally occurring day-roost and several nocturnal foraging habitats is called a 'Functional Unit' (Figure 4.21). This term was coined by Tamisier in the 1970s based on diurnal and nocturnal observations of the behaviour and movements of Common Teal in southern France and Green-winged Teal in Louisiana (Tamisier 1976,

FIG. 4.20 *Marsh Harrier forcing roosting Common Teal to take flight. Photograph by Dominique Gest.*

1978; similar patterns have been reported for Green-winged Teal in California by Euliss & Harris 1987, though Green-winged Teal there fed during 60% of the daylight hours in addition to feeding at night). Later development of radio-tracking equipment (VHF tags) allowed Tamisier & Tamisier (1981) to demonstrate that wintering Common Teal were generally faithful to their day-roost, returning there every morning.

Conversely, in winter Teal are more prone to change nocturnal foraging habitats, presumably in response to changes in food availability over time. Subsequent studies with the same VHF methodology showed that some individual Common Teal had an alternative strategy, namely to remain at or in close proximity to their day-roost to forage at night (Guillemain *et al.* 2002a). Legagneux *et al.* (2008) even observed radio-tagged Common Teal to be equally likely to remain on their day-roost at night as to leave it to forage elsewhere. Commuting flights recorded by these authors were 1–2km on average, hence not representing a major daily energy cost. Commuting distance to nocturnal foraging sites increased during the course of winter, suggesting a gradual response to food depletion near the day-roost. From an analysis of dusk and dawn hunting bags in the Camargue, Guillemain *et al.* (2008a) recorded a similar commuting flight range of 2–3km on average over the winter.

Although nocturnal foraging seems to be the rule in wintering dabbling ducks, diurnal foraging increases in periods of higher energy demands, such as after the autumn migration (Tamisier *et al.* 1995) and for birds wintering in colder areas (Guillemain *et al.* 2002b).

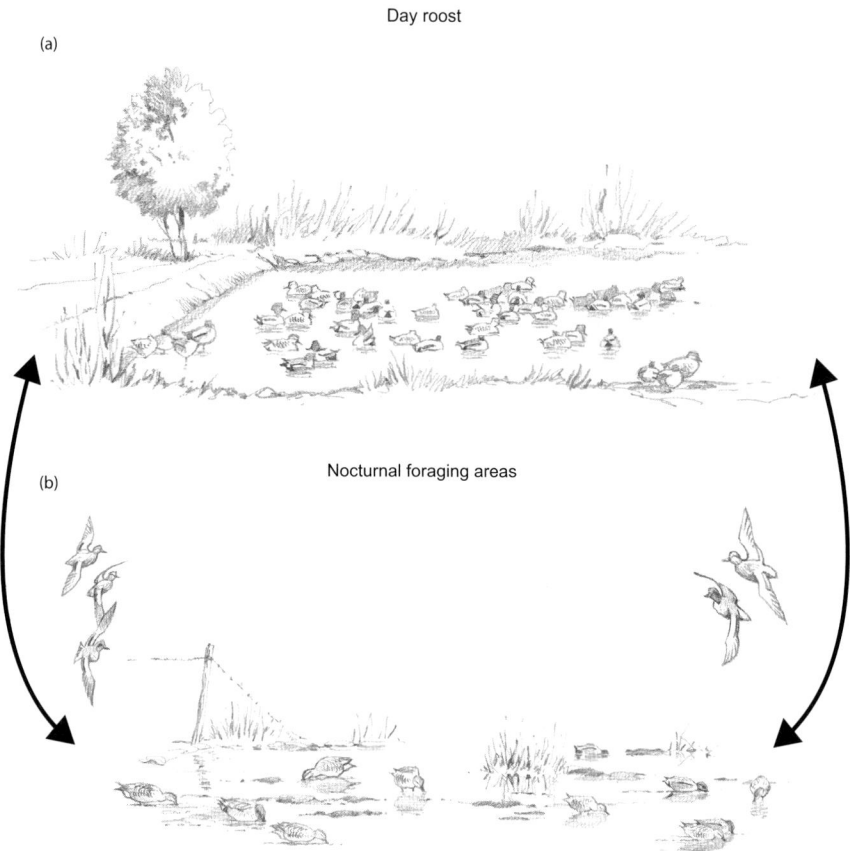

FIG. 4.21 *The functional unit of wintering Teal, composed of a daty roost (a) where birds gather in large numbers, and foraging grounds (b) towards which they disperse during the night.*

tens of metres away, within the same protected area (Guillemain *et al.* 2002a). In the Camargue, Common Teal preferentially use freshwater marshes for nocturnal foraging. They nonetheless make a greater use of cultivated areas (rice fields *Oryza sativa*) after crops are harvested in September, but switch back to more natural habitats in February, when rice fields become drier (Pirot *et al.* 1984).

The resident Aleutian Teal use marine beaches during winter, while they mostly breed on interior freshwater wetlands (Kenyon 1961). Delacour (1956) found that Green-winged Teal also use marine areas during winter.

CHAPTER 5

Feeding ecology

ENERGY REQUIREMENTS

Energy requirements of animals increase with body mass, but at a rate lower than one to one (Lasiewski & Dawson 1967): smaller species have greater relative energy requirements (the amount of energy needed for maintenance of a given body-mass unit) than larger ones. Small species therefore have to be particularly selective during the food acquisition process, and to select the food types yielding greater energy benefits.

For dabbling ducks, the very comprehensive analysis by Miller and Eadie (2006) provides an equation to compute Resting Metabolic Rate (RMR, the energy required for maintenance of a non-active individual under average environmental conditions) as a function of body mass, demonstrating this decelerating positive relationship (RMR = $457 \times BM^{0.77}$, where RMR is in kJ/day and BM is body mass in kg; Figure 5.1). Accordingly, a 350g Teal would require 204kJ/day, far less in absolute terms than the 559kJ required daily by a 1.3kg Mallard. However, while the Mallard needs 430kJ per kilogram of body mass, the equivalent value for Teal is 582kJ. To meet its daily energy requirements, a Teal thus has to ingest 35% more energy per unit mass. This can be accomplished either by ingesting a greater food volume, or by selecting more energy-rich foods. A study in Louisiana demonstrated that Green-winged Teal spent a greater percentage of time foraging (68%) than did Mallard (35%), possibly owing to such differences in relative energy requirements (Johnson & Rohwer 2000).

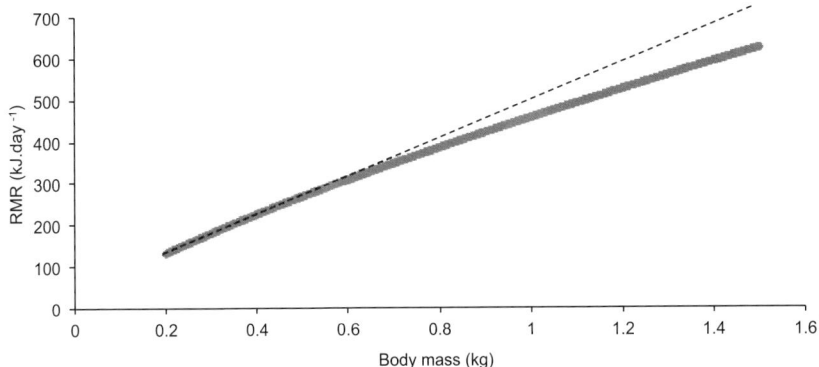

FIG. 5.1 *Relationship between species-specific Resting Metabolic Rate (RMR) and species-specific body mass among dabbling ducks (after equation in Miller & Eadie 2006). Daily energy requirements increase with increasing body mass at a decelerating rate (RMR = 457 x Body Mass$^{0.77}$). The dashed straight line shows what RMR should be if it was always linearly proportional to body mass.*

It has been calculated that in practice Teal have to ingest as much as 20–30g wet weight of natural seeds per day on average in winter (Tamisier 1971, Tamisier & Dehorter 1999). To obtain this quantity of food the birds have to spend *c.* 35% of their time foraging, which is mostly done at night. Foraging time is likely to be driving the diet of Teal in winter: if they ate vegetative parts of plants (stems, leaves) rather than fed on seeds they would have to forage for 61% of the 24-hour period to acquire the same amount of energy (Bruinzeel *et al.* 1997), which is almost impossible to accomplish over extended periods of time. Conversely, when relying on unharvested energy-rich Corn *Zea mays* kernels, Teal can apparently meet their daily requirements by spending only *c.* 20% of the day foraging (Baldassarre *et al.* 1986).

Teal energy requirements change markedly over the annual cycle. One major hurdle occurs in midwinter, when cold temperatures may freeze wetlands and food can become inaccessible for several days. In order to cope with this risk, Teal store energy reserves (lipids) from early to midwinter. Starting from a low value after autumn migration and moult to breeding plumage around October, most studies document a massive body-mass increase until December or January, in Common Teal as well as in Green-winged Teal (Baldassarre *et al.* 1986, Fox *et al.* 1992, Tamisier *et al.* 1995), except in Louisiana where temperatures are far milder (Rave & Baldassarre 1991; Figure 5.2). The same pattern of a body-mass peak in midwinter is also seen in other duck species (for example American Wigeon in Rhodes *et al.* 2006). It should be noted that such an increase has also been demonstrated in recaptured individuals weighed on several occasions during the same winter. The midwinter body-mass peak is thus not an artefact arising from sampling different groups of animals, perhaps of different origin, but rather a genuinely individual pattern (Guillemain *et al.* 2005b).

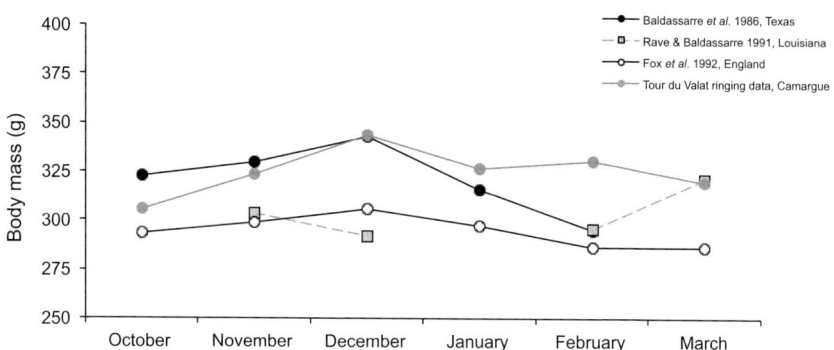

FIG. 5.2 *Change in mean body mass in adult male Teal through winter.*

Such an increase in body mass would provide Teal with insurance against adverse weather conditions caused by temporary food inaccessibility. The 20g increase recorded in Green-winged Teal wintering in Texas between October and December represents an added fasting ability of *c.* 2 days compared to the October body-mass status, assuming that the entire mass gain is in the form of lipids, that 1g lipids equals 9kcal energy, and that the daily energy requirement of an adult male is 95kcal (see details in Baldassarre *et al.* 1986).

Surprisingly, Teal do not attempt to maintain similar body reserves and 'fasting insurance' later in the season, and body mass falls to particularly low levels by February, at the onset of spring migration when one would expect a migratory bird to store energy in preparation for a long flight. However, it may simply be impossible for Teal to maintain such large body reserves despite intensive feeding if food gradually gets depleted during the course of winter. Moreover, weather conditions are usually harsher in January–February than earlier on in winter (Tamisier *et al.* 1995). Still, the potential benefit of carrying extra reserves gradually decreases over late winter as time progresses, since the likelihood of encountering very cold conditions itself decreases (Baldassarre *et al.* 1986, see also Schummer *et al.* 2010 for seaducks). Carrying extra weight has costs, especially in terms of wing loading and hence manoeuvrability, so that body weight should not be maximised when not necessary (Lima 1986, Witter & Cuthill 1993, Zimmer *et al.* 2011 for Teal, Higginson *et al.* 2012). Consistent with this hypothesis, Hepp (1986) showed that captive Black Ducks also lost body mass in late winter despite having access to *ad libitum* food.

Teal with the February body masses shown in Figure 5.2 may still have more reserves than they immediately need for spring migration: a 330g adult male Common Teal about to depart from the Camargue could theoretically fuel a flight of 4,118km (results from Flight version 1.24 software, Pennycuick 2008), while Common Teal ringed in the Camargue during winter were observed to breed only 3,000km from there on average (Guillemain *et al.*

2009a). Moreover, these birds have the opportunity to make several stopovers on their way, for instance halting in northern Italy (only a few hundred kilometres from the Camargue), where they could potentially refuel for continued spring migration (Callenge *et al.* 2010).

Some bird species store energy reserves in winter and early spring, to be used later on during the breeding season, for example for egg formation. This is typically what happens in some capital-breeding goose species (Drent & Daan 1980, Meijer & Drent 1999). Conversely, small species like Teal cannot carry such large reserves over long migration distances (Klaassen 2002), and they are more likely to acquire the energy needed for the next leg of migration through foraging activity at the present stopover, and to fuel reproduction by the energy obtained on the breeding grounds proper or in the last few weeks before reaching these, hence just before nesting (Paquette & Ankney 1998; see the section above about habitat use during late-spring migration). These birds are called 'income breeders' (Drent and Daan 1980) and, following the same logic, have also been termed 'income migrants' (Arzel *et al.* 2007b).

DAILY FORAGING TIME

Nocturnal foraging

Foraging behaviour typically refers to the process of finding food, but the actual ingestion of foods is also a fascinating process in dabbling ducks. Except for the few instances when they are grazing on fine plant parts or when picking seeds or invertebrates from plants and the water's surface, Teal and other dabblers acquire most of their food in a complicated sieving action involving the bill and tongue (see Box 5.1). Accordingly, dabbling ducks collect water or suspended sediment in the bill, and the water is expelled through lateral lamellae along the lower (mandibular) and upper (maxillary) halves of the beak. The water and suspended items are thus filtered out and foods remain in the bill. Successful feeding is therefore essentially preceded by touch foraging (Van Eerden & Munsterman 1997), which explains why Teal and other ducks can easily feed at night.

Direct assessments of the nocturnal time budget of Teal are few (see, however, Swanson & Sargeant 1972, Tamisier 1972), but extensive monitoring with a light intensifier in western France showed that 90.7% of the night was spent foraging (Guillemain *et al.* 2002b). The reasons why Teal are such nocturnal foragers are many and not mutually exclusive (Table 5.1, see also McNeil *et al.* 1992). First, Teal may feed at night because their foraging habitats are typically densely vegetated shallow wetlands, where predation risk would be too high for the birds to remain during the daylight hours (Tamisier 1976, 1978). Secondly, prey availability may peak at night either because of the emergence of insects leaving their larval stage (Swanson & Sargeant 1972), or due to day/night migration of invertebrates in the water column, both

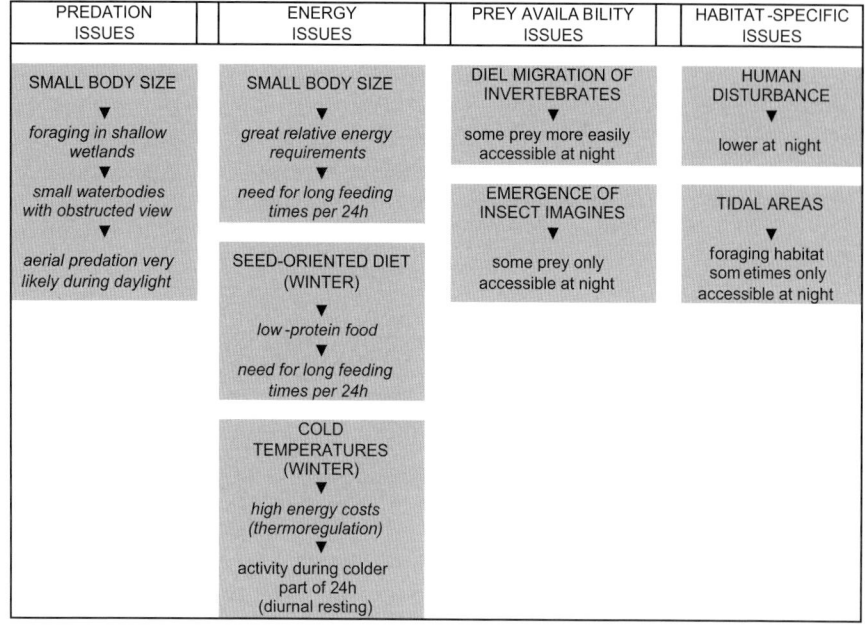

TABLE 5.1 *Main factors leading to nocturnal foraging in Teal.*

processes leading to more prey being available closer to the water's surface at night (see Guillemain *et al.* 2000b for the same process in Northern Shoveler). Finally, nocturnal foraging may be part of an energy-optimisation process, by which individuals acquire energy and produce body heat through digestion and activity during the period of the 24 hours when they need it the most; that is, long winter nights (Baldassarre *et al.* 1988b, reviews in Owen 1991 and McNeil *et al.* 1992).

In any case, nocturnal feeding seems to be the rule in Teal in winter (Box 4.4), rather than being used when they cannot meet their daily energy requirements by foraging during daylight hours only ('Preference' versus 'Supplementary' hypotheses, McNeil *et al.* 1992). Even when relying on energy-rich foods in agricultural habitats, which leads to short foraging times, Teal preferentially fly to these fields during the night, or use them temporarily at dusk and dawn, rather than forage in them in daylight (Thomas 1981, Baldassarre & Bolen 1984, Pirot *et al.* 1984).

Diurnal foraging in winter

In natural habitats, wintering Teal are thus considered to feed chiefly at night, but they can nevertheless expand their foraging activity to daylight hours when circumstances permit or when they are forced to do so. Typically, this occurs when food is only temporarily accessible, or when energy demands are high.

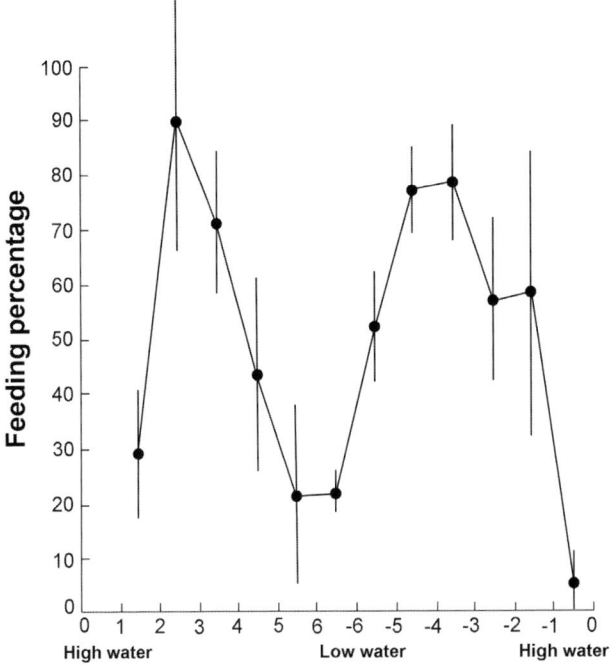

FIG. 5.3 *Mean proportion of Common Teal foraging during daylight hours over a tidal cycle in the Netherlands. Vertical bars show standard errors. From Zwarts (1976), with kind permission from the Netherlands Ornithologists' Union.*

The former situation is observed in intertidal areas, which are used by both Green-winged and Common Teal (the Netherlands: Zwarts 1976, Van Eerden 1984; Denmark: Madsen 1988; Louisiana: Gaston 1992; British Columbia: Baldwin & Lovvorn 1994; France: Blaize *et al.* 2003). Because Teal cannot feed in deep water and also avoid dry sediment, foraging on the seashore mostly occurs along the water's edge, during both rising and receding tides. One single tidal cycle during the night is unlikely to allow the birds to meet their energy requirements, so that Teal in tidal areas also forage during daylight hours (Figure 5.3).

Daytime foraging in Teal is also more common when the birds have higher energy needs. A compilation of published studies shows that Teal spend more time foraging during daylight hours if wintering in colder geographic zones, where daily energy requirements for thermoregulation are greater (Figure 5.4).

In areas where Teal have to feed more extensively in daylight during winter, they mostly do so during early morning hours, plus potentially in late afternoon if necessary, while foraging activity is generally minimal at midday (Thomas 1982, Rave & Baldassarre 1989, Gaston & Nasci 1994).

In addition to geographical differences, the diurnal time budget of Teal also shows marked fluctuations within a given wintering site, although always

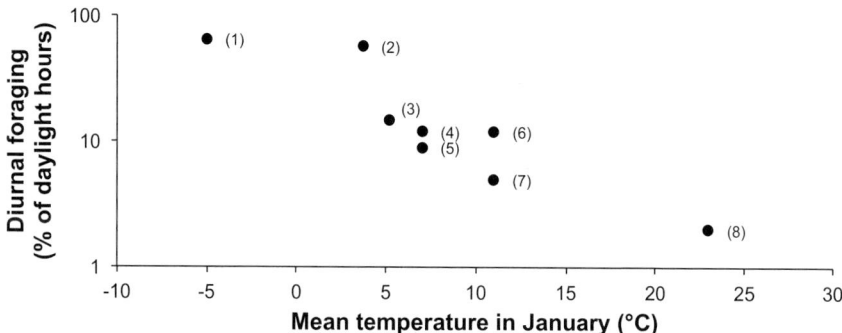

FIG. 5.4 *Mean proportion of time spent foraging by wintering seed-eating dabbling ducks (most of these being Teal) in relation to mean January temperature. (1) Jorde et al. 1983 [Nebraska], (2) Thomas 1982 [Ouse Washes, England], (3) Guillemain et al. 2002b [Western France], (4) Miller 1985 [California], (5) Tamisier 1972 [Camargue, France], (6) Quinlan & Baldassarre 1984 [Texas], (7) Tamisier 1976 [Louisiana], (8) Roux et al. 1978 [Senegal]. With kind permission from Springer Science+Business Media: Biodiversity and Conservation, 'Foraging strategies of granivorous dabbling ducks wintering in protected areas of the French Atlantic coast', volume 11, 2002, pages 1721–1732, M. Guillemain, H. Fritz & P. Duncan, Figure 1.*

following the same general pattern: diurnal foraging is particularly intense in autumn and late winter, and is at a minimum in December (Quinlan & Baldassarre 1984, Rave & Baldassarre 1989, Gaston & Nasci 1994, Tamisier *et al.* 1995, Houhamdi & Samraoui 2001; review in Paulus 1988; Figure 5.5). Tamisier *et al.* (1995) suggested that this pattern reflects a behavioural optimisation process by which Teal strategically allocate their time among different activities throughout the winter; accordingly, in autumn and early winter Teal are hypothesised to feed intensively to replenish their body reserves after autumn migration, to acquire the necessary proteins and energy for moult and to build up body stores in preparation for winter cold spells. This would be made possible by large amounts of seeds available at this time of year. In midwinter, Teal with large body stores could easily meet their daily energy requirements during the long December and January nights, so that daylight foraging would be minimal and birds would rather engage in social activities for pair formation (Tamisier 1966, Hepp 1985, see Figure 5.5). Later in the season, in February and early March, temperatures sometimes get very cold, while birds have used a part of their stored energy for maintenance and courtship. If food resources are scarce at this time, Teal are forced to increase their diurnal foraging time again.

The increase in Teal foraging time in late winter is likely to be driven by the need to meet higher daily energy demands, or by depletion that makes food increasingly difficult to find. In any case, this increase in foraging time is unlikely to result from birds trying to build up body reserves ahead of spring migration; conversely, dabbling ducks more likely lose body mass during this part of the year (see energy section above). In that sense, the process is

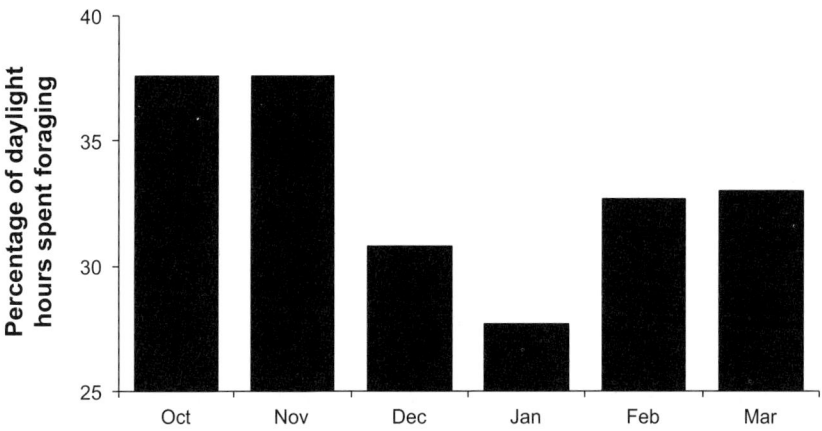

FIG. 5.5 *Mean percentage of time spent foraging during daylight hours per month by Green-winged Teal wintering in Louisiana (after data in Rave & Baldassarre 1989).*

therefore very different from spring 'hyperphagia', a behaviour observed in geese preparing for migration and breeding by foraging actively on wintering grounds during the last months of the winter season (McLandress & Raveling 1981).

Spring and summer patterns

Environmental constraints on individuals are very different in spring and summer compared with winter. As a result, the proportion of time spent foraging and the distribution of foraging time over the day and night are quite different in the more benign seasons with longer daylight hours. Although climate is milder and thermoregulatory demands are markedly lower, Teal have greater energy needs from March to September in order to fuel migrations and to breed successfully. As explained above, they fuel themselves chiefly on a daily basis – in other words they take in more energy during the demanding periods. This can only be done by extending daily foraging time (hence the terms *income migrants* and *income breeders*; see the energy requirements section above). Because most of the night is already spent foraging in winter, the only way for Teal to extend their foraging time during peak migration, breeding and moult is to feed more during daylight hours, which is exactly what is observed in the field (Figure 5.6).

Although energy needs are high during the breeding season, late spring and summer are also the parts of the year when many constraints on the foraging behaviour of dabbling ducks are largely relaxed. To some extent this is because nights are short in breeding areas (at high latitudes), and there are often fewer aerial predators (see below). In addition, the density of ducks is generally much lower than on winter and staging sites, possibly reducing

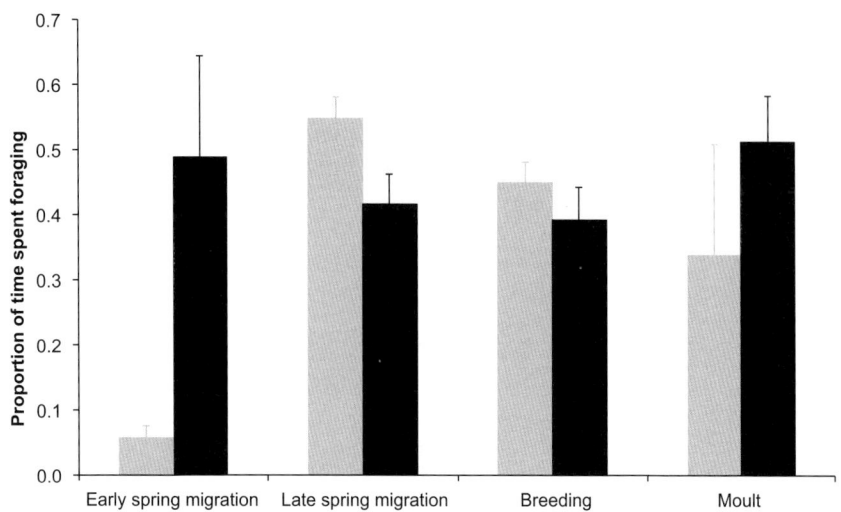

FIG. 5.6 *Mean proportion of time spent foraging by Common Teal in Europe during daylight hours (08:00–18:00, grey) and during the night (18:00–08:00, black) per period of the annual cycle. Vertical bars show standard errors. Common Teal increase diurnal feeding time in periods of higher energy demands: this was significantly lower during early migration than during late migration and breeding (after Arzel, C., Elmberg, J. & Guillemain, M. 2007b. 'A flyway perspective of foraging activity in Eurasian Green-winged Teal,* Anas crecca crecca*'. Canadian Journal of Zoology 85: 81–91. © 2008 Canadian Science Publishing or its licensors. Reproduced with kind permission).*

resource competition. Teal may therefore adopt a diel foraging regime centred more on their individual needs. For example, they can alternate periods of foraging and digestion to avoid digestive bottlenecks (Van Gils *et al.* 2005). Teal can also adjust foraging activity to when this activity is the most profitable. For example, in northern Sweden in May, the peak feeding activity of Teal occurred during the period of the day (noon) when peak emergence of midges occurred (Sjöberg & Danell 1982, Figure 5.7).

FORAGING METHODS AND POSTURES

Being dabbling ducks, Teal forage primarily on the water's surface or on land; some observations document Teal diving for food (McCanch 2012), but this is very exceptional. Foraging from the water's surface does not prevent Teal from reaching underwater prey, and they on a repertoire of methods involving a range of postures to reach food in the sediment or in the water column (Figure 5.8).

Each foraging method permits access to food at a given water depth, so that the mean foraging depth of an individual can be determined by the relative use of each method, based on morphological measurements: a Teal foraging

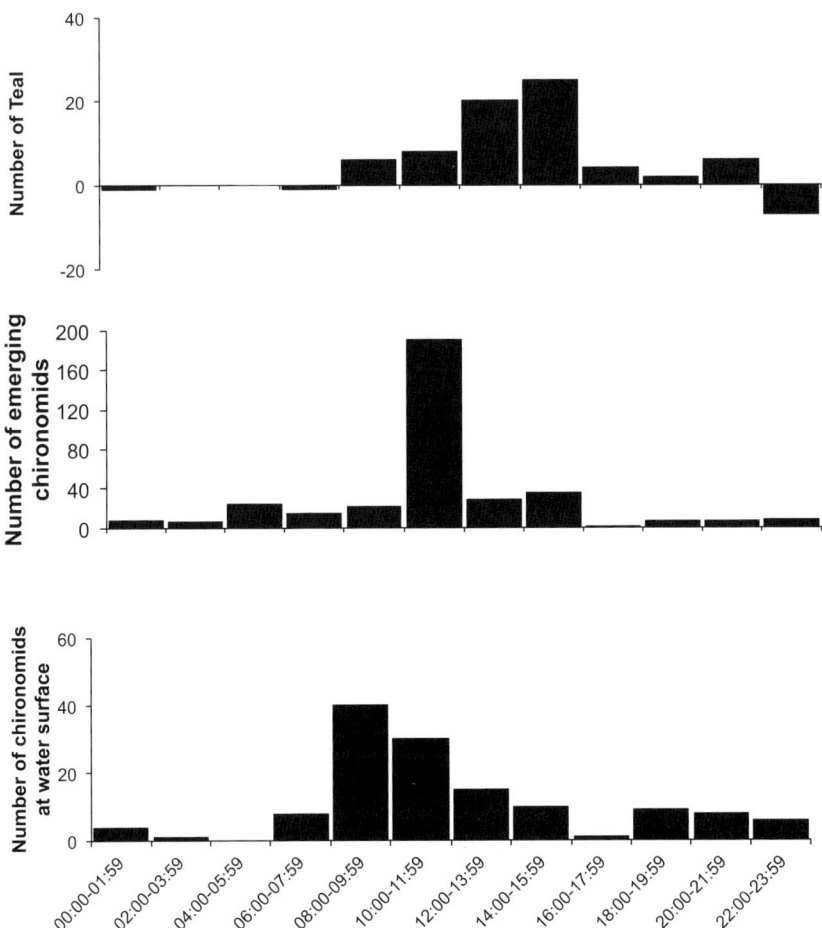

FIG. 5.7 *A 24-hour perspective on foraging activity and foraging methods in Common Teal in relation to prey availability in a boreal Swedish lake. The number of Teal engaged in surface (above axis) versus underwater foraging (below axis, indicated by a negative value) is shown (top graph). The number of emerging midges (chironomids) (middle graph) is measured as the number of pupal exuviae collected, while the number of chironomids on the water's surface refers to adults (bottom graph).* From Sjöberg, K. & Danell, K. 1982. 'Feeding activity of ducks in relation to diel emergence of chironomids.' Canadian Journal of Zoology 60: 1383–1387. © 2008 Canadian Science Publishing or its licensors. Reproduced with kind permission.

by dabbling reaches 4cm, while an upending individual forages at a water depth of up to *c.* 24cm (Table 5.2).

Teal find most of their food in shallow water

Although they could theoretically rely on a range of foraging techniques, wintering Teal show a marked preference for foraging in the shallowest zones

FIG. 5.8 *The foraging repertoire of Teal, allowing them to reach food at various depths, showing dabbling (including on foot), head under, dipping and upending, the last both to reach the sediment and to forage in the water column, in deep water.*

(preferentially by dabbling; Szijj 1965, Tamisier 1972, Euliss & Harris 1987, Johnson & Rohwer 2000). Apart from during flooding conditions that may force them to feed in deeper water (Thomas 1982), Teal will preferentially forage by dabbling during the winter, while other granivorous species gradually switch feeding methods in response to gradual food depletion and reach foods in deeper water (Guillemain & Fritz 2002; Box 5.3).

Shallow foraging enables Teal to forage with only the bill submerged, keeping the eyes above the water and permitting anti-predator vigilance while feeding (Figure 5.9, Guillemain & Fritz 2002). Dabbling ducks that forage by filtering the sediment or the water column have very wide visual fields from each eye: because they do not need stereoscopic vision in the front of the head as much as some other duck species do (for example grazing Wigeon that peck at plant leaves), natural selection has instead promoted a wider visual field towards the rear of the head. This provides species like Shoveler and Mallard (and hence presumably Teal) with total panoramic vision without the need to raise or turn the head (Martin 1986, Guillemain *et al.* 2002c).

Trading off foraging depth and predation risk

We deal at length with the potentially limiting effect of predation on Teal population dynamics in Chapter 7. We have previously pointed out that predation risk can have strong effects on foraging activity in individual Teal, for example by forcing them to forage chiefly at night and leave their feeding habitats in the early morning to fly back to day-roosts in winter. The actual predation rates on Teal during this season are extremely difficult to quantify, since such events are by essence relatively rare to observe. Furthermore, while duck remains can sometimes be collected around raptor nests or mammal dens in summer, there is no such opportunity in winter, and duck carrion is often taken by to scavengers within a few hours (Pain 1991a). The smaller size of Teal compared to other ducks makes them potential prey to a broader range of predator species (Phillips 1923). The fact that they tend to forage and rest close to the shore in shallow water also prevents them from escaping by diving in most cases, while their high density at day-roosts is very attractive

TABLE 5.2 *Maximum foraging depth associated with different foraging methods of Teal. Body measurement values from Thomas (1982).*

Method/posture	Morphological measurement	Maximum foraging depth (both sexes combined)
Dabbling	Bill length	4.1 cm
Head under	Bill + head length	8.1 cm
Dipping	Bill + head + neck length	17.1 cm
Upending	Bill tip to leg junction	23.8 cm

to predators (see Box 4.4). It is hence safe to say that Teal face high predation risk though low predation rates during winter. Only appropriate behavioural response allows them to reduce predation rates: ducks not taking off in response to raptor fly-overs will definitely be attacked individually, with very little chance of escape.

Beyond the effect of predation risk on general time budgets, the selection of a given foraging method by Teal is therefore also likely to be the result of rather dynamic and more or less momentary trade-off decisions between the need to access food and the need to maintain antipredator vigilance by appropriate foraging posture (Pöysä 1987). For example, as Teal move

FIG. 5.9 *Typical posture of a foraging Teal in winter: the shallowest zones close to the shores are generally preferred, and when adopting this foraging mode the eyes remain above the water. This permits Teal to monitor the environment for predators while feeding. Photograph by Christelle Lucas.*

towards their breeding grounds in spring they gradually encounter fewer diurnal raptors, and concomitantly increase their mean foraging depth (Guillemain *et al.* 2007b, Figure 5.10).

This pattern of change in foraging method is unlikely to reflect a simple switch from shallow- to deep-water food, but was rather hypothesised to illustrate a reduction in the predation risk constraint on vigilance as the year progresses. High predation risk during winter would constrain Teal to shallow foraging to keep their eyes above the water's surface. In response to lower predation risk in spring and summer, Teal can adjust their behaviour to more freely correspond to the spatial availability of their food, switching between deeper and shallow foraging when necessary. This is consistent with earlier studies documenting a narrower behavioural repertoire in Teal in winter, but more flexible foraging methods during the breeding season (Dubowy 1988). Teal appear to be particularly opportunistic in their use of foods and foraging methods during spring and summer (Nummi 1993).

Foraging depth over time

Such flexibility translates into changes in Teal foraging methods over several time scales during the breeding season: at the seasonal scale, Danell & Sjöberg

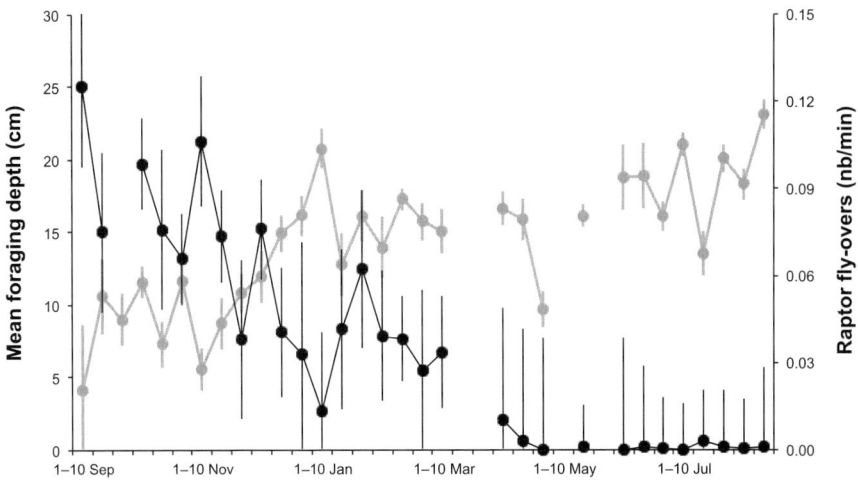

FIG. 5.10 *Mean foraging depth of Common Teal (grey) and frequency of raptor fly-overs at study sites (black, expressed as number per minute) from early winter to the end of the summer moult. Vertical bars show standard errors. Reprinted from Guillemain* et al. *(2007b), with kind permission from Elsevier.*

(1982) showed that Common Teal mostly fed at the water's surface or above it in May and June, which corresponds to the period of most intense chironomid emergence, whereas they fed underwater more intensively in August when midge larvae became more available in lake sediments (Figure 5.11).

At a finer scale, several studies have documented changes in foraging methods through the day, attributed to Teal trying to track the availability of their main prey. Figure 5.7 shows that Teal switched from underwater foraging to foraging at the water's surface around noon on a Swedish breeding lake, which was the period of the day when most chironomid emergence occurred (Sjöberg & Danell 1982).

Pöysä (1985b, 1989) also documented short-term adjustments of foraging methods in Teal, leading to gradually increasing foraging depth throughout the morning (Figure 5.12). He attributed this shift to a combination of prey being depleted in the upper strata and Teal themselves disturbing the prey animals, which moved deeper. Teal did not respond to this shift by switching to an alternate undisturbed and non-depleted patch. Instead, they lengthened patch residence time by shifting foraging methods towards deeper strata within the water column.

Finally, foraging methods change with duckling age: small ducklings aged around two weeks old were observed to forage mostly on emerging Diptera (mainly midges, Chironomidae) and Trichoptera (caddisflies), which they picked from the water's surface. Fully fledged ducklings relied more heavily on underwater invertebrates such as chironomid larvae (Danell & Sjöberg 1980), although Teal of this age also eat adult Ephemeroptera (mayflies) and

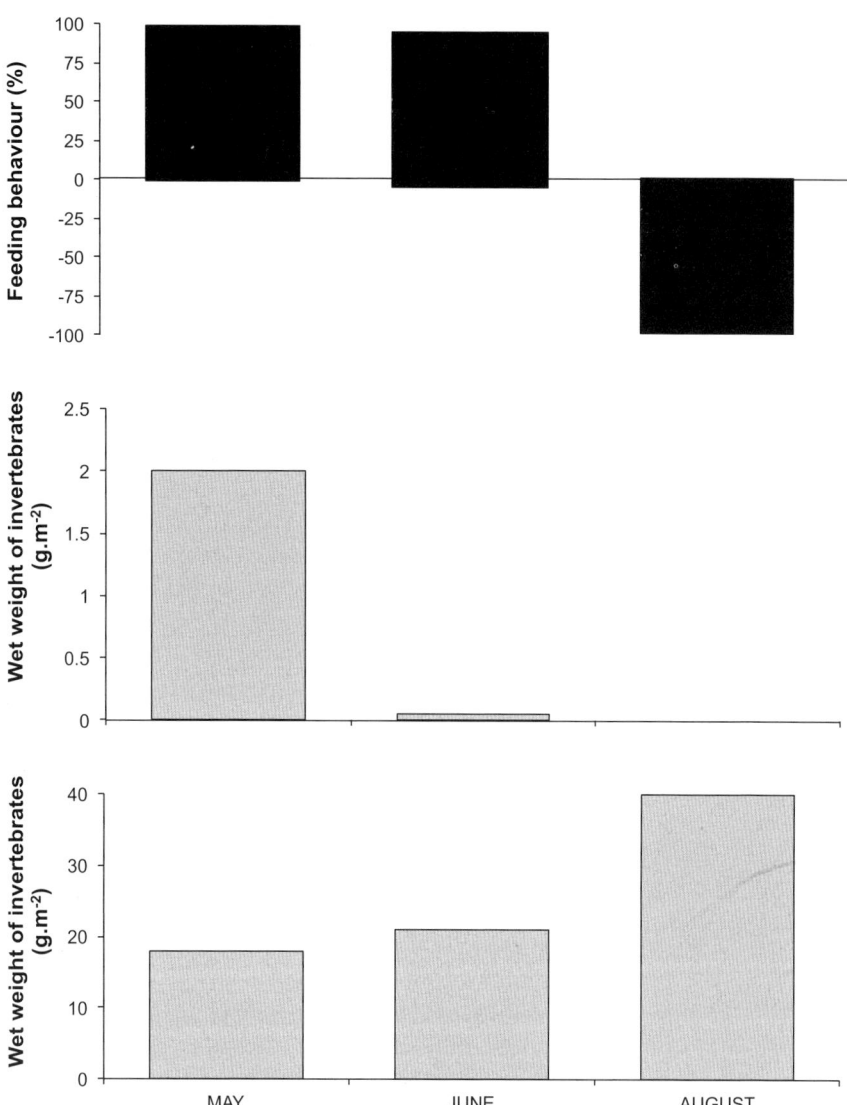

FIG. 5.11 *Feeding behaviour of Common Teal on a Swedish breeding lake (top graph: positive values indicate surface or above-surface foraging, negative values indicate underwater foraging) and availability of invertebrate food in the vegetation (middle graph: 99% of invertebrate food made of chironomid imagines) and on the lake bottom (bottom graph: 99% of invertebrate food made of benthic chironomid larvae). After Danell & Sjöberg (1982), with kind permission from John Wiley & Sons, Inc.*

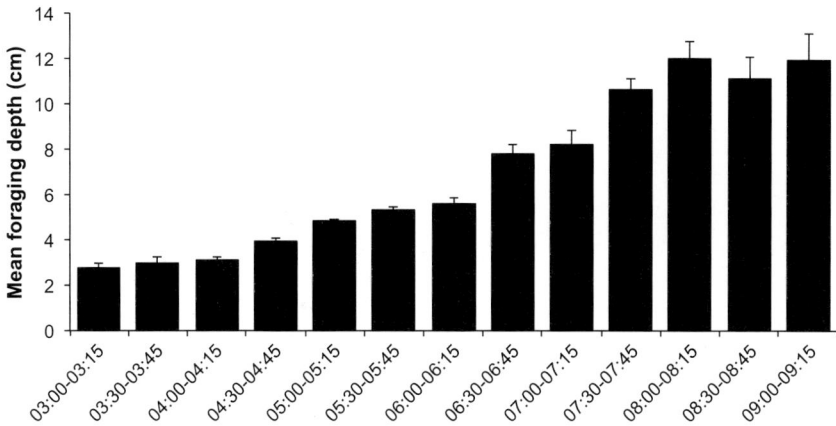

FIG. 5.12 *Mean foraging depth of Common Teal over the morning hours in Finland in June and July. Vertical bars show standard errors. The gradual daily switch to deeper foraging methods was considered to be due to prey depletion and prey disturbance by other foraging Common Teal. Redrawn after Pöysä (1989).*

Trichoptera (Nummi & Väänänen 2001). While these changes in foraging methods may partly be due to changes in relative prey availability through the summer (cf. Danell & Sjöberg 1982), young ducklings may also have difficulties foraging in the water column per se. Because of the great buoyancy provided by their down, underwater foraging is far more difficult when they are very young compared to later, when they acquire adult-like feathers.

Foraging methods and the structure of foraging behaviour

Switching from shallow (eyes above water's surface) to deep (eyes underwater) foraging methods profoundly affects the proportion of time spent head-up and vigilant: more time is spent in head-up vigilance by deep than shallow foraging Teal, probably because shallow-foraging birds can keep their eyes above the water's surface and hence already maintain some vigilance while feeding head-down, though maybe not of similar quality to head-up vigilance. Increased vigilance in deep foragers is accomplished by shortening of both feeding and scanning bouts, producing a pattern of shorter but more frequent episodes of head-up vigilance (Guillemain *et al.* 2001, Figure 5.13).

This indicates that foraging Teal aim at maintaining constant vigilance efficiency, even when deep-foraging, which is consistent with the main threat during daylight hours being sudden attacks by aerial predators (Fritz *et al.* 2000). This also indicates that deep-foraging Teal can quickly assess predation risk level from very short peeks at their surroundings. In turn, this is consistent with these birds relying on obvious signals obtained from the behaviour of their neighbours (Guillemain *et al.* 2012a), rather than scanning the environment in detail to directly detect potential predators. Increasing

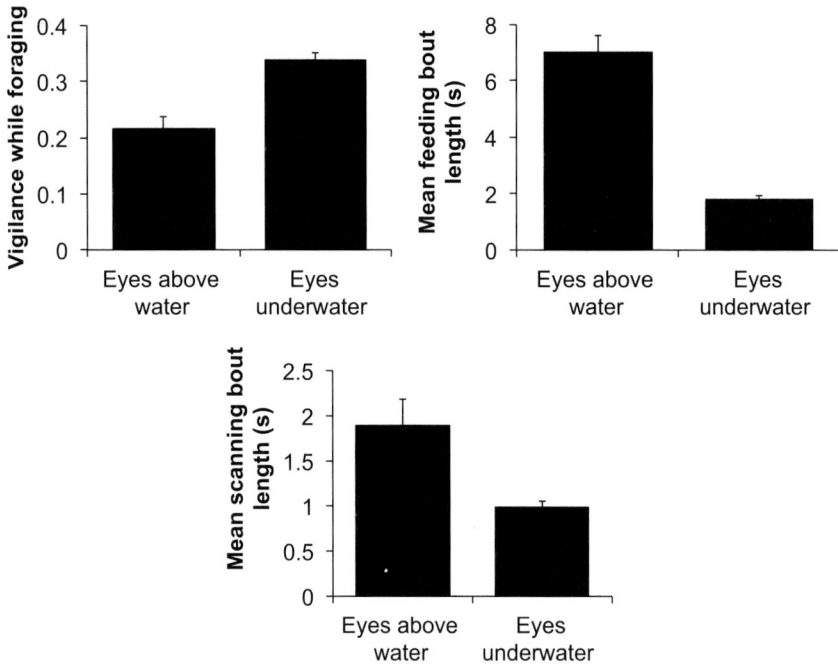

FIG. 5.13 *Proportion of time spent vigilant while foraging in Common Teal depending on whether the eyes are above (n = 47 focal individuals) or under (n = 57 focal individuals) water. Vertical bars are standard errors. Deeper foraging induces a greater proportion of time spent overtly vigilant, which is achieved through more frequent and shorter scan bouts. After data in Guillemain* et al. *(2001).*

scanning rate rather than scan duration has also proven to be less detrimental to foraging efficiency in dabbling ducks, leading to a smaller decrease in food intake rate (Fritz *et al.* 2002).

BOX 5.1. Dabbling ducks possess an efficient filtering tool: how the bill works

Teal and other dabbling ducks sometimes use their bills to strip seeds off plants, to tear pieces off fresh plants, and to pick invertebrates from the ground, vegetation and the water's surface (Olney 1963). They can also pick various large food items directly from the ground, including potatoes *Solanum tuberosum* (Thomas 1981, Baldassarre & Bolen 1984). On rare occasions they even catch flying insects using snapping motions (Sjöberg & Danell 1982). A few dabblers, notably Wigeon *Anas penelope* and *A. americana*, are elaborate grazers of grasses and even take shoots of herbaceous plants one at a time.

However, the main foraging method used by *Anas* species is the filtering of water or soft sediment to retrieve small food particles. The bill of dabbling ducks shows an extraordinary level of specialisation for this purpose: the mandible (lower half of the bill) and the maxilla (upper part) both possess lamellae used for filtration, which are most developed in the proximal part of the bill apparatus. Water and sediment trapped in the bill during feeding are sucked from the tip of the bill, using the tongue as a piston (Kooloos & Zweers 1991). The collected material is then moved through the bill cavity towards the base of the bill, where it is forced laterally and filtered through the lamellae (Figures 5.14 and 5.15). Touch feeding is made possible by abundant mechanoreceptors in the bill and on the tongue (Berkhoudt 1980), where food particles are retained during the filtering process (see also Tremblay & Couture 1986 for a detailed description).

Teal are among the ducks with the highest density of bill lamellae (19 to 20 lamellae per centimetre of bill in its proximal part, as opposed to 11

FIG. 5.14 *Close-up view of a Common Teal head, showing the bill lamellae (the 'comb'-like ridges near the edge of the inside of the bill) used to filter water and soft sediment for food items. The presence of such lamellae once led taxonomists to place dabbling ducks in a 'lamellirostre' taxonomic class (Rogeron 1903, Ternier 1922), together with other filter-feeding birds like flamingos. Photograph by Mathieu Remacle.*

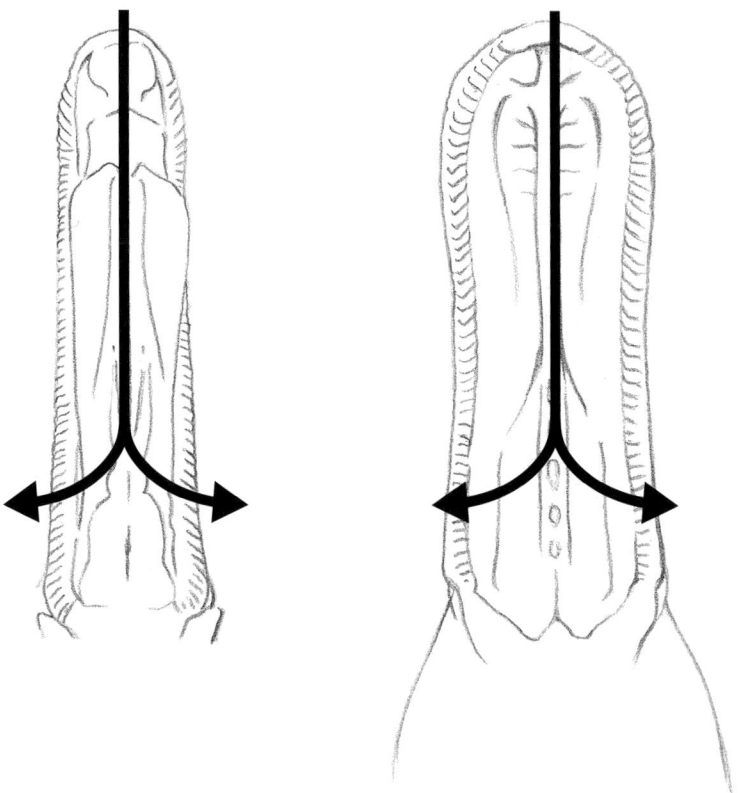

FIG. 5.15 *Schematic functioning of a duck bill, showing water and food item fluxes within the apparatus.*

on average in Mallard, Champagnon *et al.* 2010; see also Nudds *et al.* 1994 for specific values for Green-winged Teal). Despite the ability of most ducks to adjust inter-lamellar spacing by adjusting the relative angle between the mandible and maxilla (Gurd 2005), the high lamellar density of Teal represents a 0.29mm spacing between lamellae for the inner set along the mandible (Thomas 1982). Such high density allows Teal to forage on much smaller prey than, for example, coarser-lamellae Mallard (whose minimum inter-lamellar distance is 0.67mm on average, Thomas 1982). In practice, having denser lamellae provides Teal with a greater prey spectrum: just as a smaller mesh sieve allows the filtration of both small and large particles, the dense lamellae of Teal allow them to potentially forage on prey of very different individual size distributions. The coarser lamellae of the Mallard, conversely, restrict the potential diet of this species to larger prey than that of Teal (see Nudds & Bowlby 1984,

> Guillemain *et al.* 2002d). Accordingly, Thomas (1982) observed large (>1.5mm) seeds to be used to a similar extent by Teal, Mallard, Pintail and Shoveler, while Teal were more likely than Mallard to use smaller seeds. It is worth pointing out, though, that there is a trade-off between lamellar density and the risk of the bill being clogged by very fine particles, so that the species with the coarsest lamellae run the lowest risk of having their lamellae clogged in muddy waters (Tolkamp 1993).

GENERAL DIET AND SEASONAL CHANGES

The list of food items found in Teal digestive tracts is long and diverse, ranging from seeds to plant leaves, and from bryozoans to crustaceans, insects, molluscs and even fish eggs (Appendix 6). Within this extensive array of potential food types, three main features nevertheless emerge:

- a clear switch from seed-eating in late summer, autumn and winter to a diet consisting mostly of invertebrates during spring and early summer;
- the predominance of some seed (Cyperaceae) and invertebrate taxa (chironomids, especially their larvae) over other food types, whenever any of these two food types are available;
- opportunistic foraging: Teal exhibit an extraordinarily diverse diet, yet some individuals occasionally gorge on thousands of items of one food type when these are particularly abundant.

General food habits over the year

Diet studies of dabbling ducks are generally based on the analysis of digestive tract contents. However, these analyses have relied on a variety of collection methods (randomly selected, hunter-killed birds versus a scientifically targeted selection of individuals to be shot), digestive organs (muscular gizzard versus oesophagus contents) and ways of presenting results (per cent occurrence, relative dry/wet weight, relative volume, and so on). Thus, a proper meta-analysis of the many papers published has proven impossible to perform (Dessborn *et al.* 2011b, Box 5.2).

A switch from a seed-dominated diet in late summer, autumn and winter to an invertebrate-oriented diet during the rest of the year has been documented in most studies analysing the food habits of Common and Green-winged Teal (Munro 1949, Spärck 1957, Olney 1963, McGilvrey 1966, Rollo & Bolen 1969, Mazzucchi 1971, Tamisier 1971, Thomas 1982, Hughes & Young 1982, Euliss & Harris 1987, Väänänen & Nummi 2003). This pattern of change is also evident within any particular site: DeRoia (1989) showed that migrant Green-winged Teal staging at Lake Erie in spring were relying almost entirely on invertebrates, while in the same area in autumn seeds accounted for 98–100% of the gut contents.

Such a switch in diet has been related by some authors to changes in Teal energy needs over the annual cycle (Anderson *et al.* 2000), or considered by others to merely reflect changes in the relative availability of the two food types (Euliss & Harris 1987). Arzel *et al.* (2009) were able to assess invertebrate and seed food abundance throughout the year with rigorous protocols and using the same observers, avoiding common biases of flyway-scale meta-analyses. When transformed into energy density (kJ/cm^3), invertebrate food abundance increased from the wintering to the breeding areas (including post-breeding moulting sites), while the opposite was observed for seeds (Figure 5.16). It is possible that energy is not as limiting as one may believe, and that changes in diet merely reflect changes in relative availability and accessibility of seeds versus invertebrates. In any case, the seasonal decline in one food type is concomitant with the increase in the other, so that by changing their diet Teal are able to find relatively abundant food sources throughout the year.

The energy requirements of Teal also change markedly over the annual cycle (see 'Energy requirements' section above). Interestingly, the periods of greater short-term energy needs (this excludes thermoregulation during winter, which is somewhat predictable) are also those when Teal feed on invertebrates the most. First, young ducklings rely almost exclusively on

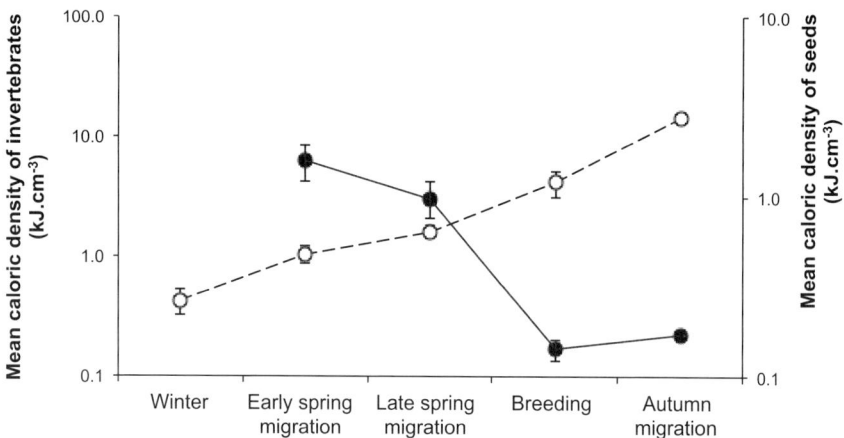

FIG. 5.16 *Mean ± SE calorific density (kJ/cm^3) of invertebrates (open circles and dashed line, left axis) and seeds (filled circles and solid line, right axis) per sampling period. The mean calorific data from the sites where Common Teal were actually present were used: the Camargue for the winter period, Normandy for early spring, Scania and Västerbotten for late spring and Västerbotten for breeding and early autumn (Log-scale). With kind permission from Springer Science+Business Media:* Journal of Ornithology, *'A flyway perspective on food resource abundance in a long-distance migrant, the Eurasian Teal (Anas crecca)', volume 150, 2009, pages 61–73, by C. Arzel, J. Elmberg, M. Guillemain, M. Lepley, F. Bosca, P. Legagneux & J.B. Noguès, Figure 3.*

invertebrates as food, particularly those they can catch on or above the water surface (Munro 1949, Danell & Sjöberg 1980, Krapu & Reinecke 1992). Insects, crustaceans and molluscs seem to be so important for downy young of dabbling ducks that it has been implied that duckling survival probability is dramatically reduced if they cannot forage on invertebrates during their first weeks of life (Sugden 1973, Street 1977).

Adult Teal are also known to rely heavily on invertebrate food during the breeding season and during moult (Danell & Sjöberg 1980, Nummi 1993, Anderson *et al.* 2000, Väänänen & Nummi 2003). It is likely that they do so during spring migration, too, but assessing Teal diet in spring is difficult because this is the period of the year when hunting is banned almost everywhere, so that collecting birds for gut analysis causes ethical concerns. The only published direct evidence comes from DeRoia (1989), who found chironomids to be the exclusive food of many birds examined. Later in the season, Anderson *et al.* (2000) showed that moulting Green-winged Teal consume invertebrates in greater proportion than their availability in the environment; that is, invertebrates are selected by foraging Teal to fuel their protein requirements during the period of feather growth.

Preferred prey – is there any?

Beyond the rather well-described but also simplistic dichotomy between seeds and invertebrate prey, it remains an open question whether Teal are opportunistic or instead select specific food types. Some authors consider Teal to eat more or less what is available in the environment (Mazzucchi 1971, Euliss & Harris 1987, Nummi 1993). Others, however, argue that Teal are selective feeders that disproportionately use some food types. For instance, Rollo & Bolen (1969) reported that Green-winged Teal in the Playa Lakes region of Texas rely on Smartweed (*Polygonum* spp.) seeds to a degree that was 3–5 times higher than their relative abundance among all seeds in the habitat. This is in line with the results of Nudds & Bowlby (1984), who considered Teal to be among the most specialised duck species examined, with a prey size distribution that differed the most from the size distribution in the environment. Similarly, Nummi & Väänänen (2001) found that Teal selected prey that were larger than the mean distribution in the environment.

Methodological inconsistencies between published studies prevent a proper comparison between what is used by the birds and what is available to them in the environment, hence making it difficult to determine if Teal should be considered generalist or specialist feeders (Callicutt *et al.* 2011). The very wide variety of food items found in Teal digestive tracts in a number of studies from widely scattered locations supports the idea that they are able to adjust their diet to what is available in their environment. Compiling all publications on Teal diet that we were able to find, we could list no fewer than 259 different prey taxa recovered in Green-winged Teal digestive tracts, and an astonishing 496 different taxa for Common Teal (Appendix 6). One should not believe, however, that Common Teal have a more diverse diet than Green-wings. Our review results could reflect a publication bias: we found 37 publications presenting detailed Green-winged Teal diets, and 53 concerning Common Teal. This is because Teal research in Europe has long had a more autoecological focus, while studies in North America have often been carried out (or presented) at the level of duck guild, or at least lumping the Teal species (Blue and Green-winged Teal until the end of the 1960s, less frequently thereafter; for example, Martin & Uhler 1939, Glasgow & Bardwell 1962). Such a problem has also been encountered by Callicutt *et al.* (2011), and in both cases this significantly reduces the number of studies that could be included in literature reviews.

The very fact that some (and only some) digestive tracts from the sample in a given area and period may contain huge numbers of one single prey type may be considered as evidence of great diet selectivity. However, such findings may also indicate that those individual birds opportunistically used superabundant prey as they happened to encounter them: Euliss & Harris (1987) report two Green-winged Teal containing 6,053 and 4,925 chironomid larvae each in California, and Tamisier (1971) similarly reports a Common Teal having 3,400 of these prey and another one having 113,000 Stonewort (Characeae) oogones in the oesophagus in the Camargue.

Because Teal use such a variety of habitats, their diet sometimes differs markedly between areas depending on what food there is in the first place; for example, when they winter in tidal areas Teal frequently feed on crustaceans (ostracods: Gaston 1992; amphipods: Munro 1949, Baldwin & Lovvorn 1994). In agricultural lands Teal frequently rely on crops, mainly rice (Tamisier 1971, Llorente *et al.* 1987), corn or wheat *Triticum aestivum* (Sell 1979, Thomas 1981, Baldassarre 1982). However, a common feature of Teal diets is that the bulk of them are often dominated by certain specific taxonomic groups: Cyperaceae seeds (and Characeae oogones to a lower extent) and chironomid insects, mainly the larval forms (Munro 1939, 1949, Coulter 1955, Spärck 1957, Olney 1963, Mazzucchi 1971, Tamisier 1971, Bengtson 1975, Danell & Sjöberg 1980, Hughes & Young 1982, Thomas 1982, Euliss & Harris 1987, Dubowy 1988, DeRoia 1989, Nummi 1993, Brochet *et al.* 2010a). Still, this may or may not indicate selectivity, since Cyperaceae (for example bulrushes *Scirpus* spp., sedges *Carex* spp.) tend to be the most abundant seed-producing plants in many wetland habitats where Teal forage, and chironomids may similarly be among the most abundant invertebrates. To cut a long story short, future studies need to be much more specific about comparing ingested and available food items if we are to understand to what extent Teal are selective feeders.

BOX 5.2. Methods of assessing the diets of Common and Green-winged Teals

A range of methods can be used to assess the diets of dabbling ducks, including Teal (see Sutherland 2004 for a review): direct observation of the birds' behaviour as they feed, faecal analysis, blood isotope analysis, and analysis of the gut contents. The last is still the most common approach by far.

Direct observation of duck behaviour: as dabbling ducks in general feed on very small food items that they collect from sediments, the bottom or shorelines of waterbodies, it is usually futile to determine from a distance which food types they are feeding on. The only exception is when they take seeds directly off plants (Madsen 1988), but this occurs infrequently.

Faecal analysis: examination of faeces is sometimes used to study duck diets, as remains in the droppings can be identified. This approach is most reliable for hard and thick structures, such as mollusc or crustacean shells, or plant stem and leaf remains. With a microscope and a good reference collection of epidermises it is possible to determine which plants were consumed, although some of these were possibly not assimilated (Owen 1975, Madsen 1988). However, vegetative parts of plants are not a common food for Teal, and invertebrates are not consumed to a similar extent as are seeds in autumn and winter, so this method in not particularly well suited to the study of Teal diets. Furthermore, faecal analysis does not provide a precise measure of the relative part of each food type in the diet, simply because digestion rates differ so much between different food types.

Isotopic signature: the components of the diet can also be very broadly categorised from the isotopic composition of some organs, or of blood; the structure of some atoms (called stable isotopes) is variable, and the ratio of the different isotopes can differ between prey types. Plant foods, for instance, differ in their carbon isotope signature (for example, depending on how they use and incorporate carbon atoms in their tissues; that is, C3 versus C4 type metabolism), freshwater prey animals differ somewhat from those in salt water, and so on. Animals build up or renew their tissues from the atoms they acquire by foraging in the environment; hence the isotope ratio in the environment is transferred to some extent into their tissues. Because blood cells are renewed very frequently, isotope ratios in the blood (especially plasma, Hobson & Clark 1993) can provide information about the assimilated diet during the previous couple of days to weeks, as long as reference values are available for the various potential prey types. Isotope composition in more permanent organs (bones, feathers) can reveal information about diet over a longer period, before the bird was sampled, when the sampled tissue was grown or partially replaced. Stable isotope analysis thus allows diet assessment at a relatively broad scale (Hobson *et al.* 1994), which can still be very useful in deducing habitat use and degree of animal matter.

Assessment of digestive tract contents: a reliable and commonly used method of assessing the diets of ducks is analysis of gut contents, even though it is also the most time consuming method and may be subject to a number of biases (see for example Campredon *et al.*'s 1982 manual of duck diet analysis). Ducks do not commonly regurgitate indigestible fragments of their food out of their digestive tract (as raptors do with pellets or seabirds do to feed their chicks), so gut content analysis in ducks has to rely on the dissection of dead individuals. It is often easy to collect dead Teal from hunters, since they are common and highly appreciated gamebirds. However, this may lead to the first of many possible biases: hunter-killed birds may be a non-representative sample of the population, especially if bait is used to increase hunting success. Though it is not practical to do so, researcher-collected birds where sampling follows a specific protocol may therefore be favoured over hunter-killed birds whenever possible (see also Heitmeyer *et al.* 1993). Hunter-killed bird collection is also restricted to some periods of the year (when hunting is open), providing little information about diet during the spring migration and breeding season, for instance.

Dissection of the digestive tract should be performed as quickly as possible after death of the animal to prevent digestion from continuing after the bird is dead. Ideally, the digestive tract should be collected and preserved (either frozen or placed in 70% alcohol) within 10 minutes after death. For this reason, too, hunter-collected birds may not always be the most reliable source of information, since rapid sample preservation is not always possible (Swanson & Bartonek 1970).

Digestion of soft items (smaller invertebrates, for example) is a very fast process, so that differential digestion rate of the various food items leads to an increasingly biased estimation of diet composition as sampling 'moves downstream' within the digestive tract. Most early studies of duck diets were based on gizzard contents, because the gizzard is an easy organ to collect. However, since the critical paper published by Swanson & Bartonek in 1970, it is considered that limiting diet analysis to gizzard contents will provide biased results, because many soft prey items are digested before the bolus reaches this organ, causing an over-representation of hard food items (especially hard seeds). The whole anterior part of the digestive tract, oesophagus, proventriculus and

FIG. 5.17 *Dissection of Teal digestive tract, showing oesophagus, gizzard and unfolded intestines.*

gizzard, should therefore always be collected, and their contents sorted and recorded separately (Figure 5.17).

When extracted, the contents of each part of the digestive tract should be sorted under a binocular magnification glass. Floatability or centrifugation techniques have been tested, but hand-sorting of the contents to separate foods from non-digestible particles (grit in the gizzard, in particular) remains the most common and efficient method of collecting prey items by taxonomic group. Due to excellent reference works (Campredon *et al.* 1982, Cappers *et al.* 2006, Legagneux *et al.* 2007), seeds are generally possible to identify to species level, while partially digested invertebrates can often be identified to the family level at best.

Methods and problems in the presentation of gut-content analysis results: when sorted, the share of each food type in the diet can be presented in several ways:

- frequency of occurrence: proportion of guts containing a given prey type;
- relative number of items: percentage of a given food type out of all prey items in the sample;
- relative weight (wet or dry weight): percentage of a given food type out of the total weight;

- relative volume: percentage of a given food type out of the total volume.

Each of these methods has advantages and pitfalls (see Swanson *et al.* 1974). For example, calculating the frequency of occurrence of a given prey item in a sample of gut contents is often very quick (one only has to detect the presence of these items in the gut contents, without necessarily individually sorting and collecting them), but such information often lacks precision regarding the importance of each type in the diet (frequency of occurrence is the same whether a bird has one or several hundred of a given food type in its digestive tract, while this does not mean the same share of the diet). Conversely, relative weight provides much more precise information, but requires far more work in the laboratory (each food item has to be separated individually from the rest of the bolus).

A major problem in the literature is that because so many different methods exist to analyse duck diet, results from different studies are very rarely comparable, and cannot even be combined in meta-analyses or be used to assess duck selectivity by comparing diet composition to the range of prey available in the environment (Callicutt *et al.* 2011, Dessborn *et al.* 2011b).

The solution: ideally, because published papers now exist that provide reference weights for a wide range of prey types (for example seeds, Arzel *et al.* 2007a), we recommend counting the relative number of each prey type per digestive tract, and providing this rough data together with the frequency of occurrence and relative weight or volume. This will allow compilation and joint analysis of studies in the future.

BOX 5.3. Teal in the dabbling duck community

Local multi-species assemblages ('communities') of dabbling ducks, which may occur in any season, are often presented as textbook examples of ecological niche partitioning, according to which each species, by virtue of its specific morphology and microhabitat preference, may reduce or avoid interspecific competition (Lack 1974). Although they may share a similar body shape, dabbling duck species differ in terms of body length and bill lamellar density, which may limit their foraging depth and ingested prey size spectrum, respectively. That such subtle morphological differences help duck species to coexist through a reduction of competition pressure is not a new idea, as Darwin (1872, page 184) commented on the role of differences in lamellar density within the duck community. To what extent ducks exhibit niche partitioning via foraging depth or prey size has long been debated (Nudds & Bowlby 1984, Nudds *et al.* 1994, Pöysä *et al.* 1994, 1996), but may actually depend on the physical attributes of the waterbody (Nudds *et al.* 2000) and may change through time owing to gradual food depletion (see Box 5.4).

Apart from ways by which dabbling ducks may cope with competition, the very existence of competitive exclusion within duck communities is debated. The fact that breeding dabbling duck communities have more species in more productive (hence, invertebrate-rich) lakes would seem to indicate that interspecific competition is otherwise a limiting factor (Elmberg *et al.* 1993). However, Elmberg *et al.* (1997) were unable to demonstrate actual competitive exclusion between breeding Mallard and Teal, since the experimental addition to breeding lakes of wing-clipped Mallard that were not able to leave the wetland did not lead to any change in density of wild breeding Teal (Figure 5.18).

It is therefore a widely held notion that interspecific competition is very limited during the breeding season, and this is believed to be possible because of the great abundance of breeding sites and of food within those sites (Pöysä 1984a,b, Nudds *et al.* 1994, Nummi & Väänänen 2001). For instance, Bethke & Nudds (1993) concluded that fluctuations in environmental conditions (for example drought) played a greater role in explaining the structure of breeding dabbling duck communities in North America than did competition.

Teal join or form mixed-species flocks very easily: 'Teal, particularly when in small parties, have no objection to associating closely with the larger surface-feeding ducks' (Phillips 1923). This is, however, not to say that interspecific competition is not occurring in subtle ways within these flocks. Overt aggression between Teal and other ducks, or even among Teal, is only rarely observed (see for example Zwarts 1976). Conversely, avoidance behaviour has been recorded (Pöysä 1986a), which may lead to individuals not interacting directly with each other but to some eventually being constrained to less profitable parts of feeding sites. For instance, Pöysä (1986a) observed increasing numbers of individuals using poorer (deeper) foraging areas in larger groups.

An alternative explanation for this pattern may be that passive interference through prey disturbance may also be at play among foraging groups of ducks, with the swimming and straining activity of increasing numbers of birds leading invertebrate prey to retreat downwards within the water column (Pöysä 1985b, 1986a). However, the Common Teal foraging niche changed little between situations in which they fed alone or in close proximity to other dabblers (Pöysä 1986b).

When Common Teal rely on seeds in winter, Guillemain & Fritz (2002) did not observe their foraging depth to be affected by the current density

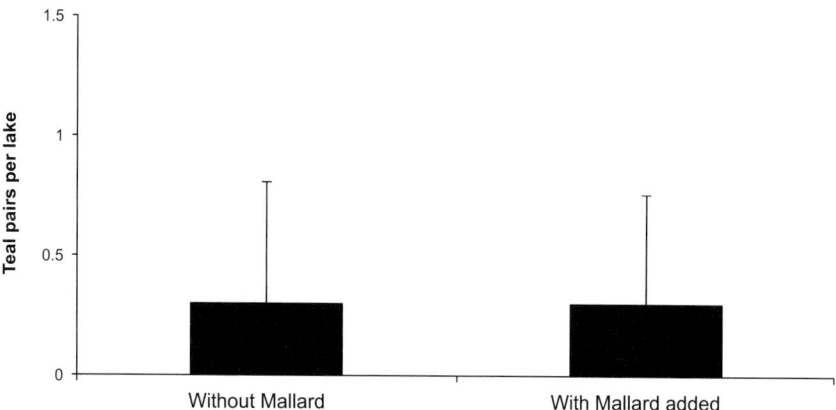

FIG. 5.18 *Mean number of breeding Common Teal pairs per lake depending on whether wing-clipped Mallard were previously experimentally introduced to the lakes or not. Sample size = 32 lakes in each case. Vertical bars show standard deviation. This experiment failed to demonstrate competitive exclusion of Teal by Mallard during the breeding season. After data in Elmberg et al. (1997).*

of foraging granivorous ducks, suggesting no despotic behaviour by which some individuals would overtly force subordinates to less profitable areas in deeper water. The seed diet of Teal, however, changes in response to depletion (see Box 5.4), suggesting that interspecific competition is more likely during winter, while it appeared to be non-existent during the breeding season (see also DuBowy 1988 for fluctuating levels of competition among ducks between seasons).

To summarise, interspecific competition may not prevent Teal from using specific sites, but at smaller scales in time and space interactions with other duck species it may affect Teal behaviour and local distribution within sites.

BOX 5.4. Teal and Mallard use different strategies when facing food depletion

Two major constraints that foraging animals face are the need to access sufficient food sources, while at the same time avoiding being prey for predators.

To avoid being killed, dabbling ducks tend to forage as much as they can with their eyes above the water's surface, so that they can monitor their environment while feeding and hence maintain anti-predator vigilance (Pöysä 1985b, 1989, Guillemain & Fritz 2002).

Laboratory experiments have demonstrated greater food-intake rates (quantity of food ingested per unit of time) when Teal were provided with larger seeds than when offered smaller ones. In a situation where seed sizes were mixed, Teal were further shown to be able to discriminate between seeds on the basis of their size and morphology (Van Eerden & Munsterman 1997).

Common Teal and Mallard wintering in western France preferentially foraged in shallow areas (less than 8–9cm deep) during early winter, where they selected larger food items (3–4.5mm-long seeds; Guillemain *et al.* 2002d). Shared preference for seeds of similar size is consistent with model predictions based on anatomical functioning of the duck bill in experimental situations; when seeds are not mixed with too much sediment or detritus, selecting larger seeds is the most profitable strategy for any dabbling duck (Gurd 2007).

Guillemain *et al.* (2002d), however, showed that when food gets depleted, Common Teal switch to much smaller prey; that is, the use of seeds less than 1.5mm long increases significantly, something they can afford to do while foraging in shallow areas due to their dense bill lamellae. This lets them experience lower predation risk at the expense of reduced foraging efficiency. Because Mallard have coarser lamellae, they cannot rely on the same strategy when facing decreasing food availability, and they are constrained to continue feeding on larger seeds. Mallard then use their longer necks to gradually increase foraging depth to reach larger seeds, this time at the expense of their safety since they cannot keep their eyes above the water's surface to remain vigilant for predators when foraging (Figure 5.19).

Interspecific competition should therefore be considered as a dynamic process in dabbling ducks, not only between seasons (see DuBowy 1988), but also within a given season, as demonstrated here. This can explain the discrepancies among earlier studies regarding food partitioning via prey size in dabbling duck communities. For example, abundant food in Finland during early autumn led Väänänen & Nummi (2003) to find no correlation between lamellar density of the different duck species and the mean length of invertebrate or plant food (all ducks could select the same food size since these prey items were very abundant). Conversely, Nudds & Bowlby (1984) found that Green-winged Teal selected specific prey sizes, leading them to state:

> Our results do not agree with those of Thomas (1982) who concluded that prey size selection is relatively unimportant in resource partitioning among coexisting dabbling duck species [...] Nor do we agree with Swanson *et al.* (1979) who concluded that the feeding niches of dabbling ducks are difficult to distinguish. Clearly, dabbling ducks do partition food resources

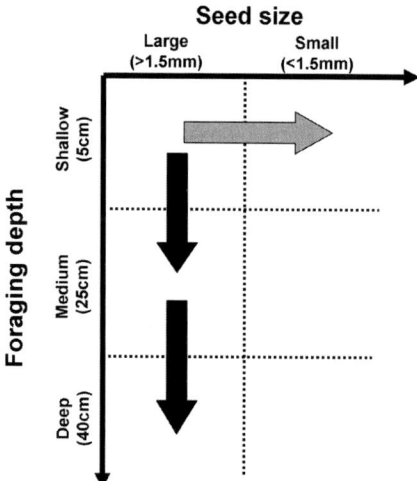

FIG. 5.19 *Divergent foraging strategies of Teal and Mallard. Both forage for large seeds at shallow depth when food is abundant, then gradually switch to alternative and diverging strategies when food starts getting depleted. Teal gradually reduce seed size (grey arrows), while Mallard gradually increase foraging depth (black arrows). Reprinted from Guillemain* et al. *(2002d), with kind permission from John Wiley & Sons, Inc.*

according to prey size, and the differences in prey sizes utilized are distinct.

Nudds *et al.* (2000) later observed ducks that partitioned food either through prey size (in North America) or through foraging depth (in Europe). Clearly, the trophic status of wetlands in which dabbling duck diet studies are carried out (oligotrophic versus eutrophic, which is associated with overall food productivity) will yield different results regarding the strength of competition between species and the way ducks adjust to this constraint.

BOX 5.5. Grit and lead poisoning

Like many other birds, ducks ingest small stones (mean diameter *c.* 1mm in Teal) which, when in the gizzard, aid digestion by taking part in the mechanical crushing of recently ingested food items. Such stones, called 'grit' or 'gastroliths', are particularly important to species feeding on hard or thick-walled food such as seeds, tubers and fibrous vegetative parts of plants. Bird species feeding frequently on these types of food generally have a greater grit mass in their gizzards than birds feeding on soft invertebrates (Figuerola *et al.* 2005). DeRoia (1989) found that Green-winged Teal had more grit in their gizzards when their diet were composed mainly of snails rather than chironomids, which suggests that individual birds could adjust their grit volumes to match their diets. In the Camargue, Tamisier (1971) showed that grit mass could represent more than 25% of the gizzard content dry weight during some periods of winter, although on average only 0.1% of the bird's total weight. Some gastroliths pass down the digestive system and have to be replaced all the time. Grit is thus sometimes sought after by the birds in some specific places (for example gritting locations of geese in Spain; see Mateo *et al.* 2000), and in any case should not be regarded as a mere by-product of feeding, incidentally ingested by the birds as they forage for proper food items.

Grit size is considered to be related to the size of sediment particles available to the ducks more than to their diets (Figuerola *et al.* 2005). Such opportunistic use of available sediment causes a major problem: when spent lead pellets from hunting cartridges are numerous in the sediment of waterbodies, birds often ingest these as grit. Gradual erosion of these pellets in the gizzard allows lead to pass into the blood circulation system, causing lead poisoning. The latter may be direct, or alternatively appear long after initial ingestion, when lead that has accumulated in fat tissues is metabolised and released into the blood (Box 7.1). In any case, lead poisoning generally causes mortality within the short to medium term even when small quantities of pellets have been ingested.

BOX 5.6. Teal give wings to plants and to flightless invertebrates

'Zoochory' is the animal-mediated dispersal of propagules; that is, seeds, rhizomes or animals (cysts, eggs or live animals). Most animal groups are suitable dispersal vectors, including invertebrates (insects, Maguire 1963), fish (Pollux *et al.* 2006), amphibians (Bohonak & Whiteman 1999), mammals (Vanschoenwinkel *et al.* 2008) and birds (Green & Figuerola 2005). Recent research confirms that waterbirds appear to be especially important vectors of propagules (Figuerola & Green 2002, Green *et al.* 2002), but this was anticipated and partly demonstrated more than 110 years ago among the earliest ecologists, notably plant sociologists attempting to understand colonisation patterns (Hesselman 1897, see also Darwin 1859, page 385).

Teal are likely to be ideal vectors for many plants and aquatic animals because they are so mobile within their wintering range, are able to fly long distances in limited time (Clausen *et al.* 2002), are restricted to wetland habitats and generally forage on very small prey items (Clausen *et al.* 2002, Soons *et al.* 2008, van Leeuwen *et al.* 2012).

Field studies in the recently revived research in zoochory have documented both external transport of propagules ('ectozoochory' or 'epizoochory') stuck onto the Teal's body or attached to feathers, and internal transport of ingested propagules ('endozoochory') that went through the digestive tract of Teal and were still viable when excreted. Intact plant seeds or stonewort *Chara* spp. oogones were found on the outsides of 18% of live Common Teal (external transport) and in 20% of Teal digestive tracts analysed (Brochet *et al.* 2010a). Similarly, 2.5% of Teal had live Cladocera or Ostracoda eggs in the posterior part of the gut, and 10.3% of birds had live invertebrate eggs in the plumage or on the legs or bill (Brochet *et al.* 2010b). Although a few individuals contained tens of live undigested seeds in the last part of the digestive tract, most birds had only a few live propagules in the gut, and generally not more than one or two in the plumag. Considering that such propagules may fall off in inhospitable habitats when Teal are flying, and may not produce adult individuals themselves later on, the probability for a given Teal individual to carry a live propagule at a given time is low. However, considering the abundance of these ducks and their frequent movements both within (between their day-roost and nocturnal foraging habitats) and among distinct geographic areas, Teal could be among the most important animal vectors for some plants and invertebrates.

FIG. 5.20 *Teal often forage in muddy environments, and may then become dispersers (vectors) of seeds, eggs and other propagules stuck on their bill or feet. Photograph by Danièle Delvart.*

CHAPTER 6

Breeding ecology

PAIR FORMATION

Mating system

Several review papers document the diverse mating systems of dabbling ducks, where Teal is shown to exemplify a typical serially monogamous duck. This means that a new pair bond is formed each year, several months before the actual breeding takes place near the female's previous breeding area or birth place (McKinney 1985, Rohwer & Anderson 1988, Oring & Sayler 1992, Baldassarre & Bolen 1994).

At least some of these features may seem relatively counter-intuitive: first, it would appear more advantageous for monogamous birds to reunite with the same partner over successive years whenever possible (Rowley 1983, Black 1996). Since this is not the case in Teal, it may seem odd that they form annual pair bonds so long before the breeding season. Second, males of several bird species do show some degree of philopatry, so the fact that only the female's decision matters as to the choice of breeding area in Teal (and other ducks) seems to be an exception in birds.

However, most aspects of the Teal mating system can readily be explained by its life-history traits: first, only the female incubates and tends the precocial ducklings. This relieves males from parental care duties, and the lack of a defended territory also means that males gain little benefit, especially compared to females, by returning to breed in an area they know. This can contribute to the explanation of why only females are philopatric (Oring & Sayler 1992).

Females pay a heavy price in being the only parent incubating the eggs, at least in terms of reduced foraging opportunities and increased predation risk during the breeding season (see below). This promotes philopatry in this sex, since successful females are more likely to come back to breed in a place they know from direct experience is safe and food-rich. Female mortality on the nest also causes a bias in the sex ratio: while sex ratio at birth is even (Bellrose *et al.* 1961, Oring & Sayler 1992, Blums & Mednis 1996), predation of incubating females leads to the adult sex ratio being strongly biased towards males (58% on average for Common Teal in the Camargue, 64% in Britain, Guillemain *et al.* 2009a). Such bias limits the opportunities for males to engage in polygyny: there is a strong competition between males to acquire and secure a mate, and the pair bond has to be reinforced and defended for a long time. This leaves few opportunities to successfully court other females. Paired males will nonetheless attempt forced extra-pair copulations on the breeding grounds when their social mate has started incubation (Baldassarre & Bolen 1994, see next section).

Due to sexual competition between males for females (the limiting sex) the former will try to form pairs as early as possible (Jonsson & Gardarsson 2001). This is also beneficial to (and hence promoted by) the females (Rohwer & Anderson 1988, Oring & Sayler 1992): early pairing in winter makes males guard their mates and protect them from harassment by other males (Ashcroft 1976, Hepp & Hair 1984). Successful pairing also provides a dominant status yielding greater foraging opportunities (Paulus 1983, Hepp & Hair 1984; paired Green-winged Teal won 83% of aggressive interactions with unpaired individuals according to Johnson & Rohwer 1998), and may also allow females to test the quality of their mate, over an extended period. The fact that female waterfowl benefit from being paired during winter, long before the actual breeding season, is also confirmed by the report that widowed Black Brant *Branta bernicla nigricans* have lower survival rates than paired females, probably because this affected their foraging, and increased predation risk and other forms of mortality (ring-recovery rate was actually higher, suggesting greater hunting harvest; Nicolai *et al.* 2012).

Rodway (2007), however, suggests that the mating system of waterfowl should not only be considered under the 'male cost hypothesis', by which females promote pairing as early as males are able. This author suggests there are also costs to females and benefits to males of being paired early, in addition to securing a partner, so that a 'mutual choice hypothesis' would be worth investigating. For instance, Nakamura & Atsumi (2000) showed that male Pintails benefitted from being paired by having longer feeding times than their single counterparts, which spent most of their time courting females.

Although most features of the Teal mating system described above are well established (see for example pairing chronology below), the statement that the same Teal do not remain mated for more than a year remains speculative. The short life expectancy of Teal reduces the probability that the same two individuals will both be alive from one winter to the next (see Chapter 8). Because it has long been considered that dabbling ducks show little fidelity to their winter quarters (Robertson & Cooke 1999), it also seemed unlikely that surviving Teal mates would be able to find each other again anyway. However, a recent study with nasal-saddled individuals demonstrated high fidelity of

Common Teal to their winter quarters (57% of the birds still alive coming back to the same site from one winter season to the next, Guillemain *et al.* 2009c). Though it remains to be shown that both members of a pair come back to the same site at approximately the same date, there is the potential that mates could find each other again on the wintering grounds. Thus, the hypothesis that the same two mates never reunite from one year to the next has never been tested in Teal due to the lack of adequate marking studies allowing individual identification during winter (a limitation already acknowledged by McKinney 1985). However, recapture of the same ringed Gadwall pair over successive winters has been described by Fedynich & Godfrey (1989). Furthermore, Mitchell (1997) demonstrated a very high remating rate in Eurasian Wigeon: from a mere 12 pairs where both members were colour-ringed, the mates of 5 pairs came back to the same place the next year, and as many as 3 pairs re-formed. The mates did not arrive at the same date, suggesting that members of the same pair migrated separately from the breeding grounds but were still able to reunite.

Because of their life-history traits (biased sex ratio, new pairs formed each year, short life expectancy), it is unlikely that all Teal (especially males) have the same likelihood of breeding in a given year. Because each individual can only expect to breed a limited number of times, owing to a short life span, the Teal is therefore among the species in which lifetime reproductive success (Newton 1989) should be expected to vary significantly among individuals, with some contributing far more to overall recruitment (and hence the gene pool) than others. This may have profound consequences for resilience of Teal populations to all sources of mortality, including harvest (see Chapter 8).

Mate selection

Although forced copulations by males exist in Teal as in other dabbling ducks (see next section), the greater investment in reproduction by females suggests that females should select males, rather than the opposite.

Mate selection is a crucial choice in female ducks, given the fact that they alone carry the costs associated with incubation and duckling attendance. That the choice of partner is important to female ducks is illustrated by the observation of a lower breeding success when they are experimentally paired with a male they had not chosen: Canvasback *Aythya valisineria* females subsequently laid no eggs (Bluhm 1985), while survival of hatched Mallard ducklings to fledging was lower than when allowed a choice of mates (Bluhm & Gowaty 2004). Mallard females experimentally widowed during winter were able to form a new pair, but had a poorer body condition and often produced fewer viable eggs, either because of the cost of re-forming a new pair late in the season or because the new male was of poorer quality (Lercel *et al.* 1999). For example, such second-choice males may have been less efficient at guarding their mates and allowing females to build up energy stores to form eggs and complete incubation. Hario *et al.* (2002) also recorded lower fecundity in widowed Common Eider *Somateria mollissima* females. Limited information exists about the mate selection process in Teal, mainly owing to a limited number of aviary studies and large-scale marking schemes in the wild. It is, however, likely that the criteria used by female Teal are similar to those described in other ducks, especially Mallard.

Pairing in ducks generally only occurs after the males have engaged in courtship behaviour for some time. Such courtship is only initiated by males when they have moulted from eclipse (basic) to breeding (alternate) plumage during autumn. Courtship displays by males largely serve to expose their bright plumage to females (see for example Sheppard *et al.* 2013 for wild Mallard; see also next section). Such secondary sexual characters may be used as a criterion by females to evaluate the quality of prospective partners under the handicap hypothesis: wearing a conspicuous bright plumage is associated with higher predation risk in waterfowl (Promislow *et al.* 1994), and iridescent feathers are less hydrophobic, hence costly for males to wear (Eliason & Shawkey 2011). Female waterfowl may hence be able to judge male quality from such signals; the ability to survive with these handicaps implies high male quality and fitness. Choosing a male with gaudy plumage could potentially enhance the fitness of a female, as her sons are likely to be chosen in the future if the bright plumage and high fitness of their father are heritable traits (Fisher's 1930 'run-away selection' principle; see also the 'sexy son hypothesis' by Weatherhead & Robertson 1979). Like many other male waterfowl, male Teal, too, possess a very intricate and gaudy plumage.

Characteristics other than plumage may also be used by female ducks to select their mates: male Mallard with brighter bills were more likely to obtain mates in aviary experiments (Omland 1996), and female Mallard invested more in their eggs (through greater egg mass and lysozyme deposition, which increases embryo antibacterial defence) when males were supplemented with carotenoid, and consequently had more brightly coloured bills (Giraudeau *et al.* 2010b). Mallard males with the brightest bills have actually been demonstrated to have semen with a greater bacteria-killing ability, which could further explain why females may use bill colour to select their mate (Rowe *et al.* 2011). Male Teal have a dull-coloured bill, so the same pattern may not occur in this species, but this illustrates how secondary sexual characters can be used by female dabbling ducks during their mate-selection process.

Female ducks may also select a mate on the basis of the intensity of his courtship activity and dominance rank during social encounters with other males. For example, female Mallard preferred individuals with a high social display activity and a high plumage status (Holmberg *et al.* 1989), as well as those that courted most actively in autumn during aviary experiments (Davis 2002b). Hepp (1989) also suggested that Black Duck females preferentially selected dominant males. Brodsky *et al.* (1988) even showed experimentally that when presented with male Mallard and Black Ducks, females of these species preferred the most dominant males, irrespective of the males' species. Preferring the most dominant and more actively courting males may again be a means by which females select males of best individual quality: in American

Wigeon, Wishart (1983) suggested that males initiate courtship only after they have reached a threshold level of body condition. The greater proportion of adult compared to juvenile Eurasian Wigeon being paired by the end of winter may in that sense reflect the fact that adults generally are in a better condition than juveniles, so that the latter engage in winter pairing behaviour to a lesser extent (Amat 1990). It is therefore not surprising that assortative pairing by age and body condition is common in dabbling ducks (Heitmeyer 1995). Again, individual studies based on marked animals' behaviour would be welcome to assess mate-selection processes specifically in Teal.

Timing of pair formation

Courtship and subsequent pair formation occur over an extended period during the non-breeding season, with differences in timing both between and within duck species (Baldassarre & Bolen 1994).

Courtship behaviour is considered to be under endocrine control (especially testosterone plasma concentration), hormone concentrations themselves being linked with changes in photoperiod (Bluhm 1988). Testosterone plasma concentration in male Common Teal increased from 34.0mg/1 in January to 747.3mg/1 in June when full spermatogenesis had occurred (Jallageas *et al.* 1978). In addition to changes in hormone concentrations, the period of increasing courtship activity is also the period of testis growth in males (Dervieux & Tamisier 1979).

Teal generally initiate courtship behaviour and form pairs later than most other dabbling ducks (Bezzel 1959). In Europe, Common Teal courtship starts in October (even late September in Iceland: Jonsson & Gardarsson 2001), increasing in frequency until it reaches maximum intensity in March (Bezzel 1959). Similarly, Lebret (1961) observed courtship to be common from November onwards in the Netherlands. As judged from Grunt-whistle activity in captivity, Dervieux & Tamisier (1979) also considered the main courtship activity to start in mid-November and to occur until May–June, with a peak in February. More recent observations based on scan sampling of wild Common Teal wintering in the Camargue confirm this pattern, although the peak courtship activity occurred in January (Figure 6.1). However, because there are communal swims (groups of birds where numerous males display around one or a few females) before and after this period, which can be considered as low-intensity courtship to some extent, pair-formation displays were considered by Dervieux & Tamisier (1979) to occur in *c.* 10 months of the year; that is, throughout the time when males wear their breeding plumage. Consistent with this idea, Lebret (1961) documented the first courtship displays as occurring as early as 10 August.

Courting apparently occurs later in Green-winged Teal, where the proportion of birds engaged in this activity during behavioural observations was, for example, 0% in November, 0.7% in December, 5.4% in January and 1.2% in February in North Carolina (Hepp & Hair 1983). Similarly, Johnson

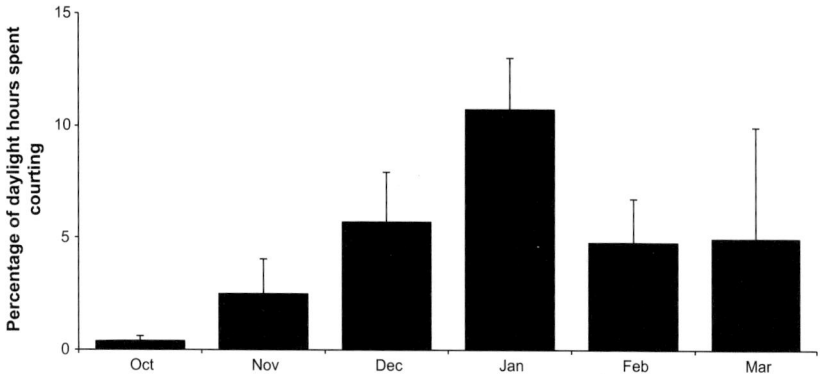

FIG. 6.1 *Percentage of Common Teal engaged in courtship activity during scan samples performed in the Camargue during winters 2002–2003 to 2004–2005. Percentages were calculated per scan sample (performed hourly, one day per week from October to March), then averaged per day and subsequently per month. Each column is the mean of the three annual values for each month. Vertical bars show standard errors. Guillemain et al., unpublished data.*

& Rohwer (1998) consider that Green-winged Teal do not start courtship before the end of December, while the main courting period was January to March in Louisiana (Rave & Baldassarre 1989, Gaston & Nasci 1994), and February–March in Texas (Quinlan & Baldassarre 1984). On the breeding grounds courtship activity decreases, especially from the period of pre-rapid follicular growth to incubation (Paquette & Ankney 1998).

Teal do gradually form pairs through the winter. At that time, paired dabbling ducks can easily be told from single ones by the close proximity of mates and their coordinated behaviour, especially in terms of movement and social interactions with other birds (Paulus 1983). However, we do not know if the birds observed in pairs during winter are necessarily those that will actually breed with each other the next spring.

Common Teal are generally observed to form pairs later than other European dabbling ducks (Figure 6.2), though Lebret (1961) recorded the earliest Teal pair on 24 October and considered Teal pairs to be 'numerous' from November onwards. The share of paired females in the Camargue was around 5% in early October, which takes us back to the question of whether some pairs reunite as soon as both pair members arrive at the wintering ground, or even whether some individuals (for example pairs that failed in reproduction) migrate in pairs to wintering areas. The percentage of paired females in the Camargue then increases exponentially throughout the winter, with half being paired by late February (Guillemain *et al.* 2007a). In the Netherlands, Bezzel (1959) observed the percentage of paired females to increase from a similar *c*. 45% in January to *c*. 95% in March. Pairing apparently occurs later in Iceland, where only 70% of females were paired in March (Jonsson & Gardarsson 2001), but pairing then quickly increased to 90% in April and reached 100% in May.

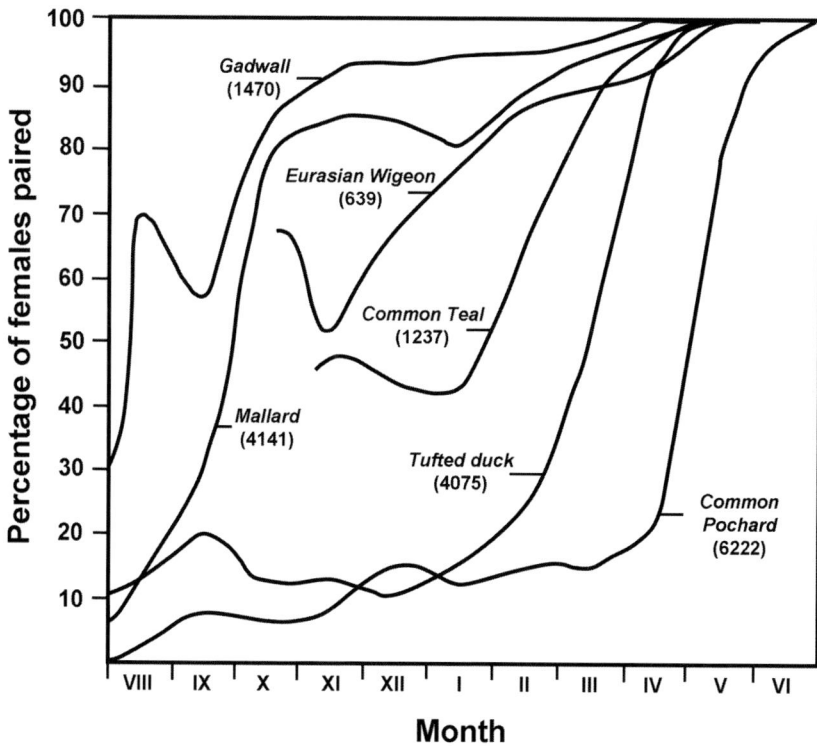

FIG. 6.2 *Annual variation in percentage of female ducks being paired at Ismaning lakes, Germany. Numbers in brackets are the total number of females examined (over several years). From Bezzel (1959), with kind permission from the Ornithologischer Anzeiger.*

Consistent with the fact that courtship is initiated later in Green-winged than in Common Teal, American birds paired later in the season than European ones. Pairs of Green-winged Teal are generally not observed before January, but the percentage of paired females then reaches 75–80% in March, just as for Common Teal in Europe (Rohwer & Anderson 1988, Baldassarre & Bolen 1994, Johnson & Rohwer 1998). McKinney (1965) considered virtually all Green-winged Teal to be paired when he studied them in mid-March in Louisiana. Hepp & Hair (1983) report the percentage of paired females to increase from 0% in December to 31.7% in January, then 79.5% in February. However, as in Europe, these authors also document a few Green-winged Teal pairs much earlier in the season (2% in November). As for Common Teal, these can be hypothesised to be experienced mates reuniting with each other for a new breeding season, although this needs to be ascertained with marked (for example nasal-saddled) individuals.

When pairs have formed, copulations (those consented to by the females, not forced within-pair copulations) are regularly observed during winter, long before the actual breeding season (Figure 6.3, see colour section). For example, in North Carolina within-pair copulations were recorded as early as January (Hepp & Hair 1983), while the first of these were observed in the Camargue in February (M. Lepley & M. Guillemain, unpublished data). Such early copulations apparently have a social rather than reproductive role: they may be used for pair-bond maintenance or for mates to test each other, but cannot lead to sperm transfer since spermatogenesis is not occurring that early in the season (cf. McKinney *et al.* 1983 for Mallard), and the penis is not fully developed.

The evolutionary factors leading to later courtship and pairing in Teal compared to other duck species has been the subject of much scientific debate. The main proximate factor put forward has often been that male Teal moult to their breeding plumage later than other ducks, and this is a prerequisite to initiating courtship (Hepp & Hair 1983, Bluhm 1988). This would be consistent with the fact that older (second calendar year or older) dabbling ducks pair earlier than first-year birds (Wishart 1983, Heitmeyer 1995), since older ducks generally moult into breeding plumage earlier than the latter. It has also been suggested that late-pairing duck species are those that breed the latest, but this hypothesis is not unequivocally supported by the data: Teal are amongst the last of the dabbling duck species to pair, but not the last ones to breed; they actually spend less time paired before the onset of the nesting season than do other dabblers (Rohwer & Anderson 1988).

It has also been hypothesised that pairing date depends on the potential benefits birds would gain from being paired. Accordingly, herbivores such as Gadwall should pair earlier because improved dominance linked with paired status would increase foraging opportunities, which is especially crucial owing to their low-energy diets (Rohwer & Anderson 1988). Similarly, Lovvorn (1989) suggested that Canvasbacks pair late because there is little benefit for them to maintain pair bonds in winter when food is generally neither limiting nor defensible (hence no benefit exists for dominant pairs). However, such hypotheses do not help to explain why, within the same guild, Teal form pairs several months later than Mallard, for example.

The current consensus is that courtship initiation is energetically constrained, which determines courtship phenology both within and among ducks: courtship displays are energetically costly to perform, and courtship is a mutually exclusive activity to foraging, so that only individuals with reduced energy requirements or those which have stored enough body reserves could start trying to acquire a mate (Wishart 1983). This would be consistent with the clear pattern of later courtship and pair formation in smaller species, which have greater relative energy requirements (Baldassarre & Bolen 1986, 1994, Rohwer & Anderson 1988). This explanation also matches observations that Common Teal wintering in Iceland (with greater energy requirements for thermoregulation) pair somewhat later than their conspecifics in continental Europe (Jonsson & Gardarsson 2001). Finally, in Black Ducks, the hypothesis of an energy-constrained courtship initiation is supported by both field and aviary studies: wild males begin courtship earlier, when temperatures are still lower, in food-rich areas (Brodsky & Weatherhead 1985). Those experimentally fed *ad libitum* paired earlier than their conspecifics on a restricted diet (Hepp 1986).

Courtship behaviour

Teal have been the subject of several studies of courtship behaviour, which has therefore been described in detail both from field and aviary observations (Lorenz 1952, Johnsgard 1960, 1965, McKinney 1965, Dervieux & Tamisier 1979, McKinney & Stolen 1982, Baldassarre & Bolen 1994).

In dabbling ducks, the courtship repertoire is broadest in those species that court on water (likely because predation risk restrains the repertoire in species courting on land), and is likely to be promoted by sexual selection since there is a correlation between repertoire size and sexual plumage dimorphism among species (Johnson 2000). Teal are sexually dimorphic and court on water, and hence rely on a wide array of courtship displays. Such displays are associated with very clear-cut postures and behaviours. These are also highly ritualised, most often performed one after the other, following the same sequence, by groups of males around one or several females.

The most common and obvious displays of the males are called 'Grunt-whistle', 'Head-up-Tail-up', and 'Down up' (Figures 6.4 and 6.5, see colour section). Various types of bill display ('Bill up', 'Bill dip'), shakes of the body or the wings, nod-swimming and ritualised preening also occur within a courtship sequence (details in McKinney 1965, from which most of the information presented below was also taken).

The 'Grunt-whistle' behaviour consists of the male raising the anterior part of its body while keeping the bill lowered (hence forming 'an arc', McKinney 1965). Typically, water droplets are then splashed in the air (presumably towards the female) (Figure 6.4 top, see colour section). During such behaviour males obviously show their iridescent head plumage to the female, but this may also serve to expose the small tuft they have at the back of the head (Figure 6.4, see colour section). The Grunt-whistle is considered the most common and characteristic display of courtship repertoire in male Teal (Johnsgard 1965, Dervieux & Tamisier 1979). A loud call is given by the male while performing the display.

'Down-up' and 'Head-up-Tail-up' are most likely used to expose the yellow lateral undertail patches of the male to the female (see also Lorenz 1952). In both cases the posterior part of the body is raised out of the water, also causing the wings to become slightly erected and open, simultaneously exposing the metallic green wing-patches (Figure 6.5, see colour section). In the Head-up-Tail-up display the head is also moved upwards. The Head-up-Tail-up behaviour is performed while the body of the male is parallel to that of the female. After this display is completed, the male invariably turns towards the female and gives a soft call. The most typical display sequence of male Teal is Grunt-whistle followed by Head-up-Tail-up, then a turn towards the female (Johnsgard 1965, McKinney 1965).

Ritualised preening (also termed 'mock preening') of the back or behind the wing is commonly practised by Teal and other dabbling ducks. This probably helps the males to show the metallic parts of their plumage (head, wing-flash marks) (Johnsgard 1960). Because the plumage of males is largely similar between Common and Green-winged Teal, it is not surprising that the display repertoires of the two species are similar, although slight differences in the relative frequency of some displays may be observed (for example nod-swim, by which ducks lower the head and swim around their partner, occurring often in Green-winged Teal but rarely in Common Teal: Laurie-Ahlberg & McKinney 1979).

In addition to displays while swimming, male Teal also perform 'jump flights' of a few metres to join females (Lebret 1958).

Female displays are less conspicuous, but may rely more on sounds and on subtle changes in posture relative to the male position. The most obvious display is 'inciting' by which the female shows 'highly ritualized, sideways threatening movements' of the head (McKinney 1965). Females are often courted by several males simultaneously, and inciting is probably used to show a male that it has been selected (by the female directing the threatening movements towards the other males). Nod-swimming is also performed by females.

Pre- and post-copulatory displays are performed by both males and females. The most common precopulatory display is head pumping, often until synchrony by the two mates is reached, followed by the female stretching out her body so as to lie on the water's surface, probably to indicate consent to copulate. As in other dabbling ducks, post-copulatory movements consist of circular swimming and flapping of the wings by the male, and mostly ritualised preening by the female. In both sexes these latter behaviours may be used to put feathers back in place after copulation, to ensure waterproofing. Postcopulatory displays are also hypothesised to act in pair-bond strengthening or to be used by the female to signal successful copulation (Johnson *et al.* 2000). The intricate anatomy of the female reproductive tract (see below) may prevent some apparently successful copulations from actually ending in sperm transmission.

Forced copulations

Male waterfowl are peculiar in that they have a proper intromittent copulatory organ (that is, a penis), while in most other birds insemination is via simple cloacal contact only (Brennan *et al.* 2007; Figure 6.6, see colour section). Such an intromittent organ may help improve sperm transmission for species copulating on the water, as waterfowl do as a rule. The organ is also considered to be involved in sperm competition between males, since its physical development is more advanced in species where forced extra-pair copulations are more frequent (Coker *et al.* 2002). Another feature of waterfowl mating systems, including that of Teal, is that forced copulations by males

FIG. 6.7. *Forced copulation is a common reproductive strategy of Teal males. Photograph by Jérôme Beziat.*

occur frequently (McKinney *et al.* 1983; Figure 6.7). In Mallard, males with more testosterone engaged in forced extra-pair copulations (FEPC) more frequently (Davis 2002a). In Teal, FEPC are generally initiated by paired males on the breeding grounds, when their mates are engaged in incubation, and targeted at females about to lay eggs or already in the egg-laying period. FEPC are therefore a secondary reproductive strategy of the males, besides social monogamy, even if such matings apparently have a poorer success with respect to subsequent second-clutch fecundity (but not in terms of sperm deposition) than copulation between pair members (McKinney & Evarts 1998; see also Dunn *et al.* 1999 for the same pattern in Ross's and Lesser Snow Geese, *Anser rossii* and *Chen caerulescens caerulescens*). In line with FEPC attempts being more common when females are the most fertile, Goodburn (1984) showed that male Mallard guard their mates (presumably to avoid FEPC) more closely when they are more fertile, during around seven days prior to egg-laying.

Females could theoretically benefit from forced copulations by males other than their social mate, because such cryptic polyandry may promote sperm competition. Surprisingly, however, 'no female of any waterfowl species has ever been observed to solicit extra-pair copulations' (McKinney & Evarts 1998). Moreover, female waterfowl are considered to always resist FEPC (Cunningham 2003). Overt resistance to FEPC may be a way for females to maintain the confidence of their social mate in their own paternity (hence leading them to maintain their mate-guarding until egg-laying is complete and incubation has started), while still benefitting from inseminations by extra males to some extent. Furthermore, McKinney & Evarts (1998) suggested that the resistance by the female may invoke attention from other males, leading to further FEPC attempts and hence increased sperm competition. At least within the pair, FEPC often leads to immediate forced within-pair copulation, likely initiated by the social mate to increase its likelihood of paternity, probably increasing sperm competition (McKinney & Stolen 1982).

However, resistance by Teal females may not be feigned, but rather be a way for them to avoid the costs of forced copulations: McKinney & Stolen (1982) described observations of female Teal dying from the stress or woundings of forced copulations in captivity. McKinney & Evarts (1998) similarly reviewed reports that up to 10% female mortality resulted from injuries linked with FEPC, though this was based on small sample sizes from aviary experiments. Another likely risk associated with multiple copulations with multiple partners may be that of sexual disease transmission, which should be especially likely given that the males have an intromittent organ (Cunningham 2003). There is, however, some controversy over the real costs of resisting FEPC in female ducks compared to the potential genetic benefit they may gain by promoting sperm competition, especially since most injury and mortality events described above were observed under captive conditions. Under such circumstances, females may have been less able to avoid harassment by males, so that in nature the real costs of FEPC may be lower compared to possible benefits (Adler 2010). Some benefits to females may be indirect, through better quality of their offspring if father quality (for example dominance status) is correlated with propensity to force extra-pair copulations. However, FEPC leads to infrequent fertilisations compared to copulations among pair members (McKinney & Evarts 1998), and offspring resulting from FEPC may be of poorer phenotypic quality, while the costs of FEPC to females are likely to be considerable (McKinney & Stolen 1982). Brennan & Prum (2012) hence suggested that female waterfowl genuinely resist FEPC to avoid fertilisation by non-preferred males rather than to promote sperm competition. Female choice (mate preference) and resistance should hence be considered as very distinct phenomena, the elaborate genitalia of female waterfowl, with several dead ends, being the ultimate way for them to resist extra-pair fertilisations. Clearly, more behavioural studies are required to better understand the underlying mechanisms of forced copulations in Teal and other waterfowl. Specifically, the relative proportion of young resulting from FEPC instead of being produced by their social father would deserve a proper assessment by genetic paternity analyses in order to better understand the importance of this behaviour to reproductive success (hence, the strength of sexual selection) in individuals of both sexes, but also at the population level.

From nest establishment to hatching of ducklings

Teal are very secretive during the nesting period, and they often nest at very low density and rely more on small forested wetlands than most other dabbling ducks (Petrula 1994). This makes Teal particularly difficult to study during this period, because locating nests is often a tedious and time-consuming activity in difficult habitats in remote locations. Phillips (1923) already noted for Green-winged Teal: 'It is very apparent from the small number of nests found, even in the great breeding areas, that the nests are better concealed and harder to find that those of the Mallard, Widgeon and Gadwall.'

As a consequence, several of the estimates or mean values provided below (for example clutch size) come from a limited number of incubating females or nests studied, and values are therefore less robust than similar data available for other duck species. For instance, the Green-winged Teal monograph in the *Birds of North America* series (Johnson 1995) only provides two estimates for clutch size, from relatively old studies both based on small sample sizes (Munro 1949, Keith 1961). Conversely, single studies providing estimates for clutch size from thousands of other dabbling duck nests are not uncommon for prairie-nesting species (Higgins *et al.* 1992).

When ducklings have hatched, the detectability of Teal broods increases considerably, so that more robust estimates of brood-rearing success are available, both for Green-winged (Toft *et al.* 1984) and Common Teal (especially so from Fennoscandia; see for example Elmberg *et al.* 2003).

Pre-nesting period

Teal arrive very early at their breeding grounds, often long before the habitats actually become suitable for nest establishment. In northern Europe, Common Teal often concentrate in coastal areas and open estuarine or riverine habitats for several weeks upon arrival, before the breeding wetlands become ice-free (Olsson & Wiklund 1999, Arzel & Elmberg 2004, Arzel *et al.* 2014) (Figure 6.8; see also the habitat section above).

A similar process has been observed in North Dakota (Higgins *et al.* 1992), Alaska (Petrula 1994), sub-Arctic Canada (Toft *et al.* 1982) and Iceland (Bengtson 1972), where Teal reached their breeding areas and waited up to three weeks before nesting was possible. Nest initiation (and even egg-laying) often occurs while the breeding lakes are not yet completely ice-free (Danell & Sjöberg 1977, see also the three references above). Elmberg *et al.* (2005) found an explanation for this pattern while evaluating long-term nesting habitat selection processes and local breeding success in Finland. earlier-settling Common Teal bred on the lakes that were the most productive on average over the long term (the best habitats) and produced more ducklings than did late-settling pairs. The cost of having to remain in unfavourable

FIG. 6.8. *Pre-breeding Common Teal in a mixed-species flock in an estuarine habitat in the boreal region (Västerbotten, Sweden). Teal arrive early in spring and stay in such areas until ice thaws on the breeding lakes located more inland. Photograph by Kjell Sjöberg*

conditions for several weeks before being able to initiate breeding would thus be counterbalanced by greater expected breeding success for early-arriving Teal that can start nesting as soon as local conditions allow.

Very little is known about time use, local movements and social interactions in Teal just prior to nesting, and especially how these factors affect subsequent breeding success. As mentioned earlier, judging by their small size, Teal are very likely to be income breeders, a fact that makes their pre-breeding activities especially relevant for research.

Nest establishment

Most female ducks are strongly philopatric, coming back to breed in the same area year after year, often very close to where they hatched (Blums *et al.* 1996). Philopatry rates of Teal in general (other than in *A. crecca* and *A. carolinensis*) seem to be much lower than in other duck species; for example, 0 to 14% in Blue-winged Teal and 1.2% in Garganey (review in Anderson *et al.* 1992), whereas philopatry was close to 100% in Northern Shoveler, Common Pochard and Tufted Duck (Blums *et al.* 1996). Unfortunately, the difficulties in finding incubating females and their nests have so far hampered a better understanding of philopatry in female Green-winged and Common Teal, for which no proper estimate of philopatry rate has to our knowledge ever been published. The only available information comes from unpublished data of F. McKinney in Johnson (1995), documenting that only one of two marked females and none of 11 marked males returned to the same breeding area in Alberta from one year to the next.

Female Teal are able to breed at one year of age (Bellrose 1976, Cramp &

FIG. 6.9. *Common Teal nest and eggs concealed in vegetation, Iceland. Photograph by Johannes B. Jonsson.*

Simmons 1977), but not all yearlings do so (Dement'ev & Gladkov 1952). On average, 7.1% of Teal females, all age classes combined, were considered to skip reproduction at Lake Mývatn, Iceland, by Bengtson (1972).

Although nests only 1m apart from each other have been documented (Cramp & Simmons 1977), nest density is typically low in Teal. Mean published values are 0.4 Common Teal pairs per km of shoreline over 12 years in Finland (Elmberg *et al.* 2003), one Green-winged Teal clutch per 243ha in pothole habitat (Drewien 1968, cited by Nudds & Ankney 1982), and 0.3 nests per 40.5ha of native prairie and 0.4 per 40.5ha of seeded grassland in North Dakota by Higgins *et al.* (1992).

Avoidance of extra-pair copulations (EPC) during the pre-laying and laying period, discussed earlier, could explain why female Teal would tend to space out from each other during these periods. However, such EPC also exist in other duck species whose nesting density is higher than in Teal. A long-term study of 52 lakes in Finland found no correlation between the number of broods or ducklings produced per pair and the number of pairs within the same area over 12 years, but did report that Teal production per pair decreased when food availability per pair decreased (Elmberg *et al.* 2003). In other words, there was no effect of pair density per se, but there was a density-dependent reduction in productivity when per capita food abundance decreased. Once again, the small body size of Teal and associated high energy costs could be responsible for this response, as the observed low nest densities could be a way to ensure sufficient food availability to incubating females and ducklings. Of course, there are alternative hypotheses to explain low nesting density, such as, for example, predator avoidance. Moreover, if the annual bottleneck regulating population size is away from the breeding grounds, there could simply not be enough Teal to fill all suitable breeding wetlands, and birds space out in a habitat selection process based more on habitat quality than on (intraspecific) competition.

Teal always build their nests on the ground and concealed under dense vegetation, with 50–85% plant cover directly above (Fox 1986), making them extremely difficult to detect. Keith (1961) measured percentage light penetration of nest cover, and observed that only 32% of the diffuse light would reach the floor of a Green-winged Teal nest, which was the lowest value (hence greatest concealment) of the 12 dabbling and diving duck species studied. Only the female Teal builds the nest: a bowl of *c.* 20cm diameter and 6cm depth is created with whichever plant material is available in the habitat (Ogilvie 1975, Cramp & Simmons 1977). Down is added to the nest by females, but only when the clutch is complete and incubation has started (Johnson 1995). Females also use down to cover their clutch when they leave the nest to forage. Nests of Common Teal are often placed under willow *Salix* spp. shrubs and in sedge tussocks, from tundra to temperate woodland, whilst breeders in the taiga often select the cover of wide conifer branches close to the ground (Dement'ev & Gladkov 1952). In forested landscapes, Teal nests are often built under windfallen trees (Sotnikov 1999), creating dense groundcover which is relatively difficult for predators to penetrate. As opposed to many other dabbling ducks, Teal favour natural habitats for building their nests. For example, no Green-winged Teal nest was ever found in tilled croplands of Woodworth Reserve, North Dakota, over 16 years of annual nest searches (Higgins *et al.* 1992), despite the species breeding in other habitats in the reserve.

Based on present and imperfect knowledge, Teal nests are invariably located close to water, rarely more than 100m from the nearest shoreline (Munro 1949, Fox 1986; see also the habitat section above). Shutler *et al.* (1998) demonstrated a significant positive relationship between duck body size and distance of the nest from water (though Green-winged Teal was not part of their dataset), which they hypothesised reflected the fact that females of smaller species would have smaller body reserves and would hence have to leave their nests more frequently to forage during the incubation period. Smaller eggs also cool faster then larger ones due to a less favourable volume/surface area relationship, so females of smaller duck species should try to minimise the time they spend off the nest. Under such circumstances placing nests closer to the water's edge would reduce the energy cost of female movement and associated predation risk. This would also limit the length of time that clutches remain unattended. Keith (1961) observed that Green-winged Teal placed their nests very close to water (less than 20m on average), but the species in the area that nested closest to water was Mallard (16.5m on average, probably because some of these birds nested on islands), thus lending little support to the above size-related hypothesis.

Egg-laying

Total clutch volume is generally limited by energy available to the female, so that there is a general negative relationship between egg size and clutch size among birds (Lack 1968). This type of resource-limitation hypothesis of clutch size determinants has been supported by recent meta-analyses incorporating phylogenetic corrections (Figuerola & Green 2006). The authors of these also demonstrated that egg mass is positively correlated with the body mass of the egg-laying female, while clutch size is not related to female mass despite the general correlation between egg mass and clutch size mentioned above. As a consequence, Teal have relatively small eggs, but relatively large clutches. Although clutch size may also be limited by specific reserve types rather than just energy (*lipid limitation hypothesis* versus *protein limitation hypothesis*; see Ankney & Alisauskas 1991, Ankney *et al.* 1991), egg-laying in Teal is in any case extremely costly to females. Green-winged Teal females did not produce larger clutches than normal when some eggs were experimentally removed from nests during the egg-laying period (Rohwer 1984), suggesting that clutch size is not only limited by energy available from the environment, but also by female reserves or deliberately by the female itself. This could be because extremely large broods cannot be adequately brooded and are hence not more productive (Dzus & Clark 1997). Another theory, termed the *viability-predation hypothesis*, proposes that an upper limit to clutch size is set because the egg-laying period otherwise becomes too long for very large clutches, increasing clutch predation risk and decreasing embryo viability of the first laid eggs (as onset of full incubation is delayed; see for example Loos & Rohwer 2004).

The average weight of a Teal egg is 27–29g (Alisauskas & Ankney 1992, Figuerola & Green 2006). A full Teal clutch (see description below) would represent 78% of the average female body mass (Lack 1968). This ratio is even more dramatic if expressed in terms of energy rather than mass: one Teal egg contains 62kcal, which is more than twice the daily energy needs of a Teal female (1 egg = 202% of Basal Metabolic Rate, which is among the greatest ratios computed for any dabbling duck; see Alisauskas and Ankney 1992). To meet the nutrient demands of egg-laying while relying mainly on ingested food to form eggs, female Teal lay one egg per day.

The energy cost of egg-laying has been studied experimentally in dabbling ducks: Mallard fed a poorer diet had reduced clutch size, smaller eggs and fewer mean nesting attempts per season, and therefore laid a smaller total number of eggs (Eldridge & Krapu 1988). Mallard experimentally fed a richer diet laid larger eggs (Pehrsson 1991). That egg-laying is energy costly to Green-winged Teal females has been clearly demonstrated in the field by Paquette & Ankney (1998). Comparing the time budgets of paired males and females, these authors showed that both members of a pair had similar time budgets during the period preceding rapid follicular growth (that is, before egg formation in the female oviduct) and that paired females greatly increased their time spent foraging during rapid follicular growth, when energy was needed for the eggs, while the time budget of males did not change (Figure 6.10). The fact that a non-negligible fraction of Common Teal females (7.1% at Lake Mývatn, Iceland, in Bengtson 1972) migrated to the breeding grounds but were subsequently not shown to nest may be related to the very high energy cost of egg production and incubation in a food-limited (or thermally demanding) environment.

Egg-laying starts in May to mid-June in most Teal populations (both Common and Green-winged Teal), with the period around 20 May often being cited as when egg-laying begins, for example in North Dakota (Higgins *et al.* 1992), Alaska (Toft *et al.* 1982), Iceland (Bengtson 1972), Finland (Väisänen 1974), western Russia (Kirov area, Sotnikov 1999), central Russia (Blinova & Blinov 1999), Amur River, eastern Russia (Babenko 2000) and Sakhalin Island (Nechaev 1991). Within this general period, however, egg-laying often occurs earlier in the south-western part of the species' breeding range, as for example between 28 April and 23 May in Wales (Fox 1986), then is later as one progresses from central Europe (mid-April) to southern Finland (early–mid May) and then Russia (late May to early June) (review in Cramp & Simmons 1977).

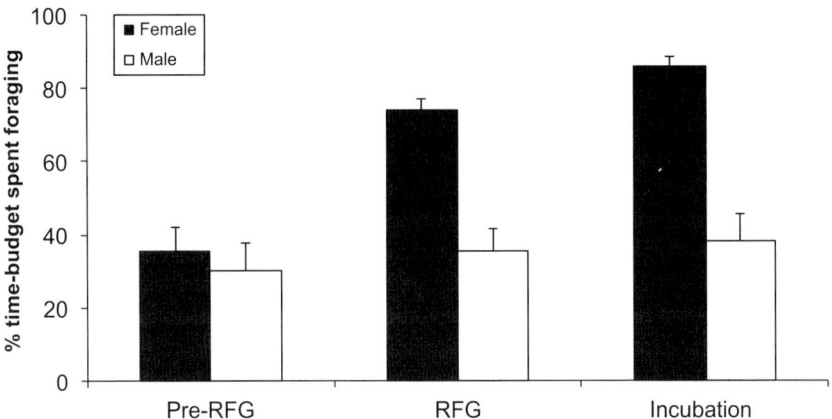

FIG. 6.10 *Percentage of time spent foraging by the two members of Green-winged Teal pairs during pre-Rapid Follicular Growth (pre-RFG, at least a week before laying), Rapid Follicular Growth (RFG, the week before laying) and incubation. While foraging remains fairly similar through the three stages of the breeding process in males, females drastically increased foraging time after pre-RFG, owing to greater energy costs for egg production (RFG) and incubation, and because of their inability to store body reserves ahead of these periods due to their small size. Vertical bars show standard errors, n = 12 in each case. Based on data in Paquette and Ankney (1998).*

Female Teal lay one egg per day on average (Sotnikov 1999). It has often been thought that the timing of egg-laying is under strong selection so that the two first weeks of ducklings' lives, when they have greater energy needs, are synchronised with peak emergence of their main insect prey (Danell & Sjöberg 1977). This was, however, not observed in the only subsequent study that directly tested this hypothesis (Dessborn et al. 2009): first, the duckling hatching peaks differed among the three species of dabbling duck studied (Mallard earliest, Wigeon second and Common Teal last), spanning two weeks. Second, insect-emergence peaks differed greatly in timing and strength among lakes in the same watershed. Finally, none of the species hatched their ducklings during the first and most prominent watershed-level burst of spring insect emergence. The hatching peak in Teal rather occurred during a period with unusually low abundance of emerging insects. Based on these findings, Dessborn and her co-workers argued that emergence patterns of midges on typical boreal lakes are neither compressed nor predictable enough to be a major selective force on the timing of egg-laying and hatching in dabbling ducks.

In many waterbird species, including ducks, the female frequently lays its eggs in the nest of another female, supposedly a strategy to decrease its parental costs while still ensuring transfer of its own genes to the next generation. Such nest parasitism was found to be almost non-existent in Common Teal when specifically studied in Iceland, where less than 1% of nests were parasitised by a female from another species, and no intraspecific nest parasitism attempt was recorded (Bengtson 1972, see also Phillips 1923). Nest parasitism has not been documented in Green-winged Teal according to Eadie et al. (1988), thus providing no support for Munro's (1949) assertion that clutches of 10 to 13 eggs could have been 'the product of more than one family'. This statement is, however, very similar to that of Dement'ev & Gladkov (1952), who considered that very large clutches up to 15 eggs could exist although these 'may be due to two females laying in one nest after one had lost her own'. Genetic analyses based on a significant number of nests will be difficult to carry out but are necessary to determine if Teal do practise nest parasitism.

Normal Teal clutches are generally composed of 8–12 eggs, with reported mean full clutch values varying from 7.5 to 10 (Keith 1961, Bent 1962, Linkola 1962, Lack 1968, Bengtson 1972, Newton & Campbell 1975, Bellrose 1976, Fox 1986, Petrula 1994, Nikolaev 1998, Gerasimov et al. 2000, Figuerola & Green 2006). It should be pointed out once again, though, that Teal have been far less studied than other duck species during the nesting period, so that many of the above values are derived from very small sample sizes, for instance only two clutch sizes reported in Johnson (1995), one from Keith (1961) and one from Munro (1949).

There seems to be a general trend for decreasing clutch size as the breeding season progresses, with mean full Common Teal clutches in Iceland being composed of 10.3 eggs if laid during the first part of the season, and late clutches being of 9.6 eggs. Re-nest clutches were even smaller, having only 8.5 eggs on average (Bengtson 1972). Exactly the same pattern has been described for Green-winged Teal in North Dakota, where mean clutch size decreases by 19% over the season (from *c.* 10 eggs in early May to *c.* 8 eggs in early July; Higgins *et al.* 1992).

Teal eggs are oval-shaped, and cream to pale green in colour (Figure 6.11). The mean size of Russian Common Teal eggs from various collections was 45.6 x 31.6mm (Dement'ev & Gladkov 1952), which is comparable to the mean value for Green-winged Teal provided by Bent (1962) of 45.8 x 34.2mm and by Petrula (1994) of 44.5 x 32.2mm. There is, however, some significant variation around those means: 46.6 x 33.7mm in Johnson (1995) in North America, 40–48 x 31–34mm in Russia in Gerasimov *et al.* (2000), 45.8 x 34mm in eastern Russia in Babenko (2000), and 45.6 x 33.5mm on Sakhalin Island in Nechaev (1991).

Similarly, mean egg mass has been given as 26.5g for Green-winged Teal (Appendix 15 in Lack 1968), 29g for Common Teal in Figuerola & Green (2006), 29.1g in the Amur area, Russia, in Babenko (2000), but only 26.1g for other Russian areas in Sotnikov (1999).

FIG. 6.11 *Common Teal egg (right) beside a Mallard egg (left). The white scale bar is 1cm. Note that the Teal egg was mistakenly labelled* 'Anas querquedula' *(that is, Garganey), while its elongated shape is clearly that of a Common Teal egg, so* 'Sarcelle d'hiver' *is correct. Collection Museum of Nîmes, France. Photograph by Matthieu Guillemain.*

Incubation

The first eggs of the clutch receive progressively more heat from the mother as she lays subsequent ones, which has been proposed to play an important role in survival of the embryos before proper incubation is initiated (Afton & Paulus 1992). Time spent on the nest by egg-laying dabbling duck females, including Green-winged Teal, increases as laying progresses, from 2–3 hours per day when the second egg is laid to to *c.* 13–16 hours per day when the tenth egg is laid (Loos & Rohwer 2004). However, duck breeders often take the eggs from captive Teal nests as these are laid, and replace them with small chicken eggs until the female starts to incubate, when the original clutch is put back under the mother. Such practice is believed to improve hatching success and to decrease the probability of malformations in captivity (J. B. Mouronval, pers. comm.). Under such circumstances the eggs are regularly turned by the farmer but do not receive any heating, suggesting that temporary warming when the female lays the next eggs is not absolutely necessary. In nature, however, this may allow females to both increase the viability of the first laid eggs and decrease clutch predation risk by their presence on the nest (see below), hence allowing them to increase clutch size following the viability-predation hypotheses described above (Loos & Rohwer 2004).

In terms of time, incubation becomes the main activity of the hens after the last egg is laid and the clutch is complete. This behaviour is generally considered to cause all eggs to receive the same amount of heat, starting at the same date (mean egg air cell temperature measured 37.9°C by Afton 1978), which would allow simultaneous hatching of all ducklings. This is particularly important in precocial species like Teal: all ducklings leave the nest shortly following hatching, and they must be able to do so in order to follow their mother and start foraging. This paradigm has, however, been challenged by the results of Loos & Rohwer (2004), which document enough incubation of the first eggs during the laying period for the embryos of these to have started development while the last eggs of the clutch are not yet laid. This raises the question of how hatching can be synchronised when all eggs have not received heat for development for a similar period of time. Firstly, ducklings soon to hatch have been shown to be able to speed up their development in order to synchronise hatching of the brood (Nilsson & Persson 2004). However, this has fitness costs in terms of later muscular and organ growth. Secondly, Eichholz & Towery (2010) demonstrated that the last eggs laid are more often in the centre than in the periphery of the clutch during incubation, hence receiving more heat and developing more quickly, enabling late-stage embryos to catch up with those in the first laid eggs. Whether the incubating hen is able to recognise individual eggs and place the last laid ones in the centre of the nest, or if those eggs end up in the centre of the clutch for other – likely physical – reasons remains to be established.

For several reasons, female physical presence on the clutch decreases the risk of nest predation: for example, brooding females offer camouflage, and they may adopt distraction behaviours against approaching predators and hence reduce the probability that the nest will be found (see below).

Females of larger duck species (Mallard, Northern Shoveler) store body reserves ahead of the incubation period (MacCluskie & Sedinger 2000), so as to maximise the time spent incubating. Once again, smaller birds like Teal may not be able to store as many reserves, even in relative terms (Gloutney & Clark 1991), which means that they have to interrupt incubation more often to forage and meet their own energy requirements (such periods are called 'recesses'). Recesses are most likely to occur around 04:00, when a clear peak is often recorded, and recess probability increases gradually from 16:00 to 20:00. Between 08:00 and 14:00, the probability of presence on the nest was estimated to be 79.8% for Green-winged Teal, which was much lower than for the other dabbling duck species studied (97.3% in Mallard, 96.3% in Gadwall, 94.8% in Northern Shoveler, 89.8% in Blue-winged Teal; Gloutney *et al.* 1993).

Studies thus far suggest that female Teal leave the nest three to four times per day, with foraging being the main activity during recesses (Afton 1978, Paquette & Ankney 1998, Figure 6.12). Precise measurements from one Green-winged Teal nest showed that the female spent 297 minutes off-nest

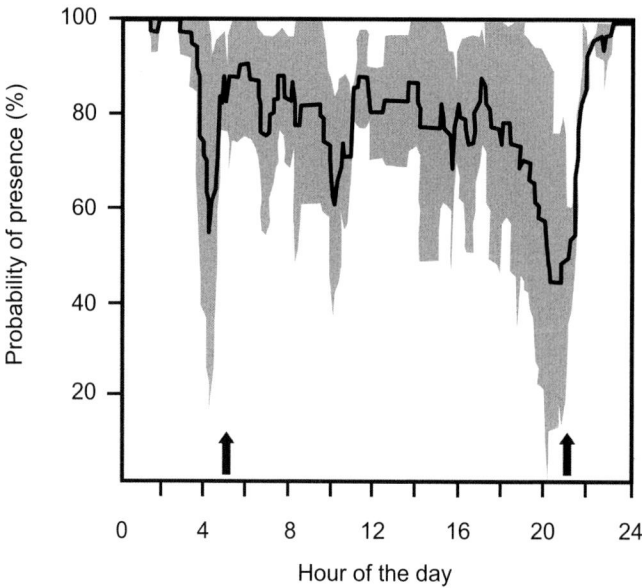

FIG. 6.12 *Probability of presence on the nest across the day for four Green-winged Teal females. The dark curve indicates the mean value, the grey area shows 2SE. The two arrows show sunrise and sunset. Redrawn from Gloutney* et al. *(1993)* Journal of Wildlife Management, *with kind permission from Wiley & Sons, Inc.*

per day, during which she spent 65% of the time feeding, equating to 193 minutes of feeding per day (Afton & Paulus 1992). This is the highest value for all ducks studied by these authors, and Afton (1978) cited Semenov-Tyan-Shanski & Bragin (1969) for similar measurements in Common Teal (note, though, that sample size was only one nest in both cases). The results presented above imply high energy demands during incubation for income-breeding species like Teal: females lose 20% of their body mass on average during the incubation period (Afton & Paulus 1992).

Incubating females typically produce some sounds (quacks) as they fly when flushed from the nest, so as to catch any predator's attention (Munro 1949). This is obviously risky for them, so that incubation and nest defence represent a trade-off between survival probability of the female and that of her embryos. Short-lived birds like Teal should theoretically be more risk-prone and hence defend their nest more than longer-lived species in which more breeding attempts can be expected. This is what Dassow et al. (2012) found: most Green-winged Teal only left the nest when the observer was within 1m. Nest defence also increased as incubation progressed (in evolutionary terms, because a date nearer hatching should mean greater hatching probability; see also Albrecht & Klvaňa 2004 for Mallard). Munro (1949, page 170) also noted that females in a more advanced stage of incubation flushed at a shorter distance.

Dabbling duck hens pay a high survival cost during incubation. For example, Sargeant et al. (1984) estimated that Red Foxes *Vulpes vulpes* take 900,000 ducks per year in mid-continental North America, most of these being dabbling ducks and the vast majority being females. In an earlier study, the lead author examined dabbling duck remains at Red Fox dens in North Dakota: while males generally outnumber females in duck populations, an average of 84% of the duck remains were from females, with an average of 75% from Green-winged Teal (Sargeant 1972). Green-winged Teal did not seem to be particularly selected by foxes, since the prevalence of Green-winged Teal remains at fox dens was found to be relative to the proportion of the species in the duck community (Sargeant et al. 1984). Greenwood et al. (1995) also found remains of Green-winged Teal in proportion to the abundance of that species within the breeding duck community, indicating no particular selection by predators or otherwise greater mortality risk for Teal compared to other ducks.

Incubation generally lasts from 21 to 23 days in Teal (Munro 1949, Lack 1968, Cramp & Simmons 1977, Fox 1986), though it may change over the breeding period: in all dabbling ducks examined by Feldheim (1997), the incubation period decreased with nest initiation date, later nests being incubated for shorter periods. Although this pattern was not tested specifically in Green-winged Teal or Common Teal, the Blue-winged Teal incubation period was reported by this author to decrease from an estimated 27 days for nests initiated on 29 April to 21 days for nests initiated on 8 June. A range of potential explanations (including milder temperatures or smaller clutches later in the season) has been proposed to explain this general pattern, but tests of these hypotheses have provided conflicting results so far (review in Feldheim 1997).

Hatching

As opposed to earlier beliefs (see for example Danell & Sjöberg 1977), hatching of Teal ducklings does not appear to be timed to chironomid emergence (which is often very erratic). Rather, the hatching peak in Common Teal was correlated to the number of days lapsed since the ice-out of the breeding wetlands, which varies considerably even in the heart of the boreal biome (Dessborn *et al.* 2009). In Dessborn *et al.*'s study from southern Finland, Teal broods hatched on average *c.* 55 days after the ice-out, which in most years corresponds to the middle of June.

On Amchitka Island (Aleutians, Alaska), newly hatched broods of *Anas crecca nimia* have been observed from 12 June to 31 July (Kenyon 1961). In Russia, the first appearance of Common Teal broods spans from 23 May (observation in the Mologa area, 36°E) to 8 August (confluence of Indigirka and Moma rivers, 144°E; Dement'ev & Gladkov 1952). Hatching in the Ivanovo region occurs around 15 June (Gerasimov *et al.* 2000), while it is mostly between mid-June and 10 July in the West Siberian Plain, further east (Vartapetov 1998). Hatching of Green-winged Teal generally occurs between 20 June and 20 July in Alaska (Toft *et al.* 1982), and between 20 June and 10 August in British Columbia (Munro 1949), as judged from the earliest and latest observations of downy young.

Causes of hatching failure are varied in dabbling ducks, including Teal, although predation of eggs or the incubating hen by birds (corvids, cf. Kalmbach 1937, Munro 1949), snakes (review in Baldassarre & Bolen 2006), and especially mammals (Higgins *et al.* 1992) is considered to be the most important reason (Greenwood *et al.* 1995). In a virtually predator-free prairie environment, Duebbert & Lokemoen (1980) recorded a 100% hatching success of 12 Green-winged Teal nests over 6 years. Although there are still unresolved issues about the scale at which predator removal is effective, dabbling duck nest success is often greater where predators have been removed (Balser *et al.* 1968, Duebbert & Kantrud 1974, Duebbert & Lokemoen 1980, Garrettson & Rohwer 2001, Drever *et al.* 2004, Chodachek & Chamberlain 2006), but still fluctuates over time, probably due to climate effects. Similarly, Mallard nest predation is lower during periods when rodents are more abundant, suggesting that predators switch between these two prey types (Ackerman 2002). Of 33 Green-winged Teal nests monitored in North Dakota that were lost, 30 were destroyed by mammals, of which 8 were identified as Red Fox, 8 as Striped Skunk *Mephitis mephitis*, 1 as ground squirrel *Spermophilus* sp., 1 as Common Raccoon *Procyon lotor* and 1 as American Badger *Taxidea taxus* (Higgins *et al.* 1992). American Mink *Neovison vison* is also often considered to be a major predator of dabbling duck nests, for example of Common Teal in Iceland (Bengtson 1972). Among less frequent dabbling duck predators, Eurasian Jackdaw *Corvus monedula* and Moorhen *Gallinula chloropus* have been documented from Loch Leven, Scotland, by Newton & Campbell (1975).

Among the other main causes of Teal nest failure are abandonment by the female and flooding of nests, especially because nests are generally placed close to the shoreline (Bengtson 1972, see also above). One of the few detailed studies of Teal nests was conducted in Iceland, and showed that 80% of clutches contained at least one unhatched egg; furthermore, among successful clutches an average of 10% of the eggs (one per clutch) did not hatch: 75% of these were non-fertile, 5% contained a dead embryo and 20% simply disappeared. Among clutches that failed completely, 56% were depredated (generally by Common Raven *Corvus corax* or mink), 19% were abandoned and 25% were flooded (Bengtson 1972). A similar study of Green-winged Teal nests in Alaska showed that 21–36% hatched, 50–57% were depredated, 0–14% were abandoned and 0–5% were flooded (Petrula 1994). In the prairie pothole region, Keith (1961) monitored the fate of 21 Green-winged Teal nests, of which 4 were abandoned, 12 were depredated (11 by Striped Skunk, 1 by an unknown predator) and 5 hatched.

A 62% hatching success (number of ducklings under 5 days of age produced compared to the number of eggs counted) was recorded in Wales (Fox 1986). Brood density in Finland (young ducklings) was 0.23 per shoreline km (Nummi & Pöysä 1995a), with 40% of settled pairs producing a brood (Elmberg *et al.* 2003). The corresponding figure was 32% of pairs producing a brood for Green-winged Teal in North Dakota (Higgins *et al.* 1992). At Lower Souris Refuge in the same state, 76.7% of nests hatched at least one egg (M. Hammond unpublished data in Palmer 1976). This is far greater than the mere 31% (32 of 104 nests) compiled by Bellrose (1976) for Green-winged Teal studies, but detection rates are unknown in all of these studies.

Another way of studying hatching success is to compare the mean brood size of downy ducklings over time and between areas (that is, ducklings under 12 days; see Box 6.1), keeping in mind that full clutches generally have 8.5–10 eggs. While clutch size is often difficult to study owing to nest crypsis, Teal broods are comparatively easy to observe when on wetlands, and brood size values have been provided by a variety of authors: 6.2 ducklings on average in British Columbia (Munro 1949), 8.5 in native prairie and 9.5 in seeded cover in North Dakota (Higgins *et al.* 1992), 5.5 to 7.1 in Finland (Nummi & Pöysä 1995a, Elmberg *et al.* 2005), 5.5 in the Canadian Northern Territories (Toft *et al.* 1984), 6.5 to 7 in Russia (Vartapetov 1998, Gerasimov *et al.* 2000) and 6.2 in Wales (Fox 1986). Whether mean brood size varies over the breeding season is still being debated, but Toft *et al.* (1984) suggested that the number of ducklings hatched per clutch decreases by 0.07 per day of delayed hatching. Broods of 10 (Nechaev 1991) or even 12 ducklings have been observed (Babenko 2000), suggesting that complete hatching success sometimes occurs.

Teal produce one brood per year. This is consistent with the great energy and time investment this represents. Following Bengtson's (1972) observations in Iceland, 75% of Common Teal females that lose their first clutch will attempt to nest a second time.

From hatching to fledging

First hours and days

At hatching, Teal ducklings weigh 15 to 17g (Koskimies & Lahti 1964, Smart 1965, Nelson 1993). Laboratory studies suggest that the newly hatched ducklings need 8.22cal per hour per gram of body mass to maintain metabolism and survive (Koskimies & Lahti 1964). As a consequence of such high energy requirements, Teal ducklings are considered to be among the most susceptible of young ducks to cold spells, which may partly explain the low survival rates recorded during their first days of life (see below).

Ducklings do hatch with substantial internal energy reserves in the form of a yolk sac remaining from the former egg. The yolk sac contents are transferred to other body stores (for example the liver) within 24 hours after hatching. Although these reserves do provide energy for the ducklings during their first days of life (Sedinger 1992), the rate of yolk-sac reduction does not depend on whether or not the ducklings can feed; that is, its function is not to buffer potentially adverse conditions at hatching. Ducklings never feed much during their very first day of life, and yolk-sac absorption is instead considered to help initiate the muscular activity of the digestive system and hence promote the initiation of foraging behaviour (Kear 1965).

Dependence on the mother

Common Teal ducklings were considered 'markedly more self-dependent' than Mallard ducklings by Dement'ev & Gladkov (1952). Nevertheless, if the nest is close to the water Teal broods will come back to it at night for several days (Sotnikov 1999), and in any case Teal hens will brood their ducklings at night until the age of about five days (Dement'ev & Gladkov 1952).

In most dabbling duck species the mother leads and takes care of her brood as the ducklings grow. Beard (1964) observed that the bond between dabbling duck hens and their broods broke up only when the ducklings were in age class III, about to begin to fly, although Green-winged Teal were not specifically included in her study. In an aviary experiment with Mallard, Boos *et al.* (2007) recorded a link between levels of prolactin in the circulatory system and parental care: the concentration of prolactin in the mother's blood decreased over time after hatching, which was associated with a gradual decline in parental care until disruption of the female–brood bond around week six, about two weeks before fledging. Unfortunately, no comparable data exist specifically for Teal.

BOX 6.1. Duckling description and ageing of Teal broods

Teal ducklings are precocial: they hatch covered with down, with their eyes open and with developed legs. This gives them good mobility and allows them to leave the nest within a couple of hours.

The general colouration and patterning of down in Teal are similar to those of most other dabbling ducks: the back and flanks are dark brown with yellowish spots, and the breast and belly are yellowish (Figure 6.14, see colour section). Similarly, the top of the head (crown) is dark brown, while the sides and lower parts are pale, though with distinctive facial patterns: a clear dark eye-stripe extends from the base of the bill to the back of the head. A cheek/ear mark is almost always visible though very variable, ranging from only a small spot to a second head-stripe below the eyeline (Figure 6.15). The bill and legs are dark grey. Very detailed descriptions of Teal ducklings are given in Fjeldså (1977) and Nelson (1993).

Although Teal ducklings are coloured and patterned much like the ducklings of the frequently co-occurring Mallard, the much smaller size of the former and the fact that the hen is often close to her brood usually allows an identification to species. Because of their small body size, Teal

FIG. 6.15 *Variability of facial patterns in Common Teal ducklings (illustration courtesy of Jon Fjeldså).*

ducklings also appear to have particularly long legs (Nelson 1993). Common Teal ducklings resemble those of Garganey, but the latter are more blackish on the back (Fjeldså 1977)

Common Teal and Green-winged Teal are illustrated by C. Harrison (1975 and 1978) with the very same drawing. In these as in other references (Ogilvie 1975, Bellrose 1976), the presence of a double eye-stripe is indicated as a distinctive feature of Teal ducklings, being shared only with Pintail ducklings (in which the pale parts of the down are light grey, rather than yellowish as in Teal). *Anas crecca nimia* ducklings are apparently darker than those of *A. carolinensis* (Nelson 1993, plate 6 and page 79). According to Nelson (1993), confusion of *A. carolinensis* ducklings with other North American Teal (Blue-winged and Cinnamon Teal) can be avoided because ducklings of the two latter species are more yellowish, with less contrasting facial patterns, and Cinnamon Teal furthermore have pale legs, while Green-winged Teal duckling legs are dark (cf. plates 6 and 7 in Nelson 1993).

Only the downy ducklings are generally illustrated in identification guides (see, however, photographs of Common Teal ducklings from 14 days to adulthood in Heinroth & Heinroth 1923), but knowledge about duckling development is also important to accurately ages the young birds. This can be useful when monitoring brood numbers (broods can then often be identified from one monitoring session to the next by the number and age of ducklings, and broods of different ages can be told apart even when occurring on the same wetland and not otherwise marked). Duckling age categories are also used to assess brood survival rates; that is, the percentage of broods surviving from duckling age class Ia to III (see main text).

Because plumage grows in a consistent way between individuals and between duck species, a common classification has been proposed to age ducklings using body size and plumage criteria. The most commonly accepted system has seven classes and subclasses, and was published by Gollop & Marshall (1954), then adapted to European species by Pirkola & Högmander (1974), and can be summarised as follows, with age ranges for Teal:

Plumage Class I: *duckling covered in down, no feathers visible*

Subclass Ia (1–4 days):
Small size, rounded body, down clearly patterned and brightly coloured, creating a 'contrasting' experience

Subclass Ib (5–8 days):
Body still rounded, down patterns less distinct

Subclass Ic (9–12 days):
Down patterns hardly distinguishable, faded colours, body more elongated with tail and neck visible

Plumage Class II: *duckling with some feathers and down*

Subclass IIa (13–17 days):
First feathers on sides and tail, but <50% of flanks feathered

Subclass IIb (18–23 days):
More than 50% of flanks feathered but still with large areas of down; first primary remiges visible

Subclass IIc (24–29 days):
Patches of down only visible in a couple of areas: nape, back or rump

> ***Plumage Class III (30–35 days):*** *duckling fully feathered but still flightless*
>
> In short, duckling plumage goes from being very crisp and contrasting as well as brightly coloured to being more worn, bleached and brownish in hue with progressing age. At the same time, body proportions change significantly: the neck and bill both become relatively longer and the body more elongated.

Habitat use by female Teal attending broods

Fox (1986) observed that Common Teal ducklings in Wales were raised close to where their nest was. However, that study was conducted in a peatland complex, with relatively permanent wetlands and homogeneous habitat. In many other situations, though, the best site to place the nest may not be the best place to raise ducklings, and the brood has to be moved by the mother from the nest site to the nearest suitable wetland, often during the first days of life. Although Teal tend to travel shorter distances than other ducks (because Teal nests are often located closer to the shoreline; see previous sections in this chapter), broods have still been recorded 175–560m (mean 260m) from the nest site 24 hours after hatching (Bengtson 1971).

Wetlands used by ducklings during the first days of life are often relatively ephemeral, forcing broods to move yet again as the habitat deteriorates and the ducklings grow. For example, Bengtson (1971) only recorded 5% of Common Teal broods to still be at their initial wetland 30 days after hatching. Such exodus from the hatching site is always a hazardous journey due to predation and other possible mortality risks as ducklings travel on dry land, leading Dement'ev & Gladkov (1952) to state that 'One of most difficult periods in life commences with forced translocation of broods to larger waters as small pools dry up, which causes greater hardship [in Common Teal] than for other river ducks owing to predilection to construct nests at smaller pools. Besides small wetlands potentially drying up, habitat requirements of ducklings may also change as they grow older. However, there is no spatial tracking study addressing habitat shifts of individual broods as the ducklings grow.

A recent Finnish study based on observational brood data collected over the entire brood season until fledging (but not temporally contiguous on the brood level; Suhonen *et al.* 2011) shows that Common Teal broods utilise lakes with less luxuriant vegetation than do Mallard broods. However, this pattern needs to be verified by further studies from a variety of settings before generality can be implied, 'the general wisdom' being that Teal hens with broods preferentially use wetlands with abundant vegetation.

Within specific wetlands Teal ducklings in age class I mostly forage near the shores, especially in shallow flooded areas and within sedge (*Carex*) stands where invertebrate prey is likely to be particularly abundant (Nummi & Pöysä 1995b). Ducklings then gradually increase their use of open water and floating vegetation towards the centre of wetlands as time progresses (age classes II and III; Bengtson 1971, Nummi & Pöysä 1995b; a similar pattern was described for Mallard from behavioural observations by Robinson *et al.* 2002). This gradual change of foraging microhabitat was considered by these authors to reflect changes in diet, but only very limited information on Teal duckling food habits is available so the hypothesis remains little tested.

Duckling diet and foraging behaviour

To our knowledge, the gut contents of no more than 22 Teal ducklings (2 Green-winged Teal and 20 Common Teal) have ever been analysed and published. Of the two downy Green-winged Teal duckling stomachs examined by Munro in British Columbia (1949), one contained 100% and the other 60% insect material, mostly midge pupae. Of the 10 Common Teal duckling oesophagi examined in Iceland by Bengtson (1975), 9 contained adult Diptera (mostly simulids), which were their sole animal food. Plant foods (mostly Cyperaceae (sedge) seeds) were also common: these represented 71% of the total number of food items, but the relative number of items may be a misleading proxy for the actual diet given the difference in weight between a Cyperaceae seed and an insect (see also Box 5.2). Of the 10 Common Teal ducklings examined in Sweden by Danell & Sjöberg (1980), 9 contained midges (Chironomidae) in adult stage, while a range of seeds (chiefly Cyperaceae) was also recorded. When these authors presented their results as relative weights, virtually 100% of the material was made up of invertebrates, while plant material was only present in trace amounts. Although very crude, these results hence confirm how heavily the young of dabbling ducks rely on invertebrate prey during their first weeks of life (see also Sugden 1973 for birds other than Teal).

We are not aware of any study detailing the foraging behaviour (other than foraging habitat use mentioned in the previous section) of Teal ducklings. Data from Mallard ducklings indicate that they initially spend *c.* 50% of their time foraging, but that foraging time decreases as they grow older. During the first 2 weeks post-hatching, 80% of the feeding time was spent catching insects at the water's surface or in the water column, while dabbling for plant food increased later on (Robinson *et al.* 2002). Mallard ducklings in wetlands with more (potentially competing) fish spent more time feeding, but their survival rate was still lower than in fish-free lakes (Hill *et al.* 1987). These authors also found fewer invertebrates in the lakes holding fish, so that the behaviour observed again highlights the importance of food availability for duckling growth and survival (see also Cox *et al.* 1998). That food availability is a factor likely to be limiting duckling survival on many boreal lakes was demonstrated by Gunnarsson *et al.* (2004) with food provision experiments: survival rate of supplementary-fed (Mallard) ducklings was significantly higher. Mortality of control birds may have been due to starvation, greater exposure to predators linked with greater difficulties in meeting daily energy requirements, or a combination of the two processes. Unfortunately, the effect of food abundance on duckling growth rate was not compared between lake types. Again, it should be noted that all the information in this paragraph relates to Mallard ducklings studied in areas where Teal are also common. Thus, a comparison between them is logical, but not verified.

Duckling growth

The growth rate of young waterfowl, inferred from studies both in the wild and in captivity, is given by the equation k = 0.31 x body mass$^{-0.22}$ (Sedinger 1992). The negative exponent indicates that smaller species grow at a faster rate than larger ones. Green-winged and Common Teal ducklings have long been known to grow faster than those of other species in their respective communities (Ogilvie 1975, Bellrose 1976). Teal reach fledging stage at 34–35 days (Bellrose 1976) as compared to 50–60 days in Mallard.

Although based on a limited number of individuals, Veselovský (1952) is the only published reference providing detailed growth measurements for (Common) Teal in captivity, together with similar data for other dabbling ducks. This direct comparison confirms the much faster growth of Teal, which attain fledging mass earlier than ducklings of the other species (Figure 6.16).

Duckling survival

The survival of ducklings is often measured as the percentage of ducklings counted at hatching (or at some stage during the first two weeks after hatching) that reach fledging age, recorded for example by following their radio-tagged mother. This is, however, based on strong assumptions relative to brood cohesion and duckling detectability, and cannot account for losses of hens or entire broods. This method tends to overestimate duckling survival

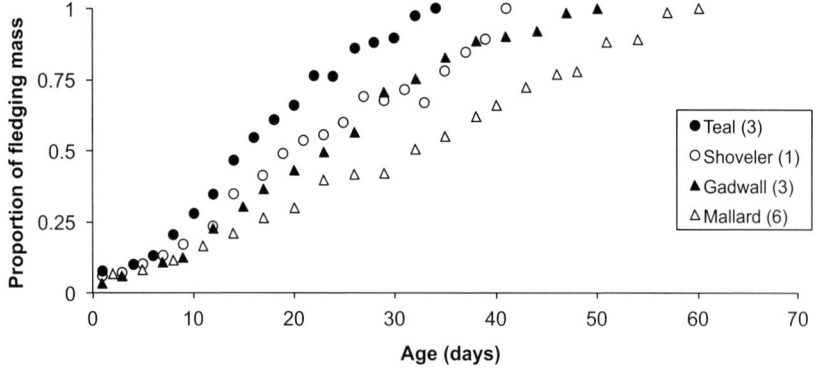

FIG. 6.16 *Time needed to attain fledging mass in Teal, Shoveler, Gadwall and Mallard, expressed as proportion of final fledging body mass, calculated from tables in Veselovský (1952). Note that sample sizes (number of different individuals given in brackets) are limited.*

(Sargeant & Raveling 1992). The recent development of lightweight radio tags allows the fitting of such equipment directly to ducklings, and hence computing individual duckling survival rate with precision and without the possible biases described above (Stafford & Pearse 2007). However, this technique has so far not been applied to Teal ducklings, so that the results presented here are only derived from changes in mean brood size over time.

Common Teal duckling survival rate from hatching to fledging has been estimated as 41% in Wales (Fox 1986), 63% in Iceland (Bengtson 1972), but only 7% in Russia by Sotnikov (1999). This illustrates the great variability in Teal brood success. Though a large part of this variability is obviously linked to differences in habitat quality and predation rate among study sites, general patterns in duckling survival also emerge that are not specific to Teal. First, most dabbling duckling mortality occurs during the first week of life, when very young birds are particularly susceptible to cold temperatures and rainy conditions (Krapu et al. 2000). In many species, the female often moves her brood over dry land, from one wetland to another, which may also be associated with high mortality risk (Keith 1961, Duncan 1986; but note that in some cases predation may be the cause rather than the consequence of movement between wetlands). By radio-tracking both Mallard hens and ducklings, Stafford et al. (2002) observed that most (83.3%) brood losses and most (64.3%) radio-marked duckling mortality occurred within the first week after hatching (see also Simpson et al. 2005 and Stafford & Pearse 2007 for increases in duckling survival with age).

The high mortality of very young Common and Green-winged Teal is obvious from the sharp decrease in mean brood size as ducklings grow (Figure 6.17).

The second common feature of duckling survival is that within a season, early duck broods are more productive than later ones, and this holds for Teal. Elmberg et al. (2005) recorded larger broods of Common Teal ducklings in age classes IIa–IIc in Finland for earlier hatching dates. This general pattern has been considered to reflect more abundant food resources and/or lower duckling predation risk earlier in the season (Dzus & Clark 1998, Dawson & Clark 2000, Blums et al. 2002, Elmberg et al. 2005).

Earlier broods are generally larger and their ducklings survive better. As previously mentioned, mean brood size was computed to decrease by 0.07 ducklings per day of delay in hatching date: in North America, Toft et al. (1984) recorded an average brood size of 6.32 class I ducklings for broods hatched before 6 July, but only 4.25 for broods hatched after 19 July. Ducklings hatched earlier also have a greater survival rate, as recorded for Common Teal in Iceland: survival rate to fledging was 60% in ducklings hatched before 1 July, but decreased to only 22% for those hatched after 1 August (Bengtson 1972). However, early clutches also have a lower nest success, which led Krapu et al. (2000) to suggest that there is a general trade-off in ducks between early nesting (low nest success, high brood survival) and late nesting (higher nest success, lower brood survival).

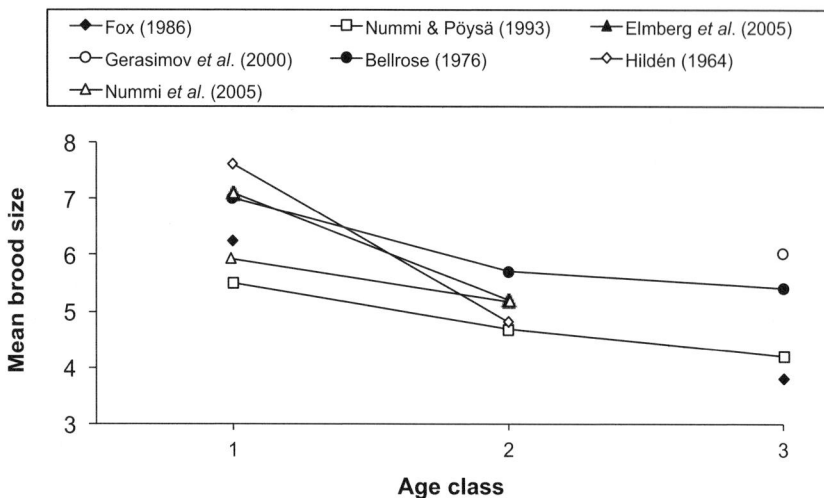

FIG. 6.17 *Mean brood size of ducklings in each age group class, based on data available in the literature. Age class I corresponds to ducklings 1–12 day old; class II: 13–29 days; class III: 30–35 days.*

Things may not always be that simple, however: Nummi *et al.* (2005) did not find any relationship between brood size of newly hatched ducklings and hatching date, while brood size of class III ducklings still decreased with advancing hatching date. This suggests that whatever factor brings a benefit to early breeders for Common Teal in Finland operates during the last phases of duckling development (classes II and III). In Mallard, Simpson *et al.* (2005) found no decrease in duckling survival with hatching date. However, they concluded that such a decrease is often observed in dynamic areas where wetland density decreases over the season, owing for example to the drying up of waterbodies, while they worked in the Great Lakes region where wetland habitats were more stable over the season. Clearly, what drives survival rates of ducklings and how this operates over time and in different habitats is an open and potentially promising field of research.

One of the likely factors affecting duckling survival to fledging is predation, though this has not been studied in detail in Teal. In general for ducks, Red Fox is considered to be particularly effective at depredating nests, as is Mink at preying on ducklings (Greenwood & Sovada 1996). In areas where predator removal is practised, Pearse & Ratti (2004) observed a 60% increase in Mallard duckling survival from hatching to 30 days of age. This was partly because fewer ducklings were depredated, but also because hatching was earlier where predators were removed, and earlier hatching per se is associated with larger clutch and brood sizes, and greater duckling survival. The link between predator abundance and duckling survival may therefore not always be direct. In line with this, Amundson & Arnold (2011) found that there was no effect on Mallard duckling survival when mammalian predators were removed. These authors showed that the rate of survival increased with the total area of seasonal and semi-permanent wetland, suggesting that food availability played a greater role in this case.

Naturally ducklings are adapted to living in a dangerous environment, and they do adopt appropriate behavioural responses to predators. Dessborn *et al.* (2012), for instance, demonstrated the innate ability of Mallard ducklings to react differently to an attacking raptor (by diving) or an attacking Pike (by running on the water's surface towards land). Temporary post-hatch brood amalgamation has also been observed in Mallard in response to predators, with families separating afterwards (Boos *et al.* 1989), although no such behaviour is known in Teal (for Green-winged Teal, see Eadie *et al.* 1988). Young ducklings were suspected of diving to *c.* 70cm underwater when frightened (as in the Dessborn *et al.* 2012 study), then grabbing underwater plants with their bills to remain there despite their high buoyancy (Sotnikov 1999).

'Final' breeding success, measured as the number of young fledged per female Common Teal per year, has been recorded to range from 1.36 (Elmberg *et al.* 2003) to 3.7 (Bengtson 1972), with Nummi & Pöysä (1995a) providing an intermediate 1.72 value. The statement of Linkola (1962) that 5.9 Common Teal ducklings fledge per brood in Finland probably concerns only those broods that survived and were counted (total brood loss was not accounted for), thus strongly overestimating duckling survival and hence annual breeding success.

BOX 6.2. Why migrate to breed thousands of kilometres from the wintering grounds? An experiment at both ends of the flyway (after Elmberg *et al.* 2009)

Understanding why birds migrate has always been a Holy Grail for ornithologists. One explanation for this behaviour, centred on the breeding grounds, postulates that birds migrate south for the winter to escape adverse meteorological conditions in northern areas, where they would otherwise be resident. It is also possible to regard the problem the other way round, and ask why birds do not breed in the mild areas where they winter. One simple answer to this question is that some of these areas become unsuitable for breeding birds (for example tidal areas where some waders winter, but could not build nests, Newton 2008; or Mediterranean wetlands that offer favourable conditions in winter but dry out every summer). However, there are wintering areas where ducks could theoretically breed, as some species like Mallard actually do, yet where breeding duck densities are low or some species are completely absent during the breeding season. This is the case in Common and Green-winged Teal in most of their wintering range. In the Camargue on the French Mediterranean coast, Common Teal winter in the tens of thousands but none attempts to breed. In contrast, species like Mallard or Gadwall do breed in this area, and encounter more abundant food resources there during spring and early summer than they would experience in most Scandinavian breeding lakes, for example (Arzel *et al.* 2009). One hypothesis to explain this paradox is that the predation risk is too high in the southern part of Teals' range for attempting to breed being an evolutionarily profitable option.

It is almost impossible to empirically test hypotheses of this type in the field, especially due to the very low density of Teal pairs in wetlands of the boreal forest (see main text), and their absence over most of the wintering range during the breeding season. In an experiment, hand-made 'semi-natural' nests filled with real duck eggs (from Mallard farms) were used instead (Figure 6.18). The idea was to build such nests in areas known to be used by breeding ducks (Mallard) in the Camargue and in northern Sweden. The crafting of these was as standardised as possible (the same fieldworker built all the nests in both countries using exactly the same techniques and the very same equipment), so that the main difference between the two sets of nests was the geographic areas and their inherent

FIG. 6.18 *Semi-natural nest of the type used by Elmberg* et al. *(2009) to compare the predation rate over time in southern France versus northern Sweden. Photograph by Johan Elmberg.*

differences rather than artefacts due to field procedures or observer effect.

Five real Mallard eggs were added to each nest; the eggs were then regularly checked to assess their survival (predation status). It is important to keep in mind that the aim of this experiment was not to assess the actual predation rate of real duck nests (on real nests, the behaviour of the incubating female allows reduced predation risk to some extent; Munro 1949, Dassow *et al.* 2012). Conversely, the goal here was, through the use of a strictly standardised field procedure, to compare how quickly semi-natural nests were depredated at the two ends of a duck's breeding range.

The results were very clear-cut: while in Sweden *c*. 15% of the semi-natural nests survived a period of 16 days, only 5% were not depredated during the same period in the Camargue (Figure 6.19). The same kind of test was carried out within Sweden, comparing the fate of semi-natural nests in wetlands in agricultural versus forested landscapes. This led to the same conclusion: duck nest predation risk is lower in lakes in the (boreal) forest (Gunnarsson & Elmberg 2008), which can explain why these are the main breeding areas for many breeding ducks in western Europe. In addition to lower nest predation, field studies also demonstrated for Common Teal that harassment by potential predators

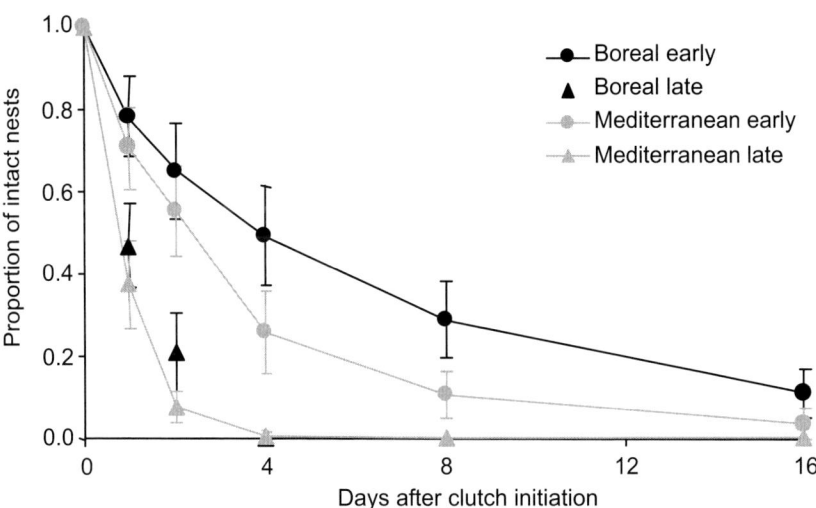

FIG. 6.19 *Mean (± SE) predation rate of semi-natural duck nests over time in northern Sweden (black symbols) and southern France (grey symbols). Mean values were computed from 14 wetlands in France and 16 in Sweden, half of which had 2 and half of which had 8 experimental nests, hence a total of 150 semi-natural nests monitored (from Elmberg et al. 2009, with kind permission from John Wiley & Sons, Inc.).*

of non-incubating ducks was clearly lower on breeding grounds compared to winter grounds (see Figure 5.10, Guillemain *et al.* 2007b).

Predation risk in general is therefore a likely factor influencing breeding habitat selection by dabbling ducks, and can at least partially explain why these birds migrate thousands of kilometres even though their wintering grounds could potentially provide otherwise suitable breeding sites. Beyond ducks, consistent results were obtained with similar experimental nests in waders, suggesting that predation risk may be invoked for the evolution of migratory behaviour in birds in general (Gilg & Yoccoz 2010, McKinnon *et al.* 2010).

Chapter 7

Mortality and limiting factors

We discuss demography at length in the next chapter, but we can already say that Teal are short-lived, at least in the wild. Most live only for one or two years, while ring recoveries indicate that they do have the biological potential to live for up to 20 years (see Chapter 8). In this chapter we describe the many factors that may explain this huge difference; that is, those that are likely to affect the individual negatively in nature, and then have repercussions on the dynamics of wild populations. Not all of these factors are necessarily lethal to Teal, but they rather affect individuals along a continuum from disturbance, pain and fitness reduction, to ultimately potential death.

DISEASES

Ducks are special among birds because so much is known about their diseases. One reason for this is that ducks are bred in captivity in most regions of the world, for ornamental purposes as well as a source of protein. In many cases, captive ducks can interact with their wild counterparts, so that disease in wild ducks may potentially have economic consequences, and in some cases even cause health problems if transmitted to humans (so-called zoonotic diseases).

Questions as to whether diseases in wild ducks threaten human health or, alternatively, if diseases in captive ducks threaten wild bird populations, are open to debate (Gauthier-Clerc *et al.* 2007). In any case, the fact that ducks (including Teal) are probable reservoirs (though not necessarily the main vectors) of Avian Influenza Viruses has put the searchlight on these birds within an epidemiological context over the last decade. In reality, however, little is known about the actual effect of most diseases on wild Teal individuals, let alone populations, and more specifically whether diseases can play any role as limiting factors in these birds. Equally, little is known about the prevalence of most diseases, simply because sick Teal in nature probably die and are scavenged, or are killed by predators, or in any case will not be discovered by humans and necropsied. In many instances the best available information is rather that a given disease (or serological evidence of previous exposure to it) has been recorded at least once in Common or Green-winged Teal (Table 7.1). Clearly, beyond a mere quantification of disease prevalence, as has mostly been done so far, more research should be carried out on the actual effect of diseases on Teal population dynamics.

Avian Influenza Viruses

Waterbirds, especially ducks and their relatives, are natural reservoirs of Avian Influenza Viruses, AIV (Webster *et al.* 1992, Olsen *et al.* 2006, Kim *et al.* 2009), a family of viruses named after the type of haemagglutinin and neuraminidase proteins found on the surface of the viruses (H1–H17 and N1–N10), from H1N1 to H17N10. Though infection by such viruses is generally benign to the birds just as annual winter flu is generally benign to humans, some Highly Pathogenic strains (HPAIV) can cause massive die-offs in wild bird populations (Olsen *et al.* 2006), and the zoonotic threat is that some viruses then become 'mammal adapted'. When viruses enter a mammalian system (for example that of livestock), the risk becomes greater that they will adapt to human-to-human transmission, causing a pandemic. Infected birds shed the virus from their digestive and respiratory tracts, over periods of time that may vary between bird species, virus strains and environmental conditions (Hénaux & Samuel 2011). Unfortunately, this does not necessarily limit the ability of wild birds, especially ducks like Teal, to spread AIV, owing to their ability to quickly cover great geographic distances during migration or winter travels (Figure 7.1, see colour section). For these reasons, and because ducks can be captured alive and large numbers of dead birds can be examined from hunting bags, ducks have received considerable attention in the context of AIV monitoring schemes. Accordingly, thousands have been sampled for AIV infection over the last decade, most often using cloacal swabs, although pharyngeal swabs have also sometimes been used owing to the possible dual routes of virus excretion (Munster *et al.* 2009).

Monitoring schemes have found Teal excreting a range of AIV strains (Appendix 7), which does support the hypothesis of their potentially significant role in the epidemiology of these viruses. However, it should be noted that all strains recorded so far from live Teal were Low Pathogenic ones (LPAIV). Outbreaks of HPAIV in wild Teal could have deadly consequences for the ducks themselves, and potentially for human health and economies. However, despite the fact that HPAIV has been recorded in many other species of waterbird (Gaidet *et al.* 2008), the potential risk that Teal and other ducks may transmit disease to poultry and humans has perhaps been emphasised too much. Some researchers argue that live birds may actually be more at risk from receiving this disease from poultry than the opposite (Gauthier-Clerc *et al.* 2007).

TABLE 7.1 Summary table of diseases known to occur in Common and Green-winged Teal (the list is not intended to be exhaustive, so that rare viral, bacterial or fungal diseases may be omitted, and not all existing references are necessarily provided for a given disease). Geographic area and season refer to what is treated in the cited studies. Most of these diseases probably have a wider geographic and temporal extent.

Type of disease	Disease or pathogen name	Pathogenicity to Teal	Geographic area	Season	Reference(s)
Virus	Paramyxoviruses (Newcastle Disease Virus – PMV-1)	Low	Iran – Caspian Sea Coast	Winter	Bozorgmehri-Fard & Keyvanfar 1979
	Paramyxoviruses (PMV-1, PMV-4 and PMV-6)	Low	Louisiana, USA	Winter	Stallknecht et al. 1991
Virus	Flavivirus (West Nile Virus)	?	Korea	Late summer to winter	Yeh et al. 2011
Virus	Herpesvirus (Duck Plague)	Low	USA		Converse & Kidd 2001
Virus	Avian Influenza Viruses	Low to high	Worldwide	Autumn to early winter	Review in Stallknecht et al. 2007
Virus	Avian Pox	?	Alaska, USA	September	Morton & Dieterich 1979
Bacteria	*Pasteurella multocida* (Avian Cholera)	High	Nebraska and California, USA		Zinkl et al. 1977, Brogden & Rhoades 1983
Bacteria	Salmonellosis	Low to medium	Oklahoma, USA		Waldrup & Kocan 1985
Bacteria	–?–	–?–	London, UK	Winter	Mitchell & Ridgwell 1971
Bacteria	*Mycobacterium avium* (Avian Tuberculosis)	?	Worldwide?		Wobeser 1997a
Bacteria	*Chlamydophila* spp. (Ornithosis)	Low	Worldwide?		Terskich 1964 in Kaleta & Taday 2003
Bacteria	*Campylobacter* spp.	?	Italy	Winter	Gargiulo et al. 2011
Bacteria biotoxin	*Clostridium botulinum* (Type C Avian Botulism)	High			Review in Rocke & Bollinger 2007
Fungi	Aspergillosis	Low			Quortrup & Shillinger 1941 in Converse 2007
Fungi	Various keratinophilic fungi	Low			Hubalek 2000

Avian botulism

Avian botulism is considered to be the most significant fatal disease for migratory waterbirds (Rocke & Bollinger 2007). Outbreaks resulting in up to millions of dead birds have occurred, especially in North America, though avian botulism has now been recorded on all continents except Antarctica, with casualties reported from a very wide range of bird species (Rocke 2006). Their general ecology puts ducks, especially dabblers, particularly at risk. Avian botulism is caused by neurotoxins produced by *Clostridium botulinum* bacteria, its toxin C_1 being the most common and deadly for waterbirds. Toxins are produced by the bacteria as they multiply, typically in freshwater wetlands at temperatures between 25 and 40°C. Though anaerobic and alkaline conditions seem more favourable to the growth of these bacteria, avian botulism has also been recorded under other abiotic conditions (review in Rocke & Bollinger 2007), and it does not seem possible so far to predict from environmental conditions where and when a botulism outbreak is most likely to occur. The most commonly accepted explanation for such outbreaks is the 'carcass-maggot cycle', almost similar to an infectious disease owing to the fact that the carcasses of birds dead from botulism may themselves become a new source of contamination for other individuals: bacteria develop on decaying organic matter, and spores can be ingested by waterfowl as they ingest such matter when foraging in shallow water. When a bird dies (for any reason), the bacteria develop in the tissues and produce toxins if they are themselves infected with a virus or bacteriophage. The toxins concentrate in sarcophagous invertebrates (fly maggots, for example), and other waterbirds may then ingest toxins at high concentration by eating the maggots, then die and thus become themselves the source of further maggot-mediated toxin transfer (Wobeser 1997b). The major role played by maggots was demonstrated by Evelsizer *et al.* (2010), who showed that Mallard mortality risk was more affected by the density of maggot-laden carcasses than of maggot-free carcasses. When ingested by the birds through the maggots, the toxin affects motor neurons, causing muscle paralysis. When infected, waterbirds often die by drowning.

The main technique used in attempts to manage avian botulism in North America is the collection of carcasses to break the carcass-maggot cycle. This, however, has proven to be inefficient, as no more than 7% of marked carcasses were found and removed by carcass-collection teams in large wetlands, leaving many carcasses undiscovered per hectare (Bollinger *et al.* 2011).

Botulism outbreaks can have devastating effects at the local and regional scales, with thousands of birds sometimes dying on a single lake (Friend & Franson 1999). Beyond this, recent studies demonstrate that botulism could be a genuinely limiting factor of dabbling duck populations, since late-summer survival at the population level was reduced by 14–44% for Mallard ringed at outbreak versus non-outbreak sites (Dufour & Bollinger 2011). Avian botulism is likely to be a limiting factor for Teal populations, at least in

FIG. 7.2 *Green-winged Teal poisoned by botulism. One of the first symptoms of the disease is paralysis of the legs and wings, which can be inferred here by the tracks in the mud left by the hanging wings. Photograph courtesy of Trent K. Bollinger (Canadian Cooperative Wildlife Health Centre).*

North America, since Green-winged Teal often represent a large share of the duck carcasses found during outbreaks (Wobeser *et al.* 1983; typically 20–30% of all duck carcasses found in Alberta according to the review in Pybus 2011). One reason for this is that the lethal consequences of toxin ingestion are reached at a lower toxin concentration compared to larger-bodied duck species: 14,000 $MIPLD_{50}$ (median mouse intraperitoneal lethal dose) in Teal compared to *c.* 45,000 $MIPLD_{50}$ in Mallard, corresponding to the ingestion of only *c.* 0.14g of infected maggots in Teal (Rocke *et al.* 2000). The exact magnitude of the mortality rate caused by avian botulism outbreaks in Teal at the population level, however, remains unknown, but Stout & Cornwell (1976) considered diseases, especially botulism, to be the second most important mortality cause for dabbling ducks in North America, after hunting.

Parasites

Just like bacterial, viral and fungal diseases, and for the same reasons of easy access to live or dead ducks to sample, Teal ecto- as well as endoparasites have been widely studied. The first list of Teal parasites was published as early as the end of the 19th century (Naumann 1897), and so far more than 200 parasite species have been recorded specifically in Teal (Appendix 8). However, many of these studies are purely descriptive, or at best quantify the prevalence of parasitism (proportion of birds infested) (Table 7.2). They show that parasite prevalence can sometimes be very high: of the 72 Green-winged Teal

TABLE 7.2. *Summary of parasites known to occur in Common and Green-winged Teal, and whose prevalence was quantified (a more comprehensive list of parasites found in Teal is given in Appendix 8).*

Parasite	Prevalence (% of birds examined)	Geographic area	Reference
Parasites in the digestive tract			
Trematodes			
Apatemon gracilis	40	Nova Scotia & New Brunswick, Canada	Turner & Threlfall 1975
	10	Oklahoma, USA	Shaw & Kocan 1980
Cotylurus brevis	14	Mexico	Martínez-Haro et al. 2012
Cotylurus cornutus	1	Nova Scotia & New Brunswick, Canada	Turner & Threlfall 1975
Cotylurus platycephalus	1	Nova Scotia & New Brunswick, Canada	Turner & Threlfall 1975
Cyathocotyle bushiensis	5	Québec, Canada	Hoeve & Scott 1988
Dendritobilharzia pulverulenta	3	Texas, USA	Canaris et al. 1981
Echinoparyphium recurvatum	13	Nova Scotia & New Brunswick, Canada	Turner & Threlfall 1975
	14	Mexico	Martínez-Haro et al. 2012
Echinostoma revolutum	70	Nova Scotia & New Brunswick, Canada	Turner & Threlfall 1975
	9	Texas, USA	Canaris et al. 1981
Hypoderaeum conoideum	2	Nova Scotia & New Brunswick, Canada	Turner & Threlfall 1975
	3	Texas, USA	Canaris et al. 1981
Microphallus primas	6	Nova Scotia & New Brunswick, Canada	Turner & Threlfall 1975
Notocotylus attenuatus	31	Nova Scotia & New Brunswick, Canada	Turner & Threlfall 1975
Notocotylus stagnicolae	19	Texas, USA	Canaris et al. 1981
Paramonostomum alveatum	7	Nova Scotia & New Brunswick, Canada	Turner & Threlfall 1975
Prosthogonimus cuneatus	2	Nova Scotia & New Brunswick, Canada	Turner & Threlfall 1975
Psilochasmus oxyurus	7	Nova Scotia & New Brunswick, Canada	Turner & Threlfall 1975
Psilostomum sp.	3	Nova Scotia & New Brunswick, Canada	Turner & Threlfall 1975
Sphaeridiotrema globulus	35	Québec, Canada	Hoeve & Scott 1988
Tracheophilus cymbius	21	Oklahoma, USA	Shaw & Kocan 1980
Typhlocoelum cucumerinum	6	Texas, USA	Canaris et al. 1981
Zygocotyle lunata	2	Nova Scotia & New Brunswick, Canada	Turner & Threlfall 1975
	21	Oklahoma, USA	Shaw & Kocan 1980
	19	Texas, USA	Canaris et al. 1981
Cestodes			
Cloacotaenia megalops	46	Texas, USA	Canaris et al. 1981
	4–28	France	Green et al. 2011

	18	Poland	Królaczyk et al. 2011
	29	Mexico	Martínez-Haro et al. 2012
Diorchis acuminata	10	Alaska, USA	Schiller 1953
Diorchis longiovum	40	Alaska, USA	Schiller 1953
Fimbriaria fasciolaris	5	Alaska, USA	Schiller 1953
Gastrotaenia cygni	10	Texas, USA	Canaris et al. 1981
Hymenolepis collaris	50	Alaska, USA	Schiller 1953
Hymenolepis gracilis	23	Texas, USA	Canaris et al. 1981
Sobolevicanthus krabbeella	6	Texas, USA	Canaris et al. 1981
	71	Mexico	Martínez-Haro et al. 2012
Acanthocephalans			
Corynosoma constrictum	26	Nova Scotia & New Brunswick, Canada	Turner & Threlfall 1975
	16	Oklahoma, USA	Shaw & Kocan 1980
	16	Texas, USA	Canaris et al. 1981
Polymorphus minutus	13	Texas, USA	Canaris et al. 1981
Pseudocorynosoma constrictum	14	Mexico	Martínez-Haro et al. 2012
Nematodes			
Amidostomum acutum	13	Nova Scotia & New Brunswick, Canada	Turner & Threlfall 1975
	51	Texas, USA	Canaris et al. 1981
	66	The Netherlands	Borgsteede et al. 2006
	54	Poland	Kavetska et al. 2011
Amidostomum anseris	9	Texas, USA	Canaris et al. 1981
Capillaria anatis	2	Nova Scotia & New Brunswick, Canada	Turner & Threlfall 1975
Capillaria contorta	6	Nova Scotia & New Brunswick, Canada	Turner & Threlfall 1975
Capillaria sp.	14	Mexico	Martínez-Haro et al. 2012
Echinuria uncinata	1	Texas, USA	Canaris et al. 1981
Epomidiostomum uncinatum	1	Nova Scotia & New Brunswick, Canada	Turner & Threlfall 1975
Tetrameres crami	21	Oklahoma, USA	Shaw & Kocan 1980
	36	Texas, USA	Canaris et al. 1981
Tetrameres ryjikovi	17	Nova Scotia & New Brunswick, Canada	Turner & Threlfall 1975
Parasites of the circulatory system			
Trematodes			
Trichobilharzia querquedulae	1	Nova Scotia & New Brunswick, Canada	Turner & Threlfall 1975
Trichobilharzia yokogawai	?	Asia	McDonald 1969
Nematodes			
Filaria sp.	3	Massachusetts, USA	Bennett et al. 1974
	5	New Brunswick, Nova Scotia & Prince Albert Island, Canada	Bennett et al. 1975
	2	Oklahoma, USA	Shaw & Kocan 1980
	3	Labrador & Newfoundland, Canada	Bennett et al. 1991

Parasite	Prevalence (% of birds examined)	Geographic area	Reference
Apicomplexa			
Haemoproteus nettionis	77	Massachusetts, USA	Bennett *et al.* 1974
	22	New Brunswick, Nova Scotia & Prince Albert Island, Canada	Bennett *et al.* 1975
	8	Oklahoma, USA	Kocan *et al.* 1979
	1	Labrador & Newfoundland, Canada	Bennett *et al.* 1991
Leucocytozoon simondi	77	Massachusetts, USA	Bennett *et al.* 1974
	17	New Brunswick, Nova Scotia & Prince Albert Island, Canada	Bennett *et al.* 1975
	18	Oklahoma, USA	Kocan *et al.* 1979
	82	Labrador & Newfoundland, Canada	Bennett *et al.* 1991
Plasmodium circumflexum	3	Massachusetts, USA	Bennett *et al.* 1974
	7	New Brunswick, Nova Scotia & Prince Albert Island, Canada	Bennett *et al.* 1975
Parasites in the muscles			
Coccidia			
Sarcocystis sp.	4	North Dakota, USA	Hoppe 1976
Ectoparasites			
Mallophaga – bird lice			
Anaticola crassicornis	76	Texas, USA	Canaris *et al.* 1981
Anatoecus icterodes	39	Texas, USA	Canaris *et al.* 1981
Holomenopon setigerum	16	Texas, USA	Canaris *et al.* 1981
Trinoton querquedulae	87	Texas, USA	Canaris *et al.* 1981
Acarina			
Epidermoptes sp.	1	Texas, USA	Canaris *et al.* 1981

examined by Canaris *et al.* (1981) in Texas, 70 had at least one parasite, while the corresponding value was 95% in 87 Green-winged Teal examined by Turner & Threlfall (1975), and 80% of 20 Aleutian Teal examined by Schiller (1953) carried at least one cestode (Table 7.2). The migratory behaviour of Teal can explain this pattern, since they are likely to be exposed to a broader range of parasites over their annual cycle than sedentary bird species (Figuerola & Green 2000). Parasite prevalence also changes markedly over time: in the Camargue, Green *et al.* (2011) found that the prevalence of the cestode *Cloacotaenia megalops* increased from 4–6% in the 1950s to 28% in the 2000s, probably due to environmental changes promoting the populations of seed shrimps (Crustacea, Ostracoda), which act as intermediate hosts for this parasite.

FIG. 1.3. *Common and Green-winged Teal males (left and right, respectively). While females of the two species are very similar, subtle but readily apparent differences exist between males, if only attention is paid to the head and flank plumage patterns (see more information in Chapter 2). Photographs by Özden Sağlam (Common Teal) and Charles McDonald (Green-winged Teal). See also Figure 2.1.*

FIG. 2.1. *Male Common (left, photo by David Lédan) and Green-winged Teal (right, photo by Charles McDonald). The latter has a vertical white bar on the side of the breast, whereas the scapulars of the former form a distinctive horizontal white line on the wing (though not always as visible as is the case in this picture). Note also that the cream-coloured lines on the face are less conspicuous in Green-winged Teal.*

FIG. 2.3. *Female Common Teal. As in many other ground-nesting species, the cryptic plumage of females is thought to be an adaptation to reduce predation risk during nest-building, egg-laying and incubation. Note the pale horizontal line on the side of the rump made up by the lateral undertail feathers, which can aid in identifying Teal from females of other dabbling duck species. Photograph by Michel Collard.*

FIG. 2.4. *Male Common Teal in active moult to breeding (alternate) plumage. Note the brown juvenile/eclipse feathers on the flanks instead of the whitish scapulars seen in males in full breeding plumage. Photograph by Marcus Wikman.*

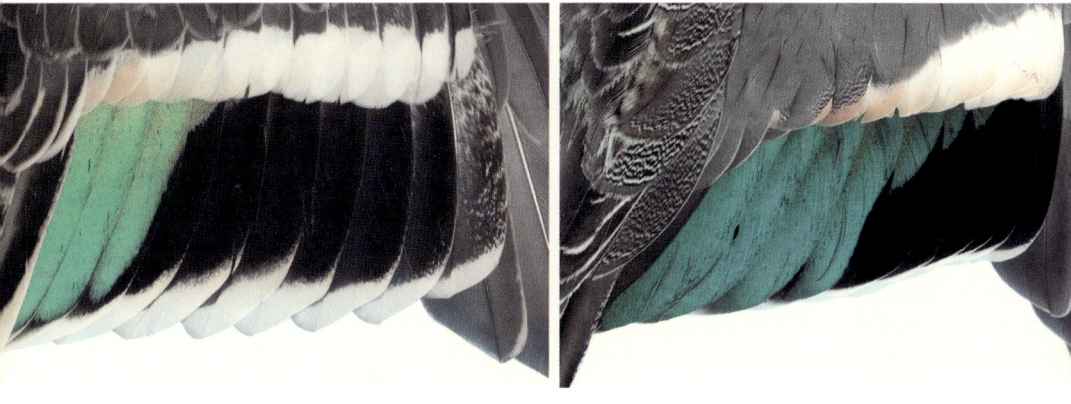

FIG. 2.5. *Female (left) and male (right) Common Teal wing speculums. While female speculums generally have fewer than four metallic green feathers, males have four or more such feathers. Photographs by Jean-Baptiste Mouronval / ONCFS.*

FIG. 2.6. *Sexing of Teal by the most distal tertial of females (left) and males (right). This feather has a wide white outer margin in females, and inside of it a dark streak that gradually grades from black (distally) to brown (proximally). In males, there is a distinct blackish outer (distal) margin, contrasting with a pearl white area inside (proximally) of it. Photographs by Jean-Baptiste Mouronval / ONCFS.*

FIG. 4.15. *A preferred breeding and moulting habitat of Common Teal: grassy shores, patches of open water and floating vegetation. Lake near Parikkala, SE Finland. Photograph by Johan Elmberg.*

FIG. 6.3. *Copulation by a pair of Green-winged Teal. Photograph by Tom Grey.*

FIG. 6.4. *Grunt-whistling male Common Teal. Note the water droplets splashed by the bill and the small tuft at the rear of the head. Photographs by Jean-Pierre Artel.*

FIG. 6.5. *Male Common Teal in Down-up position. This is mostly used during male-male interactions. When the display is directed towards a female, the head is also erected (Head-up-Tail-up display, McKinney 1965). Note the exposition of the wing flash marks and the pale lateral undertail patch. Photograph by Jean-Pierre Artel.*

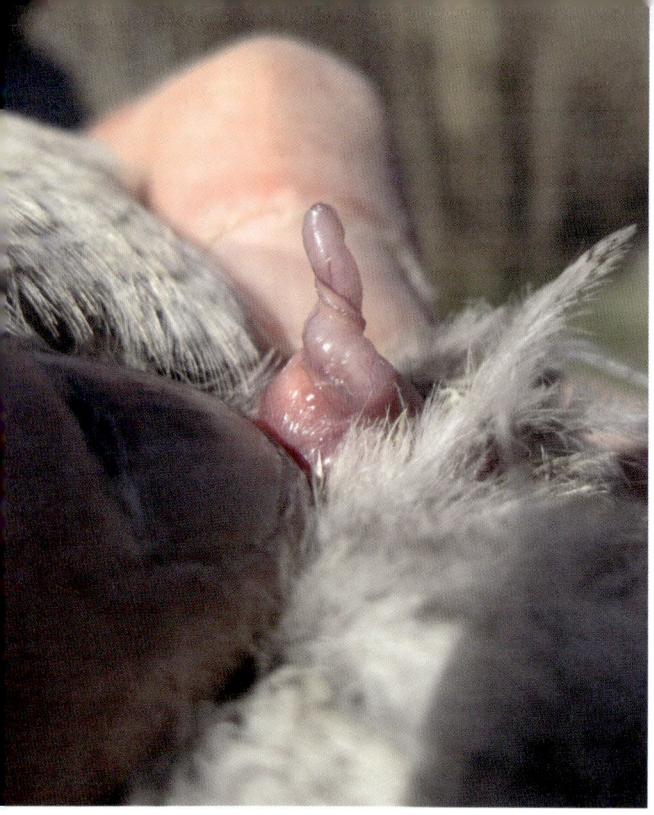

FIG. 6.6. *Penis of a first-year Common Teal in February. Waterfowl are exceptional in that males have an intromittent organ, while in most other birds sperm transmission is by simple cloacal contact. Photograph by Pär Söderquist.*

FIG. 6.13. *Common Teal female with brood (southern Sweden). Photograph by Kenneth Johansson.*

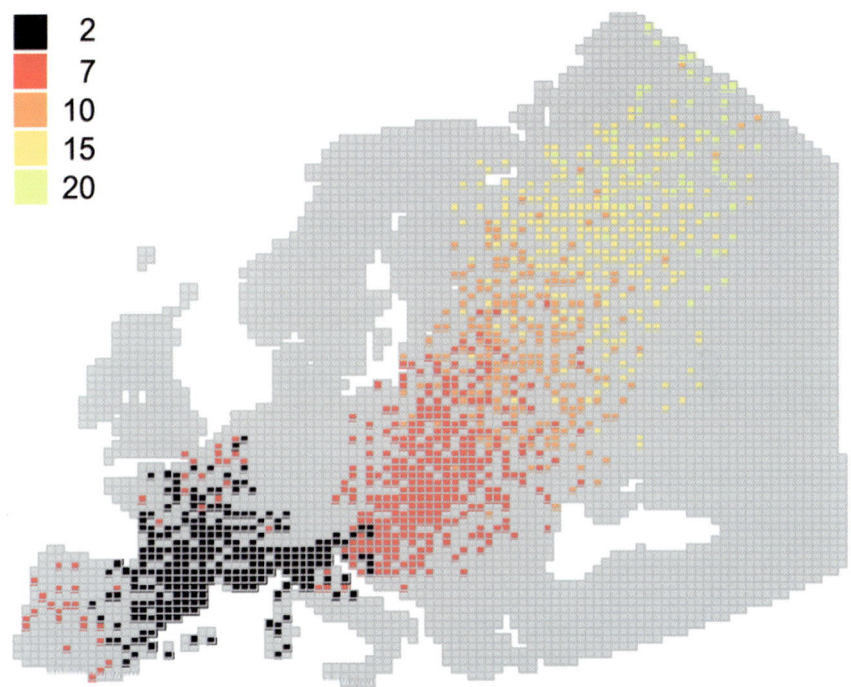

FIG. 6.14. *Downy Common Teal duckling of the youngest (Ia) age class, in other words being at the most six days old. Photograph by Romain Blanc.*

FIG. 7.1. *Modelled potential spread of AIV across western Europe and Russia by Common Teal wintering in the Camargue, southern France, depending on the duration of viral excretion duration (in days, see different colours in panel above). The model was individual-based and spatially explicit, and was developed from real Common Teal movement rates and distances inferred from direct recoveries of ringed birds. With sufficiently long virus shedding times, Teal could spread AIV anywhere along their flyway (from Lebarbenchon et al. 2009).*

FIG. 9.1. *One of the very few representations of Teal in Antique art: mosaic from the House of the Faun in Pompeii, currently presented in the Naples Museum of Archaeology. Photograph by Marie-Lan Nguyen / Wikimedia Commons.*

FIG. 9.7. *Amateur and professional breeders have produced a range of Teal varieties in captivity, which mainly differ in their coloration. Photograph by Matthieu Guillemain.*

Unfortunately, little knowledge exists about the consequences of parasites and parasite load for survival and fecundity in Teal, although high blood parasite load in waterfowl was considered a potential limiting factor at the population level (Bennett *et al.* 1974). Other accounts document the potentially lethal effects of some parasitic worms of the digestive tract in waterfowl, though not specifically in Teal (Hoeve & Scott 1988, Borgsteede *et al.* 2006). Because they recorded greater prevalence of *Cloacotaenia megalops* in heavier than in leaner Teal, Green *et al.* (2011) concluded that infestation by this parasite probably had limited deleterious effects for the birds. Clearly, the parasitology of Teal is an open field for future research. Although we already made the same statement for infectious diseases, the situation may be different for parasites, whose detrimental effects can sometimes be detected by dosing birds with an anti-parasite drug upon ringing, and comparing the fate of these birds with that of unmedicated individuals in a control group (see, for example, anthelmintic drug injection in Greater Snow Goose *Chen caerulescens atlantica* in Souchay *et al.* 2013).

Pollution

The effects of pollutants on individual bird health have been analysed in more detail in some other duck species, even addressing population dynamics in a few cases (for example, selenium in Greater and Lesser Scaup *Aythya marila* and *A. affinis*, DeVink *et al.* 2008, Ware *et al.* 2011). A range of contaminants has been found in Teal tissues, most being agricultural pesticides or trace elements originating in diffuse pollution from human activities (Table 7.3). That Teal may be contaminated by such a range of chemicals is not surprising considering that they are very mobile and live in wetlands, and hence may be exposed to run-off of a variety of contaminants from the surrounding habitats. The filtering foraging method of Teal probably makes the situation even worse, since many contaminants deposit and accumulate in wetland sediments where ducks forage and consequently consume contaminated food items.

Contamination of birds with trace elements (lead or zinc, for example) leading to death has been documented in Teal (review in Beyer *et al.* 2004). In North America, Fleischli *et al.* (2004) identified 212 Green-winged Teal carcasses of birds dead from poisoning by anticholinesterase agriculture pesticides in the U.S. Geological Survey National Wildlife Health Center mortality database between 1980 and 2000 (these compounds may have severe effects on the nervous system, including the brain). Occasional deaths of Teal linked to prolonged presence in particularly heavily polluted areas have also been described (ingestion of white phosphorus from military training activities, Steele *et al.* 1997, Sparling *et al.* 1998; waste fluids in oil pits, Trail 2006; insecticides and other pollutants in Hu & Cui 1990). The latter studies, however, probably give a conservative estimate of the true magnitude of such sources of mortality.

In contrast, most intoxication accounts concerning Teal document relatively low concentrations of contaminants (Table 7.3), or limited intoxication ('short-term cyanide intoxication' in Henny *et al.* 1994). Even Common Teal collected in Northern Ireland after the Chernobyl accident, whose radiocesium signatures (isotope ratios) indicated that they had been near the contaminated area, had radiocesium levels of 'only' 8.6–33.9Bq/kg muscle, far below the EU limit of 600Bq radiocesium/kg threshold of those days (Pearce 1995). Teal shot locally in the Chernobyl region two months after the nuclear accident, on the other hand, showed radionuclide values in their muscles that were >100 times higher than permitted in human food. Although radiocesium levels decreased over time, the measurements in Teal from the Chernobyl area were still far above the limit close to 10 years after the accident (Vyazovich 1996).

In general, reported concentrations of contaminants in Teal (Table 7.3) correspond well to levels found in other wetland birds at the same trophic level (see for example Zhang *et al.* 2011), indicating that they can be regarded as accurate and probably representative. Moreover, data from Teal fit well with other established ecotoxicological patterns, for example that fat tissue has contaminant values 100–1,000 times higher than those of feathers, and also higher than those of muscle tissue.

The fact that few accounts of massive contamination have been reported should not, however, be taken as evidence that poisoning by contaminants is not an issue in Teal (see Beyer & Heinz 2000 for a discussion on where to set threshold values for acceptable contamination of wildlife). One reason why pollution may be a bigger problem for Teal than acknowledged thus far is that many toxic and potentially widespread pollutants in wetlands have simply not been looked for in ducks. Another reason is that studies to date chiefly concern the released contaminants per se, and not the metabolites they give rise to in the duck's body after ingestion (cf. Liu *et al.* 2010). As a case in point from Table 7.3, PCB has 209 congeners and the number of potentially toxic metabolites they in turn can give rise to is extremely long, implying that only the very tip of the 'PCB iceberg of toxicity' has been studied in Teal and other ducks, with other groups of organochlorides then representing yet other groups of 'icebergs'. Another issue little explored to date in waterbirds is 'mixture toxicity', in other words the combined effects that may arise when ducks are exposed to more than one toxin, which is probably the rule in nature. On top of direct toxic effects, some of the pollutants and their metabolites may have diffuse long-term effects on, among other things, antioxidant function, hormone systems and reproduction (for example Hegseth *et al.* 2011). This constitutes another unexplored field in duck ecology in terms of the birds' longevity, behaviour and reproductive success.

In short, our understanding is still very limited about the magnitude of the effects pollution has on Teal. Nevertheless, we argue that current knowledge does not suggest that pollution is likely to be a main limiting factor for Teal populations, except locally from botulism or lead poisoning, for example (see opposite).

TABLE 7.3 *Concentrations of contaminants recorded in tissue from Common and Green-winged Teal. Among organochlorines, PCB mainly originates in industrial processes, whereas the others come from agricultural pesticides. Single numbers indicate mean measurements; otherwise maximum range of individual values is given. Doses considered lethal are indicated in the 'contaminant' column when known (expressed in (µg/g). ND: not detected; dw: dry weight; ww: wet weight; 1: Hoffman et al. 1996; 2: Blus 1996; 3: Furness 1996; 4: Thompson 1996; 5: Pain 1996; 6: Heinz 1996.*

Contaminant	Tissue analysed	Concentration (µg/g)	Study area	Reference
Organochlorine compounds				
PCB	Muscle	0.07–0.31	Sweden	Lindberg et al. 1985
200 [1]	Fat	5.15 ww	Spain	Llorente et al. 1987
	Fat	2.17 ww	Italy	Alleva et al. 2006
	Feathers	0.028	Iran	Behrooz et al. 2009
DDE	Fat	81	Iowa, USA	Johnson & Morris 1971
150–300 [2]	Fat	4.66 ww	Italy	Alleva et al. 2006
DDT	Fat	0.025	Iowa, USA	Johnson & Morris 1971
10–20 [2]	Muscles	0.01–0.05	Sweden	Lindberg et al. 1985
	Fat	8.37 ww	Spain	Llorente et al. 1987
	Feathers	0.020	Iran	Behrooz et al. 2009
HCH (Hexachlorocyclohexane)	Feathers	0.015	Iran	Behrooz et al. 2009
HCB (Hexachlorobenzene)	Feathers	0.003	Iran	Behrooz et al. 2009
Heavy metals and trace elements (from more diffuse pollution)				
Al	Digesta	260 dw	Delaware, USA	Beyer et al. 1999
	Liver	0.07–0.82 ww	Utah, USA	Vest et al. 2009
As	Liver	0.77 dw	Georgia, USA	Winger et al. 2000
	Liver	0.03–0.81 dw	Spain	Gómez et al. 2004
	Liver	0.44–2.05 ww	Utah, USA	Vest et al. 2009
	Liver	3.6–5.3 dw	Mexico	Pereda-Solis et al. 2012
Cd	Liver	0.1–0.5 ww	Texas, USA	White & Cromartie 1985
40 (liver ww) [3]	Liver	0.9 dw	Italy	Carpenè et al. 1995
	Liver	0.93 dw	Georgia, USA	Winger et al. 2000
	Liver	ND–1.76 dw	Spain	Gómez et al. 2004
	Liver	0.03–0.63 ww	Utah, USA	Vest et al. 2009
	Liver	2.0 dw	Mexico	Pereda-Solis et al. 2012
Hg	Liver	0.28	Iowa, USA	Johnson & Morris 1971
20–30 (liver ww) [4]	Liver	1.4	Spain	Baluja et al. 1983
	Muscle	0.10–0.49	Sweden	Lindberg 1984
	Liver	0.24 dw	Georgia, USA	Winger et al. 2000
	Liver	0.01–1.56	Nevada, USA	Gerstenberger 2004
	Bone	0.1121 ww	Alaska, USA	Rothschild & Duffy 2005
	Liver	0.08 ww	Italy	Alleva et al. 2006
	Liver	0.05–0.39 ww	Canada	Braune & Malone 2006
	Liver	0.41–2.16 ww	Utah, USA	Vest et al. 2009

Contaminant	Tissue analysed	Concentration (μg/G)	Study area	Reference
	Liver	4.34 dw	Iran	Zamani-Ahmadmahmoodi et al. 2009
Pb	Liver	ND–0.14 ww	Texas, USA	White & Cromartie 1985
15 (liver ww) [5]	Gut contents	4.6 dw	Delaware, USA	Beyer et al. 1999
	Liver	0.07 dw	Georgia, USA	Winger et al. 2000
	Liver	ND–2.35 dw	Spain	Gómez et al. 2004
	Liver	0.20 ww	Italy	Alleva et al. 2006
	Liver	0.04–0.19 ww	Utah, USA	Vest et al. 2009
	Liver	3.5–4.6 dw	Mexico	Pereda-Solis et al. 2012
Se	Kidney	1.0–1.5 ww	Texas, USA	White & Cromartie 1985
20 (liver ww) [6]	Liver	3.2–23.9 dw	California, USA	Ohlendorf et al. 1990
	Liver	6.74 dw	Georgia, USA	Winger et al. 2000
	Liver	1.1–8.0 ww	Canada	Braune & Malone 2006
	Liver	1.43–5.85 ww	Utah, USA	Vest et al. 2009
Zn	Liver	23–55 ww	Texas, USA	White & Cromartie 1985
	Gut contents	75 dw	Delaware, USA	Beyer et al. 1999
	Liver	166 dw	Georgia, USA	Winger et al. 2000
	Liver	54.3–276 dw	Spain	Gómez et al. 2004
	Liver	25.2–51.8 ww	Utah, USA	Vest et al. 2009
	Liver	46.4–93.7 dw	Mexico	Pereda-Solis et al. 2012

The effects of poisoning due to the ingestion of lead pellets in Teal have received most attention. Collectively, these studies convincingly demonstrate reduced survival in lead-poisoned Teal (Box 7.1), and detrimental effects on duck populations in general. These results have led to gradual implementation of bans on the use of lead shot for waterfowl hunting throughout the world (see Chapter 9).

BOX 7.1. Lead poisoning

As explained in Box 5.5, Teal and other ducks must ingest hard particles of sediment (called *grit*, or *gastroliths*) to facilitate the crushing of food items in their muscular stomach (gizzard). Unfortunately, lead pellets are sometimes taken from the sediment by the birds, causing lead poisoning after they enter the acidic gut environment and are ground by gravel and sand in the digestive tract of ducks.

Studies show that the amount of lead in the sediment in hunted areas can be remarkably high: in the Camargue in the late 1980s, densities of up to 2 million lead pellets per hectare were recorded by Pain (1991b).

The prevalence of lead shot in duck gizzards is extremely variable between geographic areas. In Britain the proportion of Common Teal with at least one lead pellet in the gizzard was 0% according to Olney

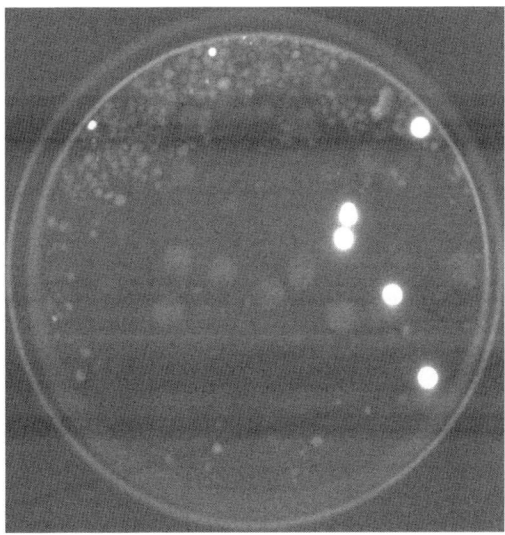

FIG. 7.3 *X-ray photography of a Teal gizzard content in a Petri dish. Five lead pellets are clearly visible to the right, plus two fragments to the top left. The other items are mineral grit. Note that the lead pellets are approximately the same size as some grit items visible at the centre of the picture. Courtesy of Jean-Yves Mondain-Monval, ONCFS.*

(1960), 0.5% after Mudge (1983) and 3.2% after Thomas (1975), while Mondain-Monval *et al.* (2002) documented 13% for the Camargue, France, and Figuerola *et al.* (2005) reported 30% for Teal for the Ebro Delta, Spain. In an early study, Bellrose (1959) found that 1.4% of Green-winged Teal carried lead pellets. Teal are not, however, necessarily the most affected birds: in the Camargue up to 50% of Pintails and 35% of Mallard had lead pellets in their gizzards (review in Mondain-Monval *et al.* 2002). Experimental studies have shown that 9% of Mallards would die from lead poisoning within two months of being force-fed a single size 4 lead pellet; when fed three pellets the percentage reached 67% (Duranel 1999). This is an ample illustration of the severity of this problem.

Lead poisoning as a result of ingestion of shotgun pellets has long been identified as a severe threat to wintering ducks (Bellrose 1951, Hoffmann 1960). The first symptom of lead poisoning is body mass loss (10% on average in Canvasback; Hohman *et al.* 1990). This is associated with a swelling of the cheeks and the production of green faeces. The bird then loses equilibrium, holding its head and neck horizontal and keeping its wings open to avoid falling. It gradually becomes more and more apathetic, its plumage deteriorates and it then loses the ability to fly (Hovette 1972, Figure 7.4). Lead-poisoned birds are easy prey to

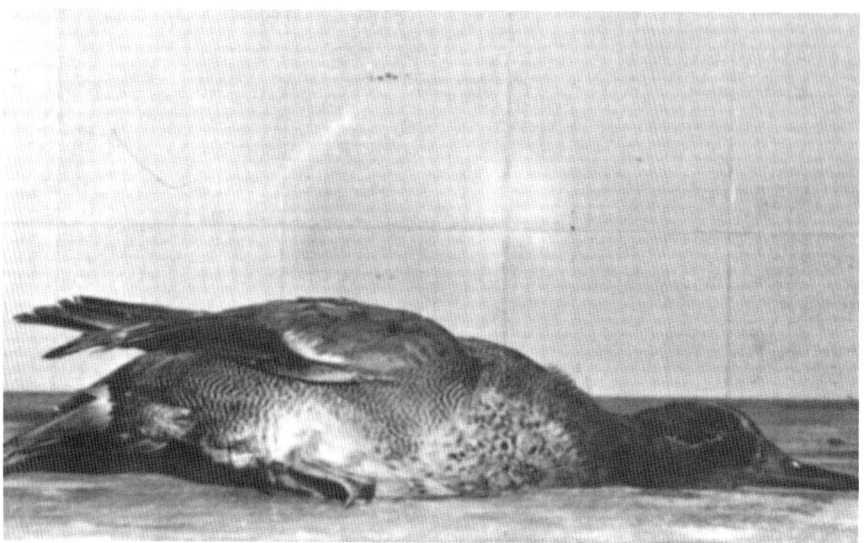

FIG. 7.4 *Lead-poisoned Common Teal. Note the general apathy and the curled toes. The bird is alive. From Hovette (1973), with kind permission.*

predators (which are also prone to becoming contaminated; Pain & Amiardtriquet 1993). Even if the bird is not killed by a predator, lead poisoning causes rapid death except if the lead pellets quickly pass through the digestive system and are evacuated naturally, which sometimes happens (Hovette 1973). Given their smaller body size, it is likely that lead poisoning after ingestion of lead shot could be more acute in Teal than in other ducks. In the Camargue, survival of Common Teal was significantly lower in birds having one or more lead pellets in the gizzard after X-ray examination, although the exact magnitude of this decrease was difficult to estimate (Guillemain *et al.* 2007c).

PREDATION

Predation plays a major role in the life history of Teal, and we have already referred to this factor many times in this book. Indeed, predation risk is considered to have promoted natural selection of female cryptic plumage (Chapter 2), to be responsible for the peculiar diel pattern of habitat use in wintering Teal (Chapter 4), and to play a major role in constraining feeding behaviour at times to specific foraging methods (Chapter 5). We have also highlighted the major role of predation as a driver of Teal breeding success, as well as the elaborate strategies female Teal have to rely on to avoid nest predation (nest habitat use, incubation behaviour, even clutch size; Chapter 6).

Teal are potential food for a diverse number of natural predators (Table 7.4). Because of their small size, Phillips (1923) noted that 'Teal are more frequently hunted by predatory birds than are the larger ducks'. It should be noted, though, that predation on Teal is mostly opportunistic for many predators: during the breeding season both Sargeant *et al.* (1984) and Greenwood *et al.* (1995) found Green-winged Teal remains at Red Fox dens, but these were proportional to the relative abundance of Teal in the duck community, indicating that Teal were not selectively targeted. Similarly, some potential predators consume Teal only very occasionally (for example Wild Boar *Sus scrofa* in Giménez-Anaya *et al.* 2008). Gulls and raptors have also been documented taking ducklings, as well as attempting to prey upon fully grown Teal in winter (Table 7.4). On the other hand, other more specialised duck predators play a more regular and thus greater role in the life of Teal. For example, Striped Skunk *Mephitis mephitis* and Red Fox are major predators of duck nests, while mink most frequently prey upon ducklings (Greenwood & Sovada 1996).

Red Fox can be viewed as a main predator on Teal throughout the year and throughout the latter's geographic range (Table 7.4). It seems probable that predation is most likely to act as a population limiting factor during the breeding season, if at all, although predator removal experiments have provided inconsistent results regarding the potential positive effect of this management action on density of breeding Teal (Duebbert & Kantrud 1974, Chodachek & Chamberlain 2006). Conversely, predation attempts on Teal in winter are more occasional, with predators mostly acting as a low-level threat at this time of year. Gulls and raptors, especially Marsh Harrier *Circus aeruginosus* in Europe, fly over wintering Teal flocks very frequently during daylight hours (up to 130 disturbances per day locally, Fritz *et al.* 2000), but very few of these fly-overs actually end up as successful predation attempts. Intensive fieldwork during an entire winter resulted in just two observations of Marsh Harrier feeding on dead Teal, without any evidence that the harriers actually killed the Teal themselves (Fritz *et al.* 2000). Similarly, Johnson & Rohwer (1996) observed Green-winged Teal to be very frequently disturbed by Northern Harrier (*Circus cyaneus hudsonius*) fly-overs, but did not witness any actual successful predation events on Teal by these raptors. Kenyon (1961) reports similar observations with Bald Eagles (*Haliaeetus leucocephalus*) disturbing Aleutian Teal.

The main exception to this 'rule of fairly unsuccessful raptors' is probably the Peregrine Falcon *Falco peregrinus*. It is not as common at winter duck roosts as harriers and hawks are, but is a very efficient Teal predator. Peregrines readily catch Teal in flight, while most other raptors cannot. Predation by Peregrines is therefore recorded regularly at wintering sites. In Asia, Vaughan & Jones (1913) stated that some Peregrines 'seem to live chiefly on Teal'. Teal is actually such a common prey species for Peregrines that Audubon (1860a) chose, among all possible prey, to illustrate them feeding on a captured Green-winged Teal (Figure 7.5). Similarly, Pike is a localised

FIG. 7.5 *Peregrine Falcon, Gadwall and Green-winged Teal, from Audubon (1860a). The fact that this influential author decided to illustrate Peregrines feeding on Green-winged Teal amply shows that Teal are a major prey item for this falcon. Photograph by Matthieu Guillemain in a private collection.*

but sometimes very efficient predator of Teal ducklings (Treasurer 1980). Solman (1945) estimated that Pike could take close to 10% of the annual duckling production in breeding areas of the Saskatchewan River Delta.

Natural predation rates are extremely difficult to quantify. They generally represent only a minute fraction of ring recoveries (1% for British-ringed Common Teal in Ogilvie 2002, 0% for Italian-ringed birds in Spina & Volponi 2008; see Table 7.5). This, however, merely reflects the fact that predation attempts generally remain unwitnessed. Based on our present knowledge, hunting is the main mortality factor in Teal, and we will get back to this in the next chapter. Predation is likely to be among the main secondary causes of Teal mortality, and as such an attempt to quantify predation rate can be made; for example, Common Teal annual mortality rate was estimated at 51.5% by Devineau *et al.* (2010), while harvest rate was simultaneously computed as 17.8%. The Teal with rings that were shot but not sent back to the ringing centres, and the birds crippled and not recovered, have to be added to this latter percentage (see next section). Still, a large though admittedly unknown proportion of the difference between these two rates is loss to natural predation.

Specific predation on females on the nest can be evaluated by computing survival rate for the breeding season, when there is virtually no hunting. Devineau *et al.* (2010) obtained an 80.7% survival rate from spring to autumn in females, compared to 95.2% in males. Most of this 15% difference is likely to be due to predation while the females are incubating or taking care of

TABLE 7.4 *Reported Teal predators in the Holarctic. While some species are regular predators (for instance mink and Red Fox), others on this list have been reported to take Teal only a handful of times. The table is not intended to be exhaustive, but to illustrate the variety of potential predators.*

Predators		Continent	References
Predators of fully grown Teal			
Mammals			
Red Fox	*Vulpes vulpes*	N America	Sargeant 1972
			Sargeant *et al.* 1984
			Greenwood *et al.* 1995
		Europe	TDV*
			Boyd 1957
Wild Boar	*Sus scrofa*	Europe	Giménez-Anaya *et al.* 2008
Domestic Cat	*Felis silvestris*	Europe	TDV & WWT*
			Boyd 1957
Domestic Dog	*Canis lupus familiaris*	Europe	Boyd 1957
Brown Rat	*Rattus norvegicus*	Europe	TDV*
Birds			
Marsh Harrier	*Circus aeruginosus*	Europe	Tamisier 1972
			Fritz *et al.* 2000
Northern Harrier	*Circus cyaneus hudsonicus*	N America	Tamisier 1976
Harris Hawk	*Parabuteo unicinctus*	N America	Miller 1925
Greater Spotted Eagle	*Aquila clanga*	Europe	Dombrovski 2010
Golden Eagle	*Aquila chrysaetos*	Europe	Platt 1951
Goshawk	*Accipiter gentilis*	Europe	Opdam *et al.* 1977
Sparrowhawk	*Accipiter nisus*	Europe	Boyd 1957
Common Buzzard	*Buteo buteo*	Europe	Kenney 1993
Gyrfalcon	*Falco rusticolus*	Europe	Lindberg 1984
Peregrine Falcon	*Falco peregrinus*	Europe	Zuberogoitia *et al.* 2013
		N America	Otnes & Otnes 1979
			Dekker 1980
			Johnson & Rohwer 1996
Eagle Owl	*Bubo bubo*	Europe	TDV*
			Boyd 1957
Yellow-legged Gull	*Larus michahellis*	Europe	Tamisier 1972
Fish			
Pike	*Esox lucius*	Europe	Treasurer 1980
Predators of eggs and ducklings			
Mammals			
Red Fox	*Vulpes vulpes*	N America	Greenwood & Sovada 1996
American Mink	*Neovison vison*	Europe	Bengtson 1972
Striped Skunk	*Mephitis mephitis*	N America	Higgins *et al.* 1992
			Keith 1961
			Bellrose 1976
Ground Squirrel	*Spermophilus* sp.	N America	Higgins *et al.* 1992
Common Raccoon	*Procyon lotor*	N America	Higgins *et al.* 1992
American Badger	*Taxidea taxus*	N America	Higgins *et al.* 1992
Birds			
Common Raven	*Corvus corax*	Europe	Bengtson 1972
American Crow	*Corvus brachyrhynchos*	N America	Bellrose 1976

Indet. gull	*Larus* sp.	N America	Higgins *et al.* 1992
Black-billed Magpie	*Pica hudsonia*	N America	Bellrose 1976
Fish			
Pike	*Esox lucius*	N America	Solman 1945

* TDV: Tour du Valat unpublished ring recovery data; WWT: Wildfowl and Wetlands Trust unpublished ring recovery data

young ducklings. The general bias in sex ratio towards males also provides some indication of female survival: while it is likely that Teal of the two sexes are exposed to relatively similar mortality rates outside the breeding season, and sex ratio at birth is even (Bellrose *et al.* 1961), Teal populations in winter typically consist of *c.* 60% males (Guillemain *et al.* 2009a). This bias indicates roughly 33% additional mortality rate in females over their whole life, probably mostly accounted for by predation during breeding. Considering that the mean life expectancy of a Teal is slightly above 2 years (see demography section, Chapter 8), this value is consistent with the additional 15% mortality of females per breeding season computed above.

Hunting

Because of their swift flight and tasty meat, Teal have always been highly valued by sportsmen and gourmets alike (for example Audubon 1860b, Phillips 1923, Kortright 1943, Bent 1962), and as such have long been exploited by hunting (see Chapter 9). Leisure hunting of Green-winged Teal nowadays occurs throughout the United States, Mexico and Canada. Common Teal are similarly hunted in all European countries except the Czech Republic, Luxembourg, Slovakia, Slovenia and the Netherlands (http://www.artemis-face.eu). In many of the countries where Teal hunting is legal, it is the most harvested duck species after Mallard (Mondain-Monval & Girard 2000 for France, Asferg 2012 for Denmark, Finnish Game and Fisheries Research Institute 2012 for Finland, Raftovich *et al.* 2012 for the USA), and as such specific harvest management for Teal is often practised (see below).

Hunting bags in North America and Europe

The annual hunting bag of Green-winged Teal in Mexico is estimated to be very limited (14,400 birds per year on average for the 1987–1993 period, Kramer *et al.* 1995), as is the case in Canada (*c.* 54,000 individuals during the 2011–2012 hunting season, 30% of which were killed in Québec alone, Gendron & Smith 2012). The bulk of the Green-winged Teal harvest in North America occurs in the United States, where an estimated 1,950,000 were shot during the 2011–2012 season (Raftovich *et al.* 2012). This harvest was not

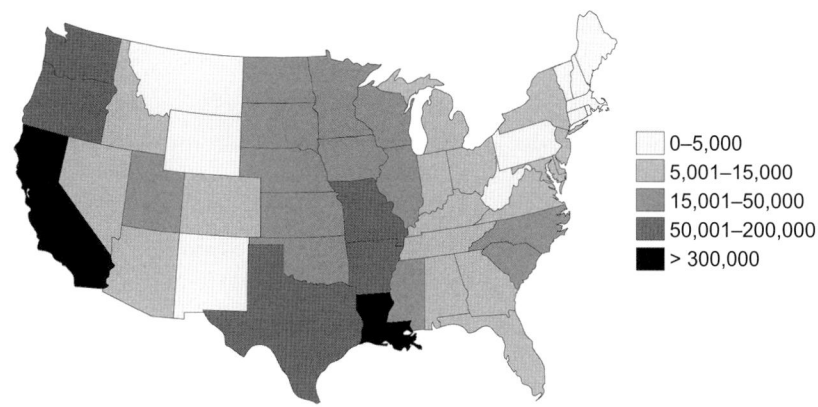

FIG. 7.6 *Estimated Green-winged Teal harvest per state in the USA during the 2011–2012 hunting season, after data in Raftovich* et al. *(2012). The estimated harvest in California and Louisiana was 311,978 and 555,781 individuals, respectively.*

evenly distributed over the different flyways: half of the birds were harvested in the Mississippi flyway alone, and on a state-by-state basis the largest harvests occurred in the main wintering areas, California and Louisiana (Figure 7.6). As a comparison, the estimated North American breeding population in spring 2011 was 2,900,100 individuals (Zimpfer *et al.* 2011). Assuming that each female produced 1.7 fledged young per season on average (see Chapter 6), the total population size immediately post-breeding would have been *c.* 5,400,000 individuals at the onset of the 2011–2012 hunting season. Harvest rate would thus have been 36% in North America that year.

Unfortunately, hunting bag statistics are not collected in a similar coordinated fashion in Europe. Harvest data are available for varying time intervals between countries, and methodologies are rarely the same. Some European countries do harvest a lot of Common Teal. For example, approximately 2.5 million birds were shot annually in the 1980s in the former USSR (back-calculated from Gerasimov & Gerasimov 1990, Priklonski & Sapetina 1990 and Yurlov *et al.* 1990, but note that this comprises a much larger area than present-day Russia, including the Asian parts). In France, 330,000 were shot during the 1998–1999 harvest season (Mondain-Monval & Girard 2000), 126,000 individuals were shot in Finland during the 2011–2012 hunting season (Finnish Game and Fisheries Research Institute, 2012), and 66,000 in Denmark in the same year (Asferg 2012). The Finnish value comprises both Common Teal and Garganey, but a previous analysis showed that Teal generally represented 99% of this total (Alhainen *et al.* 2010). The lack of consistency between hunting bag surveys among the European countries precludes any derivation of a reliable population harvest rate. In particular, the lack of comprehensive data for the number of Common Teal being produced and harvested at the flyway level (including Russia) is a major shortcoming. Mooij (2005) attempted to provide such an estimate for the countries in the European Union, suggesting that around one million Common Teal were harvested per season in this geographic area, which represented 34.7% of the total population size as estimated by the International Waterbird Counts (Wetlands International 2012). However, the author acknowledged that the quality of the data was uncertain for many of the countries under consideration. Still, Hirschfeld & Heyd (2005) provide a similar number, namely 960,000 harvested Common Teal in the European Union plus Norway and Switzerland. The ratio of the total number of Common Teal harvested to the total number counted is therefore similar to that computed in North America for Green-winged Teal.

Deriving harvest rate from ring recoveries

Harvest rate can also be derived from the analysis of ring recoveries. Virtually all ringing recoveries of Common Teal in Europe and of Green-winged Teal in North America are due to hunting (freshly dead birds reported by hunters or shot birds later found dead), none of the other causes of mortality contributing more than 1% of all recoveries (Table 7.5, see also Boyd 1957). Though this clearly indicates a heavy hunting pressure on these ducks, it must of course be kept in mind that ring recoveries only provide information on deaths reported by people, so that natural mortality causes are always strongly underestimated; for example, less than 0.8% of rings coming from predation events in any of the three datasets we analysed. Boyd (1961) noted that 'It seems unlikely that ringing [...] can make a major contribution to our knowledge of the causes and incidence of wildfowl mortality, other than the consequences of human activities'. On the other hand, that so many rings are reported by hunters provides a unique opportunity to compute harvest rate, the proportion of the population that is shot by hunters within a given time frame.

Harvest rate was computed from ring recoveries by Devineau *et al.* (2010) for both Common and Green-winged Teal, suggesting an 18% harvest rate in Europe and a 7% harvest rate in North America. This was based on the assumption that 30% of the rings recovered by hunters are reported to the ringing centre. Using a 23% ring reporting rate for most cases and the same data set, Fleming & Howell (2013) obtained a harvest rate of between 8 and 13% for Green-winged Teal, depending on age and sex class.

The discrepancy between harvest rates computed from hunting bag totals and from ring recoveries is remarkable, since the former method provides a rate that is about twice as high as the latter. One point is that the monitoring of breeding duck populations in North America and the International Waterbird Counts in Europe aim at providing robust annual estimates of population size, which can be compared across years as indices, but should not be considered as yielding a full population count. This is because land

TABLE 7.5. *Reported mortality causes of ringed Teal later recovered, expressed as percentages of total number of recoveries. Data courtesy of Tour du Valat, Wildfowl & Wetlands Trust (Abberton reservoir) and USGS Bird Banding Laboratory (North America).*

Mortality Cause	Tour Du Valat (59,187 birds ringed 1952–1978: 9,430 recoveries)	Abberton Reservoir (34,149 birds ringed 1953–1992: 5,562 recoveries)	North America (506,000 birds ringed 1914–present: 39,242 recoveries)
Hunted	92.68	91.98	97.04
Killed by predators	0.59	0.79	0.22
Deliberately captured (duck decoy, by hand, etc.)	0.41	0.00	0.03
Accidentally captured (fishing net, rat trap, etc.)	0.42	0.95	0.26
Accidental death (road casualty, wires, etc.)	0.17	0.40	0.07
Adverse weather (esp. cold)	0.49	0.07	0.02
Injured, sick, poisoned	0.10	0.13	0.47
Unknown + found dead	5.13	5.68	1.89

coverage of these schemes, though large, is far from comprehensive. The specific habitat use by Teal (low breeding density, scattered groups in winter; see Chapter 4) then probably leads to an underestimation of population size (see also Owen *et al.* 1986). The total hunting bag may hence be a fraction of a significantly larger population than estimated by these bird counts; that is, this method probably overestimates harvest rate.

The second point is related to ring reporting rate, the actual value of which is unknown. In most duck studies the reporting rate is more or less directly derived from the 30% value obtained for Mallard in North America before the 1990s by Nichols *et al.* (1991). It is known that the percentage of rings recovered in later years differs among duck species, fluctuates over time and varies between geographic areas (Nichols *et al.* 1995, Guillemain *et al.* 2011), a large part of the differences being due to varying reporting rates. Applying the Nichols *et al.* Mallard value of 30% to Green-winged Teal in different time periods and in different geographic areas may therefore be misleading. To avoid this pitfall in future analyses, a new programme has been initiated in Europe, in which a second ring is fitted to Common Teal along with the regular numbered ring (Figure 7.7). This second ring bears the mention of a significant monetary reward to the hunter reporting it. By using a reward that is large enough, a 100% reporting rate is expected for these birds, which by comparison allows estimation of the reporting rate for birds with only one standard ring (Nichols *et al.* 1991).

To summarise, it is likely that the actual harvest rate for Teal lies somewhere between the 35% computed from bag sizes and bird counts, and the 10–18% obtained from capture-recapture modelling.

FIG. 7.7 *For many reasons it is important to know the proportion of Teal that is actually harvested. A crucial first step towards this aim is to estimate reporting rate, in other words how many ringed shot birds are actually reported as recoveries by hunters. This picture shows a Common Teal fitted with both a standard numbered ring (right leg) and a reward ring (left leg). Both rings come from the same manufacturer and only their engraving differs. Reward rings bear the mention 'RECOMPENSE/REWARD 70€' and an email address plus a phone number to which the two rings of the bird have to be reported to get the reward. Only a limited number of birds are fitted with reward rings per year so as not to bias hunting pressure. The programme is advertised widely in hunting magazines and on websites so that hunters know what to do if they happen to shoot a doubly ringed Teal. Photograph by Mathieu Remacle.*

The question of crippling loss

In addition to the Teal killed and retrieved by hunters, a proportion of birds are shot down but never found, or die from their wounds some time later. Such additional deaths related to hunting are referred to as 'crippling loss' (Bellrose 1953, Anderson & Burnham 1976). Because dead ducks are notably difficult to find (cf. the low carcass-collection rate even after intensive search trials during botulism outbreaks in Bollinger *et al.* 2011), and duck carcasses generally disappear to scavengers within a very short time (Pain 1991a), crippling loss is extremely difficult to measure. Even crude estimates are difficult to obtain: French hunters interviewed while reporting a ringed duck estimated that 15% of the ducks shot dead are not retrieved, but this 'guesstimate' varied from 0% to 50% between respondents (Guillemain 2011). In addition to questionnaire surveys of hunters, crippling loss can be estimated by observing the activity of hunters directly in the field, sometimes from spy blinds. While interviews generally yield estimates of crippling loss of around 20% in ducks (17 to 20% for Green-winged, Blue-winged and Cinnamon Teal in North America, Martinson *et al.* 1966), the latter method more frequently suggests a 35–40% rate (review in Norton & Thomas 1994). Crippling loss is much lower when hunters use retrieving dogs.

The 'kill rate' *sensu* Anderson & Burnham (1976) is the proportion of birds killed by hunters during a time interval: the sum of the harvest rate (birds killed and retrieved) and the proportion of birds crippled. Kill rate is therefore a better measure of the potentially limiting role of hunting on population dynamics than is harvest rate. Considering the figures above, if the harvest rate is around 25% of the population and crippling loss is around 30% of shot birds, approximately one-third of the Teal alive at the beginning of the hunting season would die from hunting causes (other than lead poisoning) before the end of it. This is of course likely to vary among geographic areas and from year to year.

Is hunting limiting Teal populations?

Common Teal populations were believed to have considerably declined over the 19th century in Europe, probably because of the massive commercial harvest with, for example, live duck decoys and punt guns (Phillips 1923). A similar situation was described for ducks in general in North America over the same period (see accounts in Day 1949). This probably unsustainable harvest gradually led to the ban on massive commercial duck hunting on the two continents.

The question of whether current harvest, mostly practised as a hobby or sport, is still a limiting factor for Teal and other waterfowl populations has long been hotly debated (Pöysä *et al.* 2004, Sedinger & Herzog 2012, Pöysä *et al.* 2013a). Analysing the first large ringing datasets for Common Teal by the end of the 1950s, Boyd (1957, 1962) concluded that 'Heavy winter losses due to shooting are offset by reduced losses in subsequent months, [. . .] with the implication that non-shooting mortality should not be regarded as constant and independent of the kill'. Boyd hence considered that higher mortality due to hunting was largely compensated for by increased natural survival: many Teal that were shot by hunters would otherwise have died from natural causes anyway. Although he acknowledged the possible existence of such a compensation process and actually cited Boyd's work, Tamisier (1970) considered hunting mortality of Common Teal to be largely additive: a sudden halt of hunting for one season would increase the population by almost the number of birds not harvested plus the young they produced. Hunting would be a strongly limiting factor in the latter scenario. In another analysis, Moisan (1966) concluded that hunting mortality was almost completely additive to natural mortality in adult Green-winged Teal, while it was largely compensatory in first-year birds.

A considerable amount of research and monitoring has been devoted to waterfowl populations since the 1960s. Anderson & Burnham (1976) rejected the idea of a purely additive effect of hunting mortality on natural mortality, arguing that some sort of compensation by natural mortality was taking place. Alternating between more liberal and more restrictive waterfowl hunting regulations over many years demonstrated an effect on harvest (greater during more liberal years; that is, hunting regulation is effective), and has allowed the testing of the relationship between harvest and survival rates, suggesting an intermediate situation between purely additive and purely compensatory effects of hunting mortality on natural mortality (Rogers *et al.* 1979, Nichols *et al.* 1984, review in Nichols & Johnson 1996).

The specific demographic characteristics of Teal (see Chapter 8) make them particularly capable of compensating for harvest, although we suspect that such compensation is likely to be only partial. Both Common Teal in western Europe and Green-winged Teal in North America have been increasing in numbers over recent decades (see Chapter 3) despite large annual hunting bags on both sides of the Atlantic. It is therefore reasonable to conclude that hunting is possibly a limiting factor of Teal populations, which may increase even faster without harvest (see also Tamisier 1995). However, this limiting effect is probably offset to some extent by the ability of Teal to partly compensate demographically, so that harvest as currently practised does not represent a threat to these populations, albeit limiting their growth. A recent analysis of Green-winged Teal ringing data suggested that the current harvest rate was sustainable in the sense that it could still be increased without causing the population to decline (Fleming & Howell 2013). Similarly, Pöysä *et al.* (2013b) did not think that hunting is the major driver of recent population declines of breeding Common Teal in Finland.

Human disturbance

In line with their long history of harvest exploitation, Teal and other waterfowl are sensitive to human disturbance, hence the famous words of G. V. T. Matthews (1982) 'It is Man who makes wildfowl wild'. The effects of disturbance from hunting and other non-consumptive recreational activities on waterfowl have been widely studied, generally demonstrating alteration of time budgets as well as changes in temporal and spatial distribution (reviews in Meltofte 1982, Boyle & Samson 1985, Madsen & Fox 1995, Tamisier *et al.* 2003, Blanc *et al.* 2006).

Such effects of disturbance have been demonstrated specifically for

Common and Green-winged Teal, which are often considered to be among the most sensitive species. For example, in a National Wildlife Refuge in Florida, Green-winged Teal was the species whose distribution was the most extremely skewed towards the most distant areas from a drive used by tourists (Klein *et al.* 1995). Pease *et al.* (2005) similarly showed that Green-winged Teal was among the dabbling duck species responding the most to disturbance by visitors in a refuge: the birds always responded to disturbance, while some other duck species did not. Furthermore, their reaction was to take flight almost every time, while some other species only swam away. In Europe, Tuite *et al.* (1984) considered Common Teal to be among the most sensitive waterfowl species to water-related recreation. For instance, Common Teal have been documented to redistribute within British lakes and reservoirs or to completely abandon some waterbodies due to the development of sailing, fishing and other recreational activities (Batten 1977, Tuite *et al.* 1983, Bell & Austin 1985; see also Burton *et al.* 2002 for the effect of construction works on Common Teal on intertidal mudflats). Gilburn & Kirby (1992) showed that Common Teal numbers were significantly greater at British sites where human disturbance was lower, and Salmon & Fox (1994) showed that numbers within the Severn Estuary increased the most at sites protected from human disturbance.

Bregnballe *et al.* (2009b) experimentally measured the response of Common Teal to an approaching person, and showed that such a simple disturbance event influenced Teal behaviour for more than an hour. Guillemain *et al.* (2007e) showed that disturbance by ecotourists lead to Common Teal moving a great distance away from observation hides, but that they were back in their initial position one hour later. It was concluded that appropriate facilities (protection of paths to make visitors to hides invisible to the birds) could considerably reduce the disturbance effect. The most frequently visited wetland in the study area did not have a lower density of Common Teal than a sanctuary in the reserve where visitors were not allowed. This also suggests that Teal can habituate to some sources of disturbance to some extent (see also habituation to low-flying military aircraft in Conomy *et al.* 1998).

One source of disturbance to which Teal apparently do not habituate is hunting, which may be expected considering their status as a quarry species. Madsen (1988) showed that hunting disturbance displaced Common Teal from the most profitable feeding patches, which they only used at night after hunting had ceased. Tamisier & Dehorter (1999) considered hunting disturbance to be a limiting factor for the carrying capacity of the whole of the Camargue as a winter quarter, and Väänänen (2001) documented a major redistribution of Common Teal towards protected areas in response to the onset of the hunting season. This is a common behaviour of waterfowl in general (Meltofte 1982), so that the creation of adequate refuges is considered the most effective way to alleviate hunting disturbance (see Fox & Madsen 1997). Such creation of refuges has proven to be effective for Teal: the establishment of a sanctuary on the Dee Estuary in the UK led to a fivefold increase in Common Teal numbers (Hirons & Thomas 1993), while the creation of hunting-free reserves in Denmark led to a 23-fold increase (Madsen 1998, see also Bregnballe *et al.* 2004). At the scale of a country, larger numbers of Common Teal are counted in protected than unprotected sites in France (Fouque *et al.* 2009).

The establishment of a protected area can also be viewed as a scientific experiment to study the factors limiting site use by birds. The creation of a new reserve in Marais du Vigueirat in the Camargue led to a massive increase in Common Teal numbers, from a few birds to several thousand within a few years on this 1,000-ha estate (Mathevet & Tamisier 2002). Interestingly, this occurred at the expense of another protected area 10km away, suggesting that habitat features also played a major role in attracting the Teal; the birds were already wintering in the area but were previously kept from using the more attractive Marais du Vigueirat by hunting disturbance. Admittedly, though, the creation of a refuge is often not only about a hunting ban: many reserves are attractive to Teal because habitats are specifically managed to match the ecological requirements of waterfowl, and other sources of human disturbance are also limited. For example, the ban on bait digging at Lindisfarne National Nature Reserve, England, led to a sevenfold increase in Common Teal numbers within a year (Townshend & O'Connor 1993). Conversely, the use by Common Teal of a sanctuary close to Cambridge, UK, decreased when recreational fishing was also taking place (Cooke 1975).

Whether human disturbance should be considered a limiting factor for Teal populations is open to debate. Clearly, disturbance does affect Teal behaviour and distribution, sometimes profoundly. However, these are often short-term effects on the individuals, while the long-lasting impacts on individual fitness (body condition, survival and fecundity), and hence population dynamics, are far more difficult to demonstrate. This is not only true in the case of Teal, but is also rather a common feature of disturbance studies in general (Keller 1996, Robinson & Pollitt 2002, Blanc *et al.* 2006). In waterbirds this is largely due to their migratory nature and the fact that disturbance is most often studied at stopover or wintering sites, far away from the breeding sites where some ultimate consequences for individual fitness could be measured (Madsen & Fox 1995). One major exception is the research on the effects of experimental spring hunting of the Greater Snow Goose, which demonstrated that hunting disturbance had an effect on reproductive output at the population level (Mainguy *et al.* 2002). To our knowledge, the only study documenting an effect of disturbance on breeding Teal is a case study published by Reichholf (1970), who demonstrated that abandonment of two German ponds by breeding Common Teal followed an increase in angling activities. It is probably not realistic to carry out large-scale studies of breeding Teal and relate this to disturbance effects, especially winter disturbance. However, this may be done indirectly by studying body condition changes. For example, earlier studies have shown that Mallard in more disturbed situations had a lower body mass than undisturbed conspecifics, both in the field (Szymanski *et al.* 2013) and in aviary conditions (Zimmer *et al.* 2010). The latter authors considered this as a strategic decision of the birds to decrease wing loading and hence increase escape ability (disturbed and undisturbed birds were both fed *ad libitum*). Regular natural body mass changes of Teal have been widely documented, both in Europe and in North America (see Chapter 5, especially Figure 5.2). Furthermore, a general population-level relationship between late-winter female body condition and subsequent breeding success has been established (Figure 7.8), probably because Teal in better condition by the end of winter can migrate efficiently, and perhaps reach the breeding grounds earlier and with better body condition (Guillemain *et al.* 2008b). The demonstration that disturbances reduce body mass more than expected on the basis of common annual patterns may hence be viewed as evidence of long-term impact on population dynamics; that is, an indication that this source of disturbance is a limiting factor for the population.

Adverse weather

We have previously discussed the greater relative energy requirements of Teal resulting from their small body size (Chapter 5), and the fact that they preferentially forage in shallow areas (Table 5.2). The small size of Teal also makes them particularly sensitive to cold, since their Lower Critical Temperature (LCT, the temperature below which additional energy is required for thermoregulation) is around 16.5°C, and the metabolic rate of such small species is also particularly sensitive to temperature (review in Dalby *et al.* 2013). From an ecophysiological perspective, Bennett & Bolen (1978) showed that beyond temperature, it was especially winter wind velocity (hence wind chill) and relative humidity that were affecting Green-winged Teal, causing an increase in uric acid and blood urea concentration, together with a decrease in body condition. Snowfall and temperature, too, affected Green-winged Teal, although mainly sporadically during cold spells, likely through reduced access to food (deeper snow cover) or the freezing of shallow waterbodies.

That body condition in Teal is affected by adverse (winter) weather is a general finding, with cold winters leading to changes in typical temporal patterns in body mass (or body condition) through the winter (Figures 4.8 and 5.2; see also Baldassarre *et al.* 1986, Fox *et al.* 1992). If weather deteriorates, Teal soon leave their winter quarter for a milder one, generally moving south or west where their numbers rapidly build up (Munro 1949, Dobinson & Richards 1964, Wolff 1966, Saint-Gerand 1982, Ogilvie 1982, 1983, Perdeck & Clason 1983, Rüger *et al.* 1986, Monval & Pirot 1989, Ridgill & Fox 1990, Schummer *et al.* 2010, Dalby *et al.* 2013). Interestingly, the response of Teal to adverse weather greatly depends on when cold spells occur during the winter. For example, a massive departure was observed from the Camargue during the cold spell of 1963, which started in December in the UK. Conversely, the

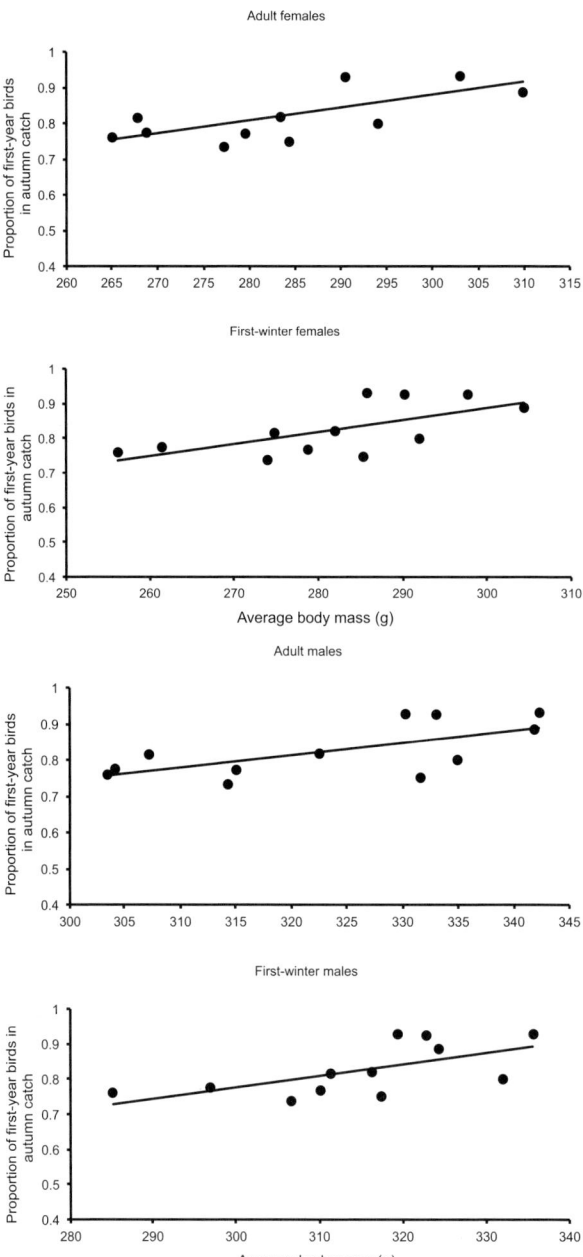

FIG. 7.8 *Relationship between the proportion of juvenile birds in the autumn catch of Common Teal in the Camargue, used as an index of breeding success, and the body mass of individuals wintering in the same area by the end of the previous winter. All regressions are significant, see Guillemain et al. (2008b) for details. Reprinted with kind permission from John Wiley & Sons, Inc.*

cold of February 1956 in the Camargue occurred at a time of the winter when Common Teal were about to leave northwards on spring migration, so the birds were likely to be more reluctant to fly 'in the wrong direction' towards the Iberian Peninsula, and more casualties due to cold were recorded locally (Impekoven 1965). Apart from such late cold spells, adverse winter weather in general translates into departure rather than massive die-offs of Teal (Boyd 1964, Harrison & Hudson 1964, Pilcher 1964). When casualties are recorded, it is more frequently male than female carcasses that are found, probably because males have a greater tendency to try to remain at cold sites further north, while females more readily retreat south (Stout & Cornwell 1976, Bennett & Bolen 1978).

This is not to say that Teal are not affected by normal midwinter cold spells: both Gitay *et al.* (1990) and Devineau (2007) recorded much lower annual survival rates in colder winters (Figure 7.10). Ogilvie (1983) also commented on this pattern, and considered that it was mainly due to increased hunting mortality in refuge countries of southern Europe during cold spells rather than being due to adverse weather per se. In addition to hunting, it is possible that some birds escaping a frozen winter quarter never make it to their expected refuge area due to depleted body reserves, or that cold stress has carry-over effects during the subsequent months in terms of survival and/or breeding success.

On the other hand, particularly mild winter conditions have also been observed to affect Teal behaviour. British-ringed Common Teal migrated earlier in spring after warmer winters (Ogilvie 1983), and similarly warm and rainy winters were associated with earlier spring arrival of migrants in Finland, likely because they were in a better body condition by the end of winter,

FIG. 7.9. *Teal sometimes have to endure periods of adverse weather during winter. Besides the physiological stress this may cause, cold periods also quickly lead to the freezing of shallow waterbodies and snow cover on the ground, making food inaccessible. Photograph by Michel Collard.*

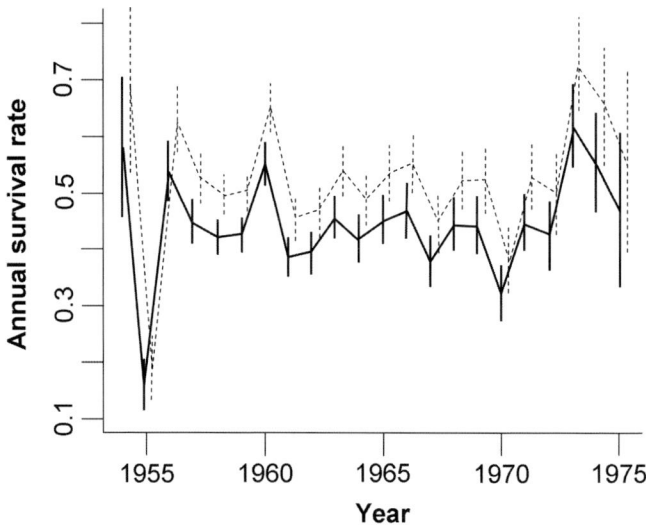

FIG. 7.10 *Annual survival rates of male (dashed line) and female (solid line) Common Teal ringed at Tour du Valat, Camargue, France. Vertical bars show 95% confidence interval. The devastating effect of the February 1956 cold spell (here year 1955, as based on calendar year at the beginning of winter) on annual survival rate is apparent. From Devineau (2007) with kind permission.*

spring was earlier, there were more tailwinds for migration and/or they wintered closer to the breeding grounds (Vähätalo et al. 2004). Krivonosov & Rusanov (1990) also documented more Common Teal wintering along the northern coast of the Caspian Sea, closer to their Russian breeding grounds, during milder winters. At the individual level, earlier arrival on the breeding grounds translates into higher breeding success in Teal (Elmberg et al. 2005). Therefore, despite the higher risks of wintering as far north as possible, there is potentially a subsequent fitness reward for those who make it successfully.

Much less is known about the effect of weather on Teal during the breeding season. Only Ogilvie (1983) documented movements within the breeding range to the north and east during drought years, which he assumed to potentially be the movements of birds seeking suitable breeding sites. Hail storms have also been reported to kill many ducks, mostly prefledged ducklings (review in Stout & Cornwell 1976), although this was not specifically in Teal.

Beyond short-term sensitivity to weather conditions in individual birds, Teal populations should also be particularly affected by long-term climate change. In Common Teal ringed in the Camargue in winter, a long-term relationship between temperature and body size has been observed; that is, they tend to be larger (have longer wings) in mild winters than in cold ones (Guillemain et al. 2005c). This was thought to reflect a decreasing need for small-bodied individuals (those with the greater energy needs) to find shelter in the south of France as the temperature gets milder. Ultimately, it was considered that this could be the first step towards a general northwards shift of the wintering geographic range of this species. Over the last 30 years, Camargue-wintering Common Teal have also become much heavier, with the increase in mean body mass being over 10% in some sex and age classes (Guillemain *et al.* 2010a). This was hypothesised to reflect a general warming of the temperature at this winter quarter, becoming less and less energy-stressful to wintering ducks, though non-mutually exclusive hypotheses linked with habitat change could not be excluded. That Gunnarsson *et al.* (2011) did not record a similar increase in body mass for migrating Common Teal caught in southern Sweden suggests in any case that the changes have mostly occurred on the wintering grounds. Such a massive increase in winter body condition provides Teal with much greater safety margins to buffer them against sudden cold spells. This also likely translates into greater breeding output (see Figure 7.8), maybe through increasingly early arrival at the breeding grounds. This is not only the case in Europe, since Green-winged Teal have also been recorded to arrive earlier at Delta Marsh, Manitoba (Canada) over the last 60 years, the arrival date being 1.6 days earlier for each degree Celsius gained (Murphy-Klassen *et al.* 2005). Migration dates in autumn have also changed over the last decades in Finland, with movement south becoming more postponed as temperature increased (Lehikoinen & Jaatinen 2012).

Finally, it is also projected that the breeding range of Teal could well change over the next several decades owing to climate change (Huntley *et al.* 2007, Figure 7.11). However, because Common Teal already breed as far north as the northern coast of the European continent this may translate into a shrinking of the breeding geographic range rather than a physically impossible northwards shift. Whether this will cause increases in Teal density large enough to reduce breeding success in the remaining breeding area through density-dependent processes remains speculative.

To summarise, weather conditions can be a limiting factor for Teal populations, in particular given the effect of cold spells on individual survival and, possibly, on breeding success via decreased body condition. Climate change may potentially make wintering conditions more benign for ducks, and hence promote the individual fitness of Teal in the future. The consequences of long-term climate change on the breeding grounds, apart from potential geographic redistribution, are so far unknown.

Habitat loss and change

Besides hunting, textbooks list habitat loss and change as the most severe threats to waterfowl, with wetland loss to agriculture in turn being the main cause (Bellrose 1976, Owen *et al.* 1986, see also Long *et al.* 2007). The same has been claimed specifically for Common Teal, for example in the former USSR (Flint & Krivenko 1990), in Chinese lakes of the Yangtze and Han

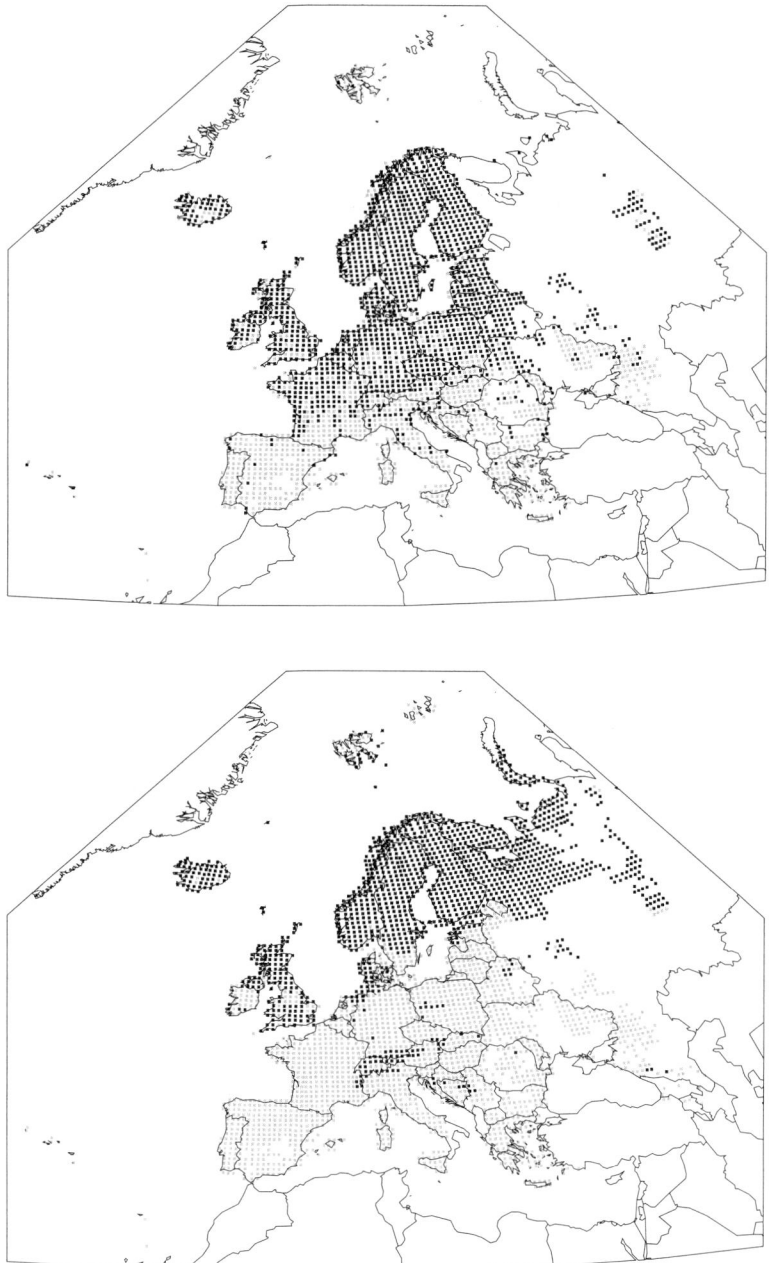

FIG. 7.11 *Present (top) and predicted (bottom, for the end of the 21st century) distribution of Common Teal under current global climate change scenarios. The pale area is the wintering range, the dark area is the breeding range. Note that the overall breeding range is predicted to shrink drastically, as there is no possibility for Teal to expand further north. From Huntley* et al. *(2007), with kind permission.*

basins (Hu & Cui 1990), and in Europe as a whole (Rüger *et al.* 1986, Fox 2005b).

Some data do exist on long-term habitat changes in wetlands, especially those used by Teal in winter, and these can be compared with bird population size estimates at the same spatial resolution. At small geographic scales, this has proven a useful method for studying and demonstrating a negative link between these two variables; for example, the dramatic crash of Common Teal numbers in the Baie de l'Aiguillon, once the second most important wintering site for waterbirds in France, coincided with the drainage of surrounding wet grasslands used for nocturnal feeding (Duncan *et al.* 1999). At broader geographic scales, such statements are obviously speculative based on correlational studies alone, although it is very likely that major large-scale habitat changes linked with the mechanisation of agriculture, for example, do have an effect on the birds using these former wetlands.

However, several specific features of Teal ecology make these birds special among waterbirds when it comes to the potential effects of habitat loss and change. First, Teal breed mainly in remote northern areas, both in the Palearctic and the Nearctic (see Chapter 3). This may to a large extent have saved them thus far from massive habitat change linked to human activities, in contrast to more densely inhabited zones at lower latitudes. In North America, for instance, Green-winged Teal breed mainly in the boreal zone rather than in the prairies, which have been more dramatically affected by agricultural changes over the last century than boreal forests. Breeding further north has also largely allowed Green-winged Teal to escape the major losses suffered by prairie-nesting ducks during long drought periods such as those in the 1930s and 1950s (Bellrose 1976, Williams 1990, Bethke & Nudds 1995, Johnson 1995). Conversely, Green-winged Teal may have been particularly affected by the collapse of Beaver populations in the 1930s and the resulting disappearance of Beaver ponds (Batt 2012), which are a major breeding habitat for Teal on both continents (Baldassarre & Bolen 2006, Nummi & Hahtola 2008; see Box 4.3), but data are lacking to precisely assess the magnitude of this effect. In fact, both Common Teal and Green-winged Teal numbers have globally increased since waterfowl counts were initiated in the second half of the 20th century, so that it is difficult to argue that these species have been negatively affected by habitat degradation and loss over that period. Many authors actually refer to habitat improvements over the species' wintering range, especially through the creation of protected areas with adequate habitat management, as a likely explanation for population increases (Rüger *et al.* 1986, Deceuninck 1997, Fox 2005b). Similarly, albeit at a more restricted geographical scale, the increase in numbers and maturation of gravel pits in the UK may have contributed to the rapid increase in the number of wintering Common Teal (Gilburn & Kirby 1992, Kirby *et al.* 1995). Although habitat loss and degradation may be a theoretically limiting factor for Teal populations, rather the opposite process has been observed over the last few decades: increases in Teal numbers linked with habitat improvement.

CHAPTER 8

Demography

We have already touched on Teal demography in several parts of this book. We dealt with population size and trends in numbers, as well as the distribution of individuals within their geographic range, in Chapter 3. We also considered some demographic parameters such as fecundity (Chapter 6) and mortality (Chapter 7). In this chapter these parameters are put in perspective by compiling the available information on Teal vital rates. This allows us to see how these are responsible for the composition of populations as indexed by variation in age structure and sex ratios. Based on this information, we have computed expected population dynamics, and compared these with estimates of population trends.

TEAL SURVIVAL RATES

Since pre-fledging mortality patterns and rates have already been dealt with at length in Chapter 6, we here consider only survival rates after fledging. Over the years, a relatively large number of annual survival rates have been estimated for Teal (Table 8.1), ranging from 26.5% for juvenile birds in Québec (Moisan 1966) to 66% for adult females ringed in Alaska (Lake *et al.* 2006). Changes in analytical methods over time and differences in sample sizes likely explain a large part of the differences among studies. In particular, most of the early estimates are derived from composite life tables (cf. Hickey 1952), while estimates published after 1970 are more generally based on Brownie models of ring recoveries (Brownie *et al.* 1985), sometimes also encompassing live recaptures as in Devineau *et al.* (2010). However, there are general patterns, often discussed in many of the papers listed in Table 8.1: i) annual survival rates may vary dramatically between years for a given ringing site (see differences between time periods for British-ringed birds in Gitay *et al.* 1990), ii) adults generally have higher survival rates than juveniles, and iii) males generally have higher survival rates than females.

That Teal survival rates vary from year to year is not surprising given the fluctuating nature of wetland ecosystems, whose area or even mere presence is profoundly dependent upon environmental conditions, especially rainfall (see changes in the number of May ponds depending on drought conditions in USFWS 2012). Similarly, very adverse winter weather can cause survival rates to drop, as was observed for Common Teal in the Camargue during the exceptionally cold period of February 1956 (Figure 7.9; see also Figure 7.10).

It also seems natural that less experienced juveniles have lower survival rates than adults (Boyd 1957, Owen & Black 1990, Johnson *et al.* 1992); in addition to having a poorer knowledge of their environment, first-year birds may also have a less efficient immune system (hence being more prone to potential effects of diseases). It should be kept in mind that survival rate has always been derived from winter ringing in Europe, and in many cases also in North America (Devineau *et al.* 2010). Hence, this provides only an estimate of juvenile survival from the first winter to the next, while the survival rate during the first months of life, from fledging to arrival at the wintering grounds, is generally unknown (Owen & Black 1990). This problem has recently been overcome by the use of indirect methods based on wing examination of hunted Teal, which demonstrate considerable mortality in juveniles during the first few months after fledging (Box 8.1).

BOX 8.1. Survival of juvenile Teal during the first months after fledging

Teal ringing in Europe, and to some extent in North America, has historically relied on autumn and winter catches (Devineau *et al.* 2010). This is because Teal occur in greater concentrations and are easier to catch for ringing during these seasons. Table 8.1 lists annual survival rate estimates obtained from winter ringing. Unfortunately, this does not provide any information about survival of juveniles from fledging to arrival at the wintering grounds, since only birds surviving long enough to be ringed on wintering sites are included. This will inflate survival estimates, and is problematic for several reasons.

The first few months after fledging represent a crucial part of the life cycle, when young individuals have to make their first migration flight while still being especially naive about predators, food finding, disease (naive immune system) and hunters, and not necessarily knowing the migration route, and not yet being experienced foragers. However desirable, it is nevertheless generally difficult to obtain survival estimates

TABLE 8.1 Annual survival rate estimates for Common Teal and Green-winged Teal based on analysis of ring recoveries. SD denotes standard deviation and SE standard error.

Geographic area[1]	Period of ringing	Sex[2]	Age[3]	Annual survival	Reference
Common Teal					
United Kindom	Winter 1949–1955	M	All	50.7 ± 1.9% SD	Boyd 1957
				43.0 ± 2.1% SD	Boyd 1957
United Kingdom	Winter – years unknown	All	All	49 ± 3% SD	Boyd 1962
Finland	Summer 1949–1963	All	J	41%	Grenquist 1965
United Kingdom	All seasons 1949–1966	All	All	55%	Wainwright 1967
Camargue	Winter 1955–1968	M	All	46.8%	Tamisier 1970
		F	All	43.7%	Tamisier 1970
France	Winter 1955–1968	M	All	47.2%	Tamisier 1970
		F	All	44.5%	Tamisier 1970
Italy+Spain+Portugal	Winter 1955–1968	M	All	44.5%	Tamisier 1970
		F	All	40.4%	Tamisier 1970
Central Europe	Winter 1955–1968	M	All	55.2%	Tamisier 1970
		F	All	50.7%	Tamisier 1970
Former USSR	Winter 1955–1968	M	All	51.1%	Tamisier 1970
		F	All	46.2%	Tamisier 1970
Britain	All seasons 1954–1970	All	A	59.2 ± 1.4% SE	Gitay et al. 1990
		All	J	53.0 ± 3.2% SE	Gitay et al. 1990
	All seasons 1974–1985	All	A	52.5 ± 2.1% SE	Gitay et al. 1990
		All	J	53.2 ± 5.4% SE	Gitay et al. 1990
	All seasons 1968–1985	All	A	54.1 ± 1.2% SE	Gitay et al. 1990
		All	J	56.6 ± 3.5% SE	Gitay et al. 1990
	All seasons 1968–1978	All	A	57.6 ± 3.1% SE	Gitay et al. 1990
		All	J	56.0 ± 4.2% SE	Gitay et al. 1990
Abberton, England	All seasons 1954–1970	All	A	57.9 ± 1.6% SE	Gitay et al. 1990
		All	J	49.2 ± 3.3% SE	Gitay et al. 1990
		M	All	54.5 ± 0.9% SE	Gitay et al. 1990
		F	All	40.7 ± 2.5% SE	Gitay et al. 1990

Camargue, France	Winter 1954–1976	M	A	52.5 ± 10.8% SE	Devineau et al. 2010
		F	A	44.5 ± 9.2% SE	Devineau et al. 2010
Green-winged Teal					
Utah, USA	Summer–Autumn 1929–1948	All	All	52 ± 2% SD	Boyd 1962[4]
North America	Summer–Autumn up to 1961	All	All	37%	Moisan 1966
Québec, Canada	Summer–Autumn up to 1961	All	J	26.5%	Moisan 1966
		M	A	49.3%	Moisan 1966
		F	A	37.7%	Moisan 1966
Saskatchewan, Canada	Summer 1955–1961	M	A	49.5 ± 4.1% SE	Martin et al. 1979
North America	Jan–Feb 1950–1989	M	All	55%	Chu et al. 1995
		F	All	51%	Chu et al. 1995
Alaska, USA	Summer 1989–2000	M	A	64 ± 6% SE	Lake et al. 2006
			J	50 ± 7% SE	Lake et al. 2006
		F	A	66 ± 4% SE	Lake et al. 2006
			J	61 ± 9% SE	Lake et al. 2006
North America	Jan–Feb 1960–1998	All	A	54.5 ± 1% SE	Devineau et al. 2010
North America	Summer 1970–2008	M	A	58 ± 9% SE	Olson 2013
			J	52 ± 6% SE	Olson 2013
		F	A	50 ± 11% SE	Olson 2013
			J	41 ± 7% SE	Olson 2013

[1] this refers to where the birds were initially ringed, except in Tamisier (1970) where all birds were ringed in Camargue, southern France, but geographic area refers to where the birds were later recovered.
[2] M: males; F: females; All: both sexes combined or differences not specified in reference.
[3] A: adults; J: juveniles (first 12 months); All: age class combined or differences not specified in reference.
[4] from data in Van Den Akker & Wilson (1949).

for the late summer period, mostly because at this time of year Teal occur at low densities in remote areas (see Chapter 5).

To overcome this problem, an indirect method was recently used (Guillemain *et al.* 2010b): Common Teal hunting bags were examined throughout the western European flyway, from the breeding grounds in northern Finland to wintering areas in western France. Because the adult survival rate is known to be *c.* 83.5% for the three months of the autumn migration (from the ringing of these birds during previous winters, after Devineau 2007), the idea was to monitor the percentage of first-year birds in the bags of hunters throughout the autumn migration period, across the flyway. A constant age-ratio would indicate that the juvenile survival rate is similar to that of adults. Conversely, if the age-ratio changed along the flyway then the specific survival rate of juveniles could be computed using adult survival as a reference.

This approach showed that the percentage of first-year birds in the hunters' bags clearly decreased along the autumn migration flyway, from 88.7% in northern Finland to 58% in western France three months later (Figure 8.1).

This indicates that a theoretical population of 1,000 Common Teal shot in northern Finland would be composed of 887 juveniles and 113 adults. The survival rate of adults being 83.5%, only 94 would survive and reach western France three months later, where they would then represent 42% of the population. Hence, this winter population would comprise 224 individuals, of which 130 would be first-year birds. The survival rate of those first-year birds over the study period would therefore be 130/887, or 14.7%. Accordingly, less than 15% of the fledged juveniles are predicted to survive their first autumn migration. Given the uncertainties inherent in this type of data, the 15% percentage should be considered a conservative and mainly informative value. However, another analysis based on wings in hunters' bags suggested that the autumn survival rate of juvenile Eurasian Wigeon was only *c.* 50% of that of adults (Guillemain *et al.* 2013a). Taken together, these studies suggest that low juvenile survival may be a general pattern in dabbling ducks.

Even when considering winter ringing (which underestimates first-year mortality, see above), it is very clear that Teal are short-lived birds. Most mortality occurs during the very first years of life. Figure 8.2 shows the annual proportion of the total number of ring recoveries for birds ringed as juveniles in the Camargue, which drops very quickly: more than 40% of the recoveries are obtained during the very winter of ringing and an additional 31% during the next one; 7 out of 10 birds are recovered before reaching 2 years of age (see also Lebret 1947).

In addition to being different between age classes, survival rates also differ

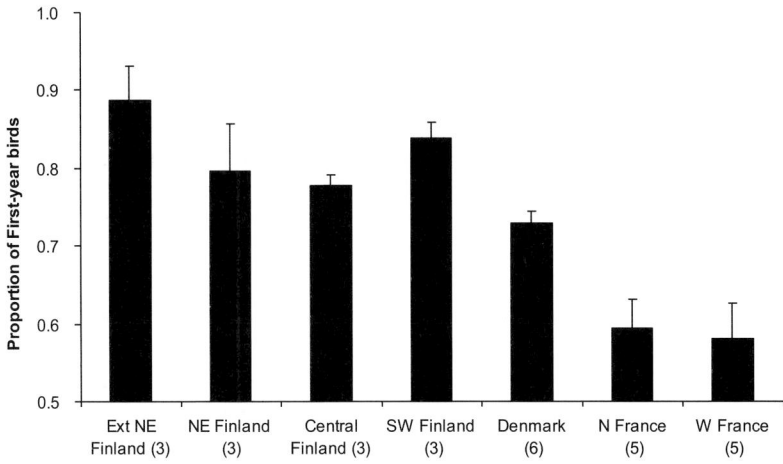

FIG. 8.1 *Percentage of first-year birds among Common Teal in the bags of hunters throughout the western European flyway. Vertical bars show standard errors and the numbers in brackets indicate the number of hunting seasons from which the values were computed in each case. With kind permission from Springer Science+Business Media:* Journal of Ornithology, *'How many juvenile Teal* Anas crecca *reach the wintering grounds? Flyway-scale survival rate inferred from wing age-ratios', volume 151, 2010, pages 51–60, M. Guillemain, J. M. Bertout, T. K. Christensen, H. Pöysä, V. M. Väänänen, P. Triplet, V. Schricke & A. D. Fox, Figure 2.*

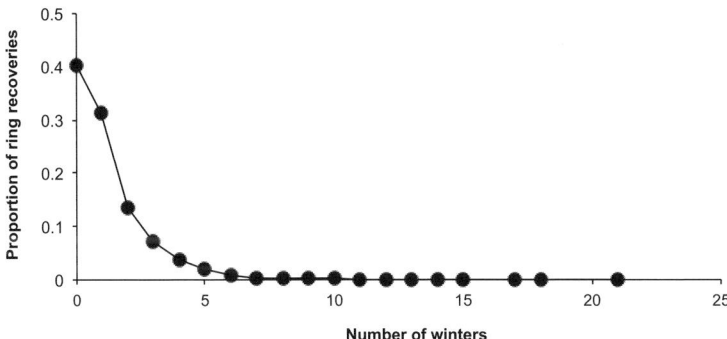

FIG. 8.2 *Proportion of the total number of ring recoveries obtained per number of winters elapsed since ringing. Data for Common Teal ringed as first-year birds in winter in the Camargue. Most birds were recovered during the very winter of ringing (as juveniles) or during the next winter (Tour du Valat, unpublished data).*

markedly between the sexes in Teal, as in other dabbling ducks (see for example Olson 2013). This is generally thought to result from greater predation risk in females during incubation or when taking care of the ducklings, a cost of reproduction that male dabbling ducks do not pay (Sargeant & Raveling 1992). Indeed, Devineau *et al.* (2010) compared

Common Teal survival estimates between sexes by season, and only observed differences between males and females for the hunting-free spring-to-autumn period, when males had a very high survival rate (95.2% ± 0.3% SE) whereas female survival was much lower (80.7% ± 1.8% SE).

Such differences in survival rate between the sexes gradually translate into unbalanced sex ratios. While the sex ratio is generally close to one at hatching, at fledging and in juvenile dabbling ducks (primary, secondary and tertiary sex ratios), skewed proportions towards males arise after the first breeding season (Bellrose *et al.* 1961); these authors analysed a sample of nearly 7,000 Green-winged Teal shot in Utah, and observed that the sex ratio increased from 58% to 72.9% males from juvenile age to adulthood (see also Figure 8.3). Considering the whole population (juveniles and adults combined), most studies have found an excess of males compared to females. This is true for samples of trapped or shot Teal, as well as visual observations of Teal flocks, with the ratio ranging from as low as 51% males in the bags of North Carolina hunters (Bellrose *et al.* 1961) to as high as 79% males in trapped birds throughout North America (Lincoln 1932, data provided in Petrides 1944). Intermediate values were provided by Furniss (1935), McIlhenny (1940), Homes (1942), Johnsgard & Buss (1956) and Guillemain *et al.* (2009b). Sex ratio estimates may be biased by assessment methods (for example, due to duck traps preferentially catching males), but the existence of a bias towards males is therefore itself likely to be true (Petrides 1944).

The annual survival rates in Table 8.1 can also be used to compute average life expectancy (after Brownie *et al.* 1985, page 39): the most recent survival

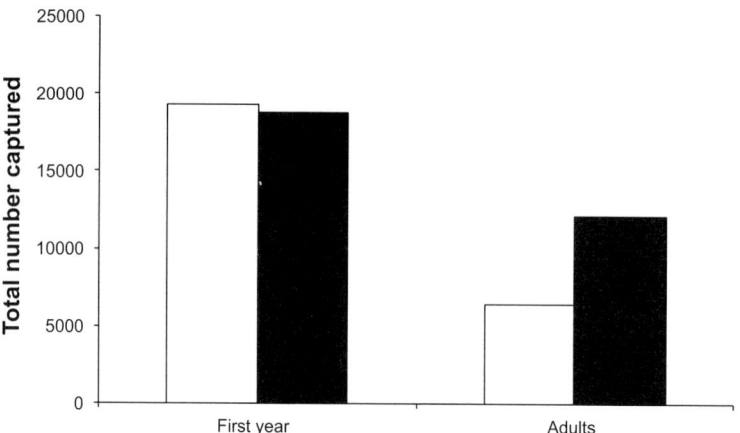

FIG. 8.3 *Total number of Common Teal of each age and sex class (white bars: females, black bars: males) among the 56,604 individuals ringed in the Camargue between 1952 and 1978. Sex ratio does not depart from unity in first-year birds (Chi^2 = 3.00, P >0.05), while it is strongly unbalanced in favour of males in adults (Chi^2 = 870.8, P <0.0001). Given the different ratios of the two age classes it is unlikely that the unbalanced sex ratio of adults only reflects capture methods biased towards catching males (Tour du Valat, unpublished data).*

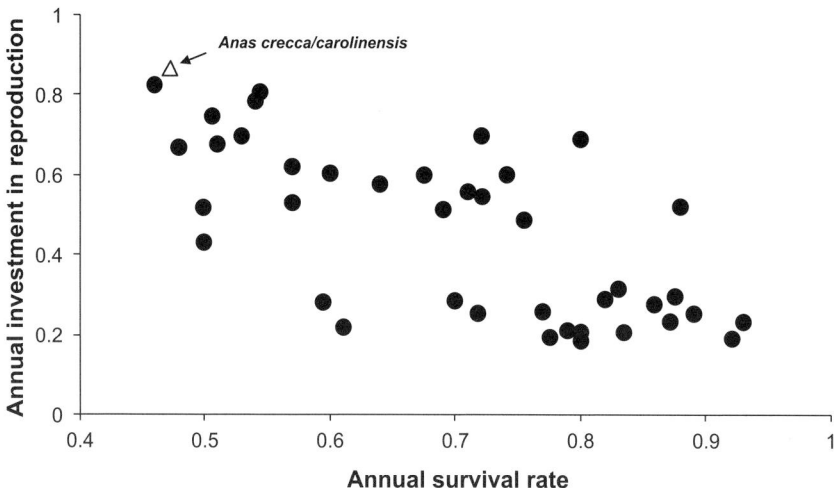

FIG. 8.4 *Annual investment in reproduction (expressed as the ratio between mean clutch mass and mean female body mass) plotted against mean adult survival rate in Anatidae (each dot represents a species of duck, geese or swan). Teal are represented by a triangle. Species in the top left are more 'rapid' in the sense that they invest a lot in reproduction annually but can only expect a limited number of breeding seasons owing to short life expectancy ('r-selected' species), whereas species in the bottom right are the opposite ('K-selected' species). Reproduction and body mass data from Figuerola & Green (2006), survival data from monographs in Kear (2005). Survival rates were computed as the mean of the two extreme range values when more than one value was available. We used annual survival of adult females when available, or mean adult survival otherwise.*

data suggest that mean life span ranges from 2.24 years in female Common Teal (Camargue-ringed birds in Devineau *et al.* 2010) to 2.84 in male Green-winged Teal (Olson 2013), both of which are much lower than those of most other waterfowl (see adult annual survival rates in Figure 8.4). These numbers are markedly lower than the highest recorded age in nature based on ring recoveries, 27 years and 1 month for a Common Teal ringed in the UK (Staav 1998, see also Bønløkk *et al.* 2006). The corresponding value for Green-winged Teal is 20 years and 3 months (USGS 2013). Such differences between potential and realised life spans again highlight the significance of the limiting factors listed in Chapter 7. In both Common and Green-winged Teal, most birds are thus likely to breed only twice in their lifetimes: once at one year of age, plus once during the following year.

FECUNDITY AND AGE STRUCTURE

We have already dealt with the breeding ecology of Teal in Chapter 6. The various constraints imposed on breeding Teal result in each breeding female producing 1.7 fledged juveniles annually on average. However, this is likely to vary greatly from year to year, and also to depend on the individual quality of females. As summarised by Nudds & Cole (1991), events during the non-breeding season (including hunting mortality) should be of relatively less importance than breeding success in determining population size in dabbling ducks. Unfortunately, the factors that determine breeding output are among the least well studied, understood and quantified in Teal.

First, breeding propensity varies with environmental conditions and with age in dabbling ducks. Second-year and older birds are generally more likely to breed regardless of environmental conditions than are first-year birds (review in Baldassarre & Bolen 2006). Unfortunately, due to the paucity of Teal studies focusing on the breeding season no quantitative data are available for these species specifically. However, given the skewed sex ratios in Teal populations (see above), females should have no difficulties in securing a mate, even late in the season. It seems likely that virtually all second-year and older females, and most first-year females (70% in ducks in general according to Blums *et al.* 1996), attempt to breed in a given year, but this remains to be further studied.

Among females attempting to breed, reproductive performance has been observed to increase with age before later decreasing due to senescence (review in Fox 2005a). Again, no data on the relationship between age and fecundity are available specifically for Teal, so that only the mean value (1.7 fledged juveniles per breeding attempt) can be used, although it is likely that first-year female Teal are less productive.

Beyond such potential intra-sexual differences, the most striking characteristic of breeding Teal is the limited number of possible breeding attempts, due to short life expectancy, coupled with a great annual investment in reproduction compared to other waterfowl (see Figure 8.4). Laying a clutch of 8–10 eggs is not uncommon in waterfowl, but what makes Teal very special is that their clutch represents an enormous energy investment, more than 85% of the female's body mass if one considers a mean clutch size of 9.5 and a mean egg mass of 29g (Figuerola & Green 2006). For comparison, the full clutch of a female swan, despite being composed of much larger eggs, only represents *c.* 20% of the bird's body mass (Figure 8.4).

Among waterfowl, Teal can be considered as 'fast' or 'r-selected' species with a rapid individual turnover, which favours annual breeding at the expense of adult survival. This was noted already by Boyd (1962), who stated that 'large waterfowl survive better, and produce fewer offspring, than small ones'. Interestingly, the Anseriformes are also among the 'fastest' bird species in terms of age of first reproduction, fecundity and adult life expectancy (Gaillard *et al.* 1989); Teal are therefore special among birds in general when it comes to their breeding strategy, as they are among the fastest of a group of already very fast species.

One immediate consequence of the large investment of female Teal in annual reproduction is the skewed age-ratio of their populations, where first-year birds generally represent the majority of the individuals, despite the very high autumn mortality rate of juveniles described above: in England and southern France, for example, wintering populations comprised 62% (±2% SE) juveniles (Guillemain *et al.* 2009b, see Figure 8.3). In the Mississippi flyway, Bellrose *et al.* (1961) even recorded 3.7 juveniles per adult female (that is, 78% juveniles) in hunters' bags, after correcting for differential susceptibility of the two age classes to hunting.

Another main feature of Common and Green-winged Teal is their very short generation time, which can be computed from the age at first reproduction, adult survival rate and population growth rate (see Appendix 1 in Niel & Lebreton 2005). Considering that age at first reproduction is one year (actually 11 months) in Teal, that adult survival rate equals 48.5% (Devineau *et al.* 2010) and that population growth rate is 1.0395 (obtained from a mean population increase of 3.95% over the period 1987–2009 in France: Fouque *et al.* 2009), Common Teal have a generation time of 1.53 years. As a comparison, generation time is about 7 years in Barnacle Goose *Branta leucopsis*, and generation times over 10 years are common in seabirds (Niel & Lebreton 2005). As discussed in the next section, such a demographic property has major consequences for maximum potential growth rate in Teal, and hence for their ability to sustain harvest.

LIFE CYCLE AND POPULATION TRENDS

The demographic parameters of Teal can easily be visualised as a schematic life cycle (Figure 8.5, see also Devineau 2007). This figure is relevant to the population immediately post-breeding; that is, for a fledged juvenile female to produce fledged juvenile females herself she has first to survive one year (S_j), then to attempt breeding (α). Her net production is then given by the

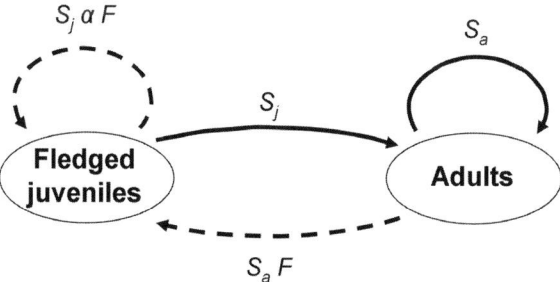

FIG. 8.5 *Formal model of Teal life cycle. Note that the population is considered as closed: there is neither immigration nor emigration of individuals to/from the system. The system is represented as if being composed of only females. Right-pointing plain arrows: ageing; that is, survival of the individuals. Left-pointing dotted arrows depict production of new individuals; that is, reproductive output. The life cycle is represented as a post-breeding pulse system: 'Adults' indicate birds that have just bred and are aged 12 months or older.*

population's mean fecundity rate (F). In this model, all females in their second year or older are considered to breed. Survival rates for first-year birds (S_j) and adults (S_a) are given in Table 8.1 for both Common and Green-winged Teal. Values for F and α are given in Chapter 6 and in the previous paragraphs of this chapter.

From these demographic parameters it is possible to derive the population growth rate, or the multiplication factor between population sizes from one year to the next (see Caswell 2001). For the Common Teal population wintering in the Camargue, the population growth rate was derived by Devineau (2007) and was found to be 1.014; that is, the population increased by 1.4% per year. This is less than what was recorded in the national winter duck counts since the mid-1980s (3.95% in Fouque *et al.* 2009, see above). Furthermore, this 1.014 value was based on very conservative parameters, such as no crippling loss and high ring return rate. With more realistic parameter values (some of which we are able to estimate only poorly), population growth rate should be lower, even below 1, hence indicating a declining population, while duck counts show just the opposite (Devineau 2007). Devineau's principal hypothesis to explain this apparent paradox is that the population under study is not 'closed', but rather receives immigrants from other populations or unstudied parts of the present population where demographic parameters could be different: the population would be spatially heterogeneous. Abmigration leading to porosity between populations has been demonstrated for Teal, both in Europe and America (see Chapter 4), so Devineau's hypothesis seems plausible.

Another non-mutually exclusive explanation has been addressed recently: rather than all individuals in the population having the same demographic parameters, some individuals may have a lower survival rate or breeding success, thus contributing less to the next generation (lower 'reproductive value'); the population would be demographically heterogeneous. Accordingly, the harvest of individuals of low reproductive value would have less negative impact on population growth rate than harvest of more productive birds; modelling the population in this way, Guillemain *et al.* (2012b) obtained a population growth rate >1, whereas a corresponding population with homogeneous demographic parameters was predicted to decline. The development of analytical software opens new avenues for research on the effect of individual heterogeneity within populations (Pradel 2009) and, in particular, the extent to which this increases the ability of populations to sustain harvest. The next steps in research should link such individual heterogeneity to an individual's 'quality', morphological features or behavioural decisions such as those associated with habitat selection.

DEMOGRAPHIC FEATURES AND RESILIENCE TO HARVEST

As explained above, demographic heterogeneity among individual Teal may greatly contribute to the capacity of populations to sustain harvest. However, this is a recent finding in this species and the extent to which demographic heterogeneity contributes to harvest compensation needs to be further studied.

What is already more clearly established is the impact of generation time on susceptibility to harvest. Niel & Lebreton (2005) worked on 'demographic invariants', and demonstrated a constant negative relationship between a species' generation time and its maximum population growth rate. Fleming & Howell (2013) demonstrated that, despite already being heavily harvested, Green-winged Teal with a generation time of *c.* 1.5 years sustain maximum population growth rates well above 1. This means that a significant additional part of the population could still be harvested without leading to a population decline. This is quite consistent with the observed long-term increases of many Teal populations worldwide in the presence of high harvest levels.

Compensation of harvest is often considered to arise through the release from density-dependent limiting factors on survival after removal of some individuals from the population. Under this scenario, non-harvested individuals have greater natural survival after harvested conspecifics are removed from the population (Anderson & Burnham 1976, Nichols *et al.* 1984). Density-dependent non-hunting mortality has been demonstrated in waterfowl (Gunnarsson *et al.* 2013). However, the strength of density-dependent feedback on survival is generally not likely to provide more than a weak compensation of harvest in most bird species (Lebreton 2005; see, however, Sedinger & Herzog 2012). Conversely, Lebreton (2005) demonstrated a greater potential for compensation of harvest by density-dependent control of reproduction (see also Pöysä *et al.* 2004). A 10% increase of harvest in a duck like Teal, with a 1.5 year generation time, would be compensated by only a 15% increase in fecundity. It is thus clear that the demographic characteristics of Teal make their harvest particularly likely to be compensated for, although this compensation is probably only partial. For instance, it is probably impossible that fecundity increases by 45% to fully compensate for the 33% kill rate of Teal mentioned in Chapter 7 (see also Péron 2013). However, a somewhat lower density-dependent increase in fecundity, combined with demographic and spatial heterogeneity among individuals, begins to provide a robust potential for compensation in Teal in response to harvest (see also Sedinger & Herzog 2012). This is not to say that harvest is a minor factor in Teal population dynamics, as it is clear that a very large proportion of individuals is shot by hunters each year. However, the very specific demographic characteristics of Common and Green-winged Teal allow these species to sustain harvest at current levels, and even to show increasing populations despite large annual bags.

CHAPTER 9

Management, harvest and conservation

TEAL POPULARITY WITH HUMANS

Genetic analyses suggest that Common and Green-winged Teal started to diverge genetically during the Pliocene-Pleistocene transition, *c.* 2.6 million years ago (Peters *et al.* 2012). The earliest known Teal fossils date to the Pleistocene, and evidence of consumption by humans has been unearthed at various prehistoric sites, both in Europe and in North America (review in Johnson 1995; see also Lydekker 1891, Ericson & Tyrberg 2004).

As opposed to Pintail, Mallard and geese, Teal were very rarely represented in Egyptian art, even though Common Teal were probably common along the Nile in winter. Only two occurrences of Common Teal in tomb paintings are known (Khnumhotep III, dynasty XII at Beni Hasan and Kenamun, dynasty XVIII at Thebes; Arnott 2007). Teal mummies have been found, probably due to Teal's function as both sacred birds and funerary gifts (Lortet & Gaillard 1909). Interestingly, these authors document the finding of immature or flightless Common Teal, which may indicate some kind of captive breeding by ancient Egyptians. Common Teal are not frequent in Roman art either, although a male is very nicely illustrated in a mosaic from the House of the Faun in Pompeii (Arnott 2007; Figure 9.1, see colour section).

Despite an intensive search, we have not found explicit mentions or illustrations of Teal in ancient cultures in North America, northern Europe or Asia, and Teal have featured only rarely in art compared with other ducks.

FIG. 9.2 *Green-winged Teal was selected as the emblem species for the youth education programme of Ducks Unlimited, Inc. and Ducks Unlimited Canada. This is likely to be because Green-winged Teal is the smallest of all dabbling ducks in North America, but also to be due to the commonness and popularity of this species with the general public. Illustration courtesy of Ducks Unlimited Canada.*

This is very surprising given how common Teal are over vast geographic areas, how highly valued they are as game and food (see next section), and the aesthetic artistic potential of the male's elaborate breeding plumage and colour pattern. Still, that Teal are nowadays common and popular is illustrated by the selection of this species by Ducks Unlimited, Inc. and Ducks Unlimited Canada, among the most influential private waterfowl habitat conservation organisations worldwide, as the emblem of their youth education programme (Figure 9.2).

TEAL ON THE TABLE, THEN AND NOW

Virtually all classical authorities on natural history have written about the high quality of Teal meat, and the excitement that Teal's swift flight can provide to hunters (spelling as in the original references):

> *This bird for the delicate taste of its flesh, and the wholsom nourishment it affords the body, doth deservedly challenge the first place among those of its kind* (Ray & Willughby 1678)

> *Some persons consider the Green-winged Teal the most desirable duck as far as flavor is concerned. […] This duck is also very attractive from a sportsman's standpoint; indeed many hunters consider it their 'best' duck* (Grinnell *et al.* 1918)

> *They certainly deserve to rank among the finest of the shoal-water ducks, the only disadvantage being their small size* (Phillips 1923)

> *It stands high on the list of table birds and is a game duck of distinction – the more discerning appreciating the quality of speed in flight that tests the hunter's skill* (Munro 1949)

Relatively small size compensated for by excellent flavour of flesh and extreme plumpness in satisfactory seasons (Dement'ev & Gladkov 1967)

Teal's highly appraised tastiness, plus the fact that they once gathered in very large numbers in some areas during autumn and winter, led to the development of a massive commercial harvest and market sale of Teal, especially during the 18th and 19th, up to the mid-20th century.

In North America, commercial harvest was then mostly practised with regular guns (though often low gauge; that is, large diameter), or with huge guns set on small boats, which could kill tens or hundreds of birds per shot (punt-guns; see for example Bland 1950). In Europe, commercial harvest was mostly by means of live catching in large traps named 'duck decoys', which were especially efficient at catching Teal (Box 9.1).

BOX 9.1. Duck decoys: from game trade to scientific duck ringing

From at least the 16th century in the Netherlands and the 17th century in Britain, large duck traps called 'duck decoys' were built to catch birds for the purpose of selling game meat at local markets. Several hundred of these traps were in operation during the 18th century and the first half of the 19th century.

The decoys were made of long funnels built around a central pond. The funnels had overarching nets (hence their name, 'pipes') and a curved shape so as to prevent ducks from seeing the end when entering the trap. Birds were attracted to the opening of the trap by captive live birds, seed bait or more frequently a trained dog (traditionally called a 'piper' in English). In the last case, benefit was taken from a peculiar behaviour of ducks towards terrestrial predators: when the latter approach a waterbody, ducks tend to come closer and swim along the bank, while keeping a short safety distance. By doing so, they indicate to the predator that it has been detected, while remaining out of its reach (Kear 1990; see also the description of such behaviour of ducks towards Coyotes *Canis latrans* in Keith 1961). In such a duck-trapping system, the *decoy man* let his dog gradually move towards the end of the decoy, so as to get the ducks to follow it inside the trap. In Europe dogs used for this purpose were specifically bred for their russet colour to resemble a fox, and were trained to move in between screens so as to be visible to ducks only intermittently, which was considered more attractive to the birds. When the ducks were far enough into the pipe, the decoy man himself appeared to flush them towards its end, where they were collected from a small funnel trap (Figure 9.3).

Some of these decoys were extremely effective, with catches of up to *c.* 10,000 ducks per year (Matthews 1969, Stott & Mitchell 1991), and a total

FIG. 9.3 *A duck decoy in operation: the decoyman, hiding behind a screen of reeds to the left in the illustration, is sending his dog to herd the ducks towards the end of the pipe. Line drawing by Sir Peter Scott, with kind permission of Dafila Scott.*

of 300,000 ducks caught in decoys for commercial food purposes in the Netherlands during 1952 (Karelse 1994). Teal were among the species most frequently caught in duck decoys, Phillips (1923) mentioning 44,568 Common Teal being caught out of a total of 96,000 ducks in Ashby Decoy (North Lincolnshire, UK) between 1833 and 1866, and an astonishing 33,000 Common Teal out of 35,490 ducks for the year 1877 in the decoys of Föhr, a German island in the North Sea (Figure 9.4).

Changes in agricultural practices, in particular the draining of wetlands, reduced the catch by duck decoys dramatically, especially in Britain, leading to gradual abandonment of this practice (Heaton 2001). Some decoys have, however, been preserved and are still in use, mainly for educational and ringing purposes (Kear 1990, Bub 1991). Duck ringing based on decoy catching was particularly intense in Britain from the mid-1950s to the 1990s, with tens of thousands of ducks caught per year, mostly Mallard, Common Teal and Eurasian Wigeon (Mitchell & Ogilvie 1996). More than one hundred decoys, catching more than 3,000 Common Teal in total annually, were still in use in the Netherlands in the early 1990s (Karelse 1994). In 2001, around 10 decoys were still operated for ringing purposes in Britain (Heaton 2001).

FIG. 9.4 *Label of a can of Teal meat from the Island of Föhr, Germany. The central part of the label shows the decoyman by the end of the pipe, beside the final funnel trap. Catches of Teal were so great along the North Sea in the 19th century that a canning factory was built to preserve the Teal meat for commercial sale far beyond local markets. Photograph kindly provided from the archives of Dr Carl Häberlin Friesen Museum, Wyk auf Föhr, Germany.*

FIG. 9.5 *Commercial harvest of Teal by means of duck decoys has been discontinued. However, a few are still on display or even operated to catch ducks for ringing purposes. Photograph courtesy of Abbotsbury Swannery in Dorset, England.*

As early as 1555, Belon stated that because its meat was so appreciated, Common Teal could be sold for the price of a goose or a capon (castrated rooster), even in Roman times. More recent accounts instead indicate that a dozen Green-winged Teal were sold for around half the price of a dozen Mallard (Grinnell *et al.* 1918), hence the term 'half fowl' for Teal and Wigeon, as opposed to 'whole fowl' for larger duck species (Ray & Willughby 1678). Still, given the difference in body size between a Teal and a Mallard, this highlights how highly valued the former have always been compared to other waterfowl.

By the turn of the 20th century the game trade had become a huge market, with tens of thousands of Teal being sold in markets locally (more than 82,000 Green-winged Teal in the markets of San Francisco and Los Angeles in the 1895–1896 season alone, Phillips 1923). Similarly, Hume & Marshall (1881) estimated that hundreds of thousands of Common Teal were captured and sold annually in India. Because of the difficulties in keeping game meat fresh in hot climates, birds were often kept alive in aviaries until being eaten, such Indian 'tealeries' being described in detail by these two authors.

The waterfowl game-meat trade grew so much that it gradually became evident it was decimating some wild populations, for example in California (Grinnell *et al.* 1918). For this reason, the sale of game birds was banned early in North America (through the Migratory Bird Treaty Act of 1918, see Day 1949). Sale of hunted ducks continued in other geographic areas, though, and was still considered a threat to game-bird populations in western and southern Europe as recently as in the 1980s (Lampio 1982). Based on a questionnaire made by the Hunting Rationalization Research Group of the International Waterfowl Research Bureau for the 1969–1970 hunting season, as much as an estimated 75% of harvested waterfowl was sold in some western countries (Denmark, the Netherlands, France and Spain; Lampio 1974).

Hunting for market sale has since then been banned in most European

FIG. 9.6 *Common Teal and other live birds for sale in a market, Damietta, Egypt, January 1980. The board at the back lists the birds for sale and their prices: raptors, Moorhen, Northern Pintail, Common Teal and Northern Shoveler. Photograph by Wim Mullié.*

countries, but continues to be legal in Britain, for example (England, Wales and Scotland, but not Northern Ireland), from 1 September to 28 February (Wildlife and Countryside Act 1981, amended). There are no restrictions or conditions on such sale (J. Harradine, pers. comm.). No precise accounts are available about the number of ducks being sold annually in the UK, but these are likely to be in the thousands, and *c.* 15,000 dead wild ducks 'in feathers' were exported annually from the country in the mid-1980s (Kear 1990). The number of Teal killed and sold at local markets today could be considerably greater in other parts of the world (Figure 9.6), though precise estimates are generally lacking. A survey conducted in the markets of three cities in northern Iran by Balmaki & Barati (2006) recorded the monthly number of Common Teal being sold to gradually increase from 15,659 in November 2001 to an astonishing 75,465 in January 2002, most of which were caught and sold alive. Mullié & Meininger (1983) reported an annual commercial catch of *c.* 10,000 Common Teal on Lake Manzala, Egypt, alone (see also Goodman *et al.* 1989 for more information about commercial harvest of waterbirds in this country). The consequences of continued commercial harvest for the dynamics of the local populations, and the importance of this practice compared with non-commercial leisure harvest remain completely unknown.

On a lighter note, commercial trade has led to ring recoveries being made in local markets (for example in the city of Arles for Common Teal ringed in the Camargue), or in even more unexpected places: Wainwright (1967) documented how two rings he himself had fitted onto Common Teal were recovered 'by diners in London restaurants. [...] a Teal was served up on a plate with the ring, 28 days old, still on its leg!'

TEAL IN CAPTIVITY

We have already mentioned that wild Common Teal were caught and kept in Indian tealeries before being killed and eaten, and archaeological remains suggest that ancient Egyptians may have bred Common Teal in captivity (though this speculation is based on indirect evidence). Teal have also been kept for a long time in bird collections for aesthetic reasons; according to Phillips (1923) this has been the case since the Middle Ages, and he provides a reference stating that the Prince of Condé had some captive Common Teal in the Chateau de Chantilly in France as early as 1663.

One major problem faced by aviculturists is that Common and Green-winged Teal captured in the wild never adopt tame behaviour and are difficult to entice to breed (Delacour 1956). This problem, however, quickly disappears in subsequent generations of captive-born individuals, which are then very easy to keep and breed (Vietti *et al.* 2010). Several of the results presented in this book, especially those on mating behaviour, were actually obtained from Teal kept in aviaries (see McKinney's work in Chapter 6).

It should be noted that some amateur duck breeders also cross duck species and types, producing captive Teal varieties whose plumage ranges from pure white to glossy black (Figure 9.7, see colour section). Hybrids between Teal and other dabbling ducks have been deliberately produced but also occur naturally, especially with Pintail (both for *A. crecca* and *A. carolinensis*; Howell 1959, Sage 1960, Harrison & Harrison 1971), but also, for example, Blue-winged x Green-winged Teal (Strubbe 1967). Hybrids between duck species are often produced by owners of live decoys where these are allowed for hunting. It has for a while been forbidden in France to use European duck species as live decoys, while exotics are still legal, which led has to a growing market for Green-winged Teal, either pure or hybrids.

HARVEST MANAGEMENT

Like most other dabbling ducks, Common and Green-winged Teal have benefitted from the general measures taken over the years to protect waterfowl, such as the ban on commercial hunting in North America through the Migratory Bird Treaty Act of 1918, and the implementation of closed hunting seasons in most countries (see also Box 3.3). In Europe, the Common Teal is generally not subject to specific harvest management; the hunting rules are the same as for most other dabbling duck species. The most significant change in policy from which this species has benefitted at the continental scale is the gradual ban on toxic shot currently being implemented in most countries (Box 9.2), and the general trend towards shorter hunting seasons over recent decades (Figure 9.8).

In North America, the teal species (Green-winged, Blue-winged and Cinnamon Teal) are subject to specific legislation, with an additional special hunting season compared to other ducks in some areas. Because Blue-winged Teal migrate earlier than other species and their populations seemed to be in a favourable conservation status, a specific September Teal hunting season was initiated in some US states during the 1960s, then expanded to other states in several flyways over the following decades. The Teal season consists of a number of possible additional hunting days in September compared with the regular waterfowl season, with a set daily maximum bag limit. The Green-winged Teal was included in this scheme because it was felt that confusion with Blue-winged would lead to the shooting of both species anyway. Since then, the legislation for specific Teal seasons has changed over time in several areas, a development summarised in Fleming & Howell (2013). According to these authors, the specific Teal season now only represents *c.* 2% of the annual Green-winged Teal harvest in North America. As is the case for all other waterfowl, Green-winged Teal harvest regulation is derived from an Adaptive Harvest Management scheme (which is devised mainly for Mallard; see for example Nichols *et al.* 2007). Through this scheme, bag limits and season lengths are adapted each year to the long-term trends of waterfowl

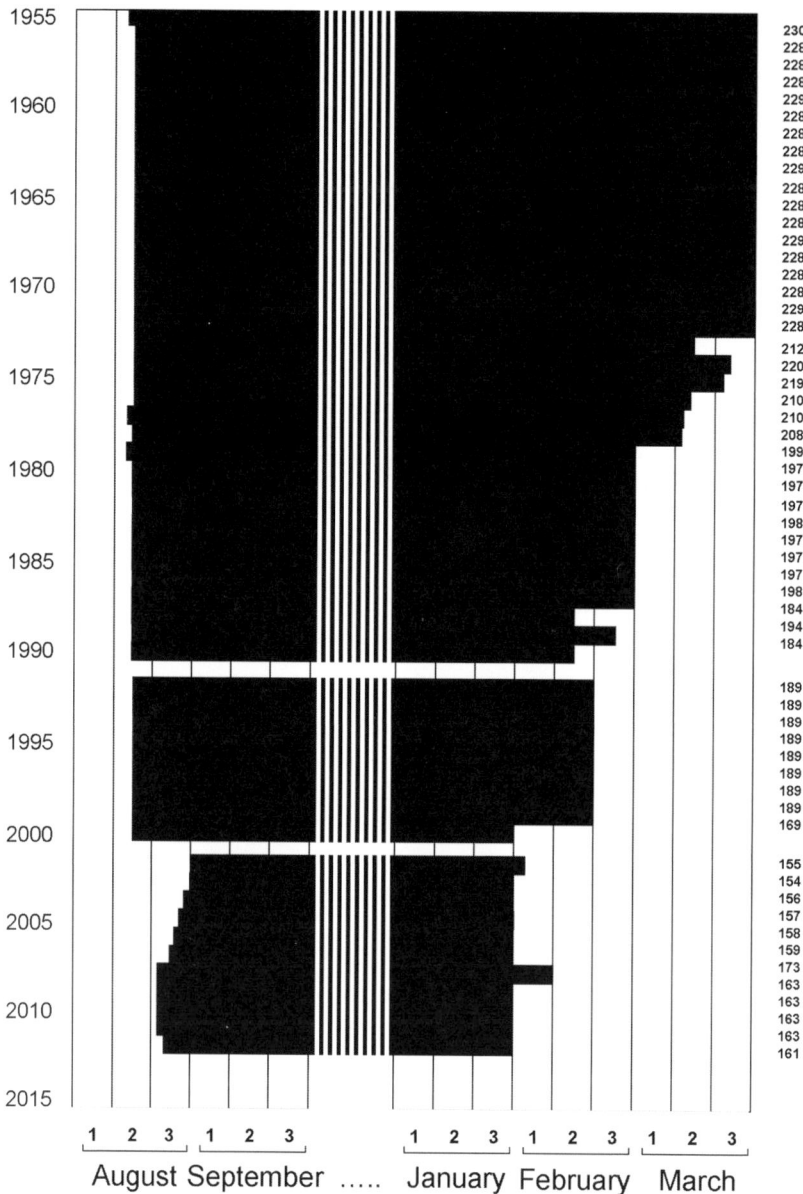

FIG. 9.8 *Opening and closing dates of Common Teal hunting season in the Camargue, southern France, over the period 1955–2012 (data missing for seasons 1991 and 2001; numbers on the x axis indicate 10-day periods). Mostly as a consequence of the advancement of the closing date, the hunting season was gradually shortened from 230 to c. 160 days (numbers to the right). Graph from a compilation of annual local hunting legislation leaflets and ministry acts.*

> **BOX 9.2. Lead poisoning and the ban on toxic hunting shot**
>
> Because of the threat represented by lead poisoning to wild waterfowl populations (Box 7.1), the use of lead shot has gradually been banned for wildfowling (in 1991 in the USA, 1997 in Canada and mostly in the 2000s in Europe; Figure 9.9). Wetlands that have been heavily hunted over long periods sometimes retain a very high density of residual lead pellets in their upper sediments. These remain accessible to foraging birds for long periods: an experimental study demonstrated that 90% of lead pellets spread at the surface of sediment were still within the uppermost four centimetres after four years, and were hence accessible to dabbling ducks (Pain 1991b). However, a ban on toxic shot rapidly has positive effects: blood lead in adult American Black Ducks decreased by 44% between two measurements, one done two years before the ban in the USA and one done seven years after (Samuel and Bowers 2000). Similarly, a decade after lead shot was banned in North America, no lead shot residue was found in 98 Green-winged Teal gizzards examined, while some non-toxic shot residues were found instead (Garrison *et al.* 2011). The gradual ban on lead shot in Europe should eventually increase individual waterbird survival rates, including in Common Teal.

numbers and their population size compared to long-term goals set in the North American Waterfowl Management Plan. For Green-winged Teal, the mean breeding population estimate over the 2002–2011 period was 2.8 million birds, well above the 1.9 million long-term objective (NAWMP 2012), and the long-term trend is clearly a continuous increase in numbers (see Chapter 3).

Habitat management

Implementation of habitat-management procedures that alleviate the limiting factors listed in Chapter 7 should benefit Teal. Therefore, the creation of hunting-free, disturbance-free protected areas is the most obvious option, and one that is considered to have contributed significantly to the increase of Teal populations over recent decades, at least in Europe (Rüger *et al.* 1986, Deceuninck 1997, Fox 2005b). The creation of a protected area on wintering grounds often readily translates into a clear and rapid increase in Teal numbers (Chapter 7). Like other dabbling ducks, Common Teal wintering in France shifted from unprotected marine coastal areas to more inland nature reserves as these were established between the 1970s and 1990s (Appendix A in Guillemain *et al.* 2002a). Such protected areas are especially beneficial to Teal, and may translate into a long-term increase in population size. A rapid increase in numbers at newly protected sites, however, often results from

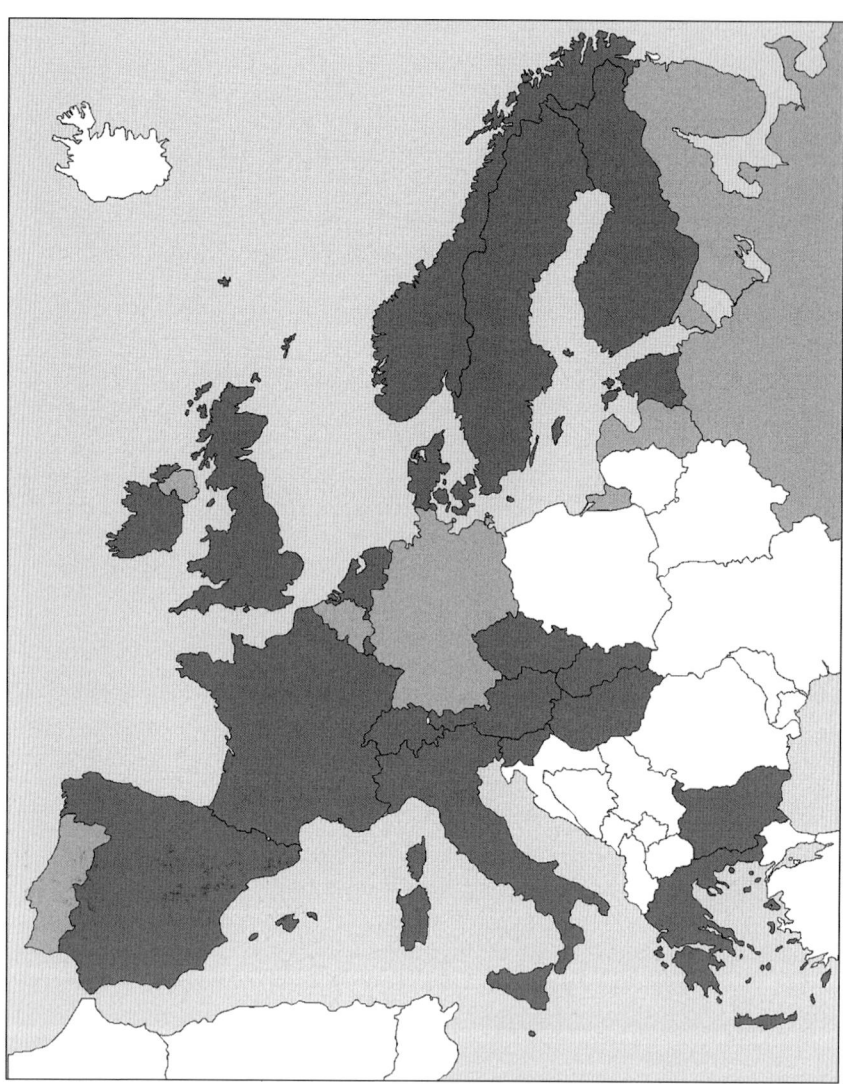

FIG. 9.9 *European countries in which lead shot is completely banned for hunting waterfowl or hunting in wetlands (dark grey, including countries in which the ban was to be implemented as of June 2013), countries with a partial ban in some wetlands (light grey) and countries where the use of lead shot is still legal in all wetlands (white). From information in Lehmann (2007), Avery & Watson (2009) and personal communication by Cy Griffin (FACE).*

short-term redistribution of local birds, sometimes from other nearby protected areas. This suggests that site-specific habitat features also play a major role in attracting the Teal. In the Camargue, the creation of a 1,000-ha protected area led to a sudden increase from virtually none to *c.* 10,000

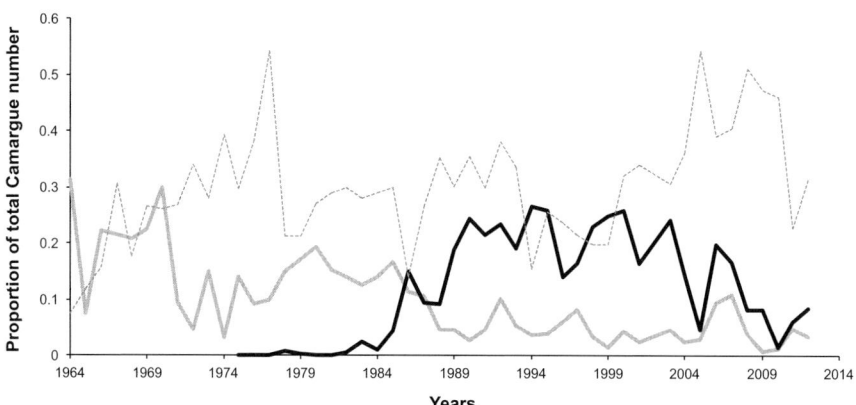

FIG. 9.10 *Changes in Common Teal distribution within a winter quarter: Common Teal numbers declined at Saint Seren (bold grey) as numbers increased in Vigueirat (bold black) following the protection of the latter site in the mid-1980s. The increase in Vigueirat was only transitory, as the numbers then decreased during the second half of the 2000s, to the benefit of the main western Camargue private estates (dotted grey). All values expressed as the mean proportion of total number of Common Teal in the Camargue each winter (based on counts in December, January and February). 1964 stands for the 1964–1965 winter, and so on. Data courtesy of Alain Tamisier (Centre National de la Recherche Scientifique, seasons 1964–2002) and Michel Gauthier-Clerc (Centre de Recherche de la Tour du Valat, seasons 2003–2013).*

Common Teal in the mid-1980s (Mathevet & Tamisier 2002, Brochet *et al.* 2009), at the expense of another nearby protected site. These large Teal numbers were, however, only transitory, and despite relatively constant local habitat management the main Teal roosts later moved towards other estates (Figure 9.10). Gradual succession of vegetation, with the development of terrestrial plants eventually including trees (for example *Tamarix* spp. or *Salix* spp.) could be among the factors contributing to some wetlands only being temporarily attractive to Teal.

The provision of such protected areas is beneficial to Teal populations because some individuals become highly faithful to these sites. Only 60 to 70% of nasal-saddled Common Teal remained within a set of four neighbouring day-roosts in the Camargue from week to week in winter, indicating high mortality and/or emigration rates. However, most of these resident Common Teal consistently used the same day-roost: only 2–6% of the individuals switched from one roost to another from week to week (Guillemain *et al.* 2010c). This suggests that two strategies coexist within the wintering Teal population: resident birds faithful to one given day-roost versus more mobile birds moving around within the winter range. This is in line with Fedynich's (1987) result obtained from the study of lake use by wintering birds in the southern High Plains of Texas, which led him to conclude that 'In all [duck] species, some individuals appeared relatively sedentary while others were never seen again'. Baldassarre *et al.* (1988a) also documented a similar coexistence of potentially distinct habitat-use strategies among Green-winged Teal wintering in the same area.

Another Camargue study showed that resident Common Teal in such protected day-roosts were heavier than birds shot nearby in unprotected areas (Guillemain *et al.* 2007f). It is thus likely that the demographic heterogeneity we mentioned in the preceding chapter is actually related to such different habitat-use patterns, with resident birds in protected day-roosts having a greater survival rate than more transient, naive, poor-condition individuals. This would indicate that the protection of Teal day-roosts contributes to improving individual fitness, which translates into positive effects on population growth rates.

In addition to safety from humans, management of day-roosts should also consider safety from predators, by providing large waterbodies with unobstructed visibility during daylight hours (cf. Box 4.4). Control of vegetation to increase the area of open water has proven to be an effective method to promote local Teal numbers during migration, for example in Germany (Biologische Station 'Rieselfelder Münster' 1982). A similar approach combining managed flooding and grazing by cattle was used to limit stands of tall reeds and bulrush in the Camargue, creating extensive shallow shores that led to a significant increase in wintering Common Teal numbers very locally through redistribution to these new areas (Duncan *et al.* 1982).

Beside safety and resting requirements at day-roosts, managers may also promote Teal by improving foraging habitats so as to increase food abundance and accessibility. Regular drawdown creating shallow waters to concentrate invertebrate food items and promote seed production can be a successful option for habitats hosting migrating and wintering Teal (Heitmeyer & Vohs 1984, see Baldassarre & Bolen 2006 for a review). Flooding of otherwise dry areas (for example saltmarshes) will also make seeds available and readily used by Teal if water levels remain low (Isola *et al.* 2000, Therkildsen & Bregnballe 2006, Bregnballe *et al.* 2009a), and freshwater input will promote the use of otherwise brackish coastal areas (Fitzsimmons *et al.* 2012). In addition to natural habitats, flooding of agricultural fields can be most attractive to waterfowl, including Teal, and can relatively easily be practised in rice fields between harvest and the next seeding season, for wintering birds (Tamisier 1972, Pirot *et al.* 1984, Twedt & Nelms 1999). It is now increasingly recommended that management efforts be targeted at improving and securing nocturnal foraging habitats rather than protecting additional day-roosts during winter (Bregnballe *et al.* 2009a, Brochet *et al.* 2009). This recommendation echoes the results of Anderson & Smith (1999), who suggest that appropriate water management of playa wetlands in the south-western United States can significantly enhance nocturnal use by waterbirds already present in the region, including Green-winged Teal.

Common and Green-winged Teal breed mainly in remote wetlands of the boreal forest, where general policies affecting habitat conservation

FIG. 9.11 *Artificial damming of a creek in Evo, southern Finland, to imitate Beaver activity: (a) 'before Beaver' (1984), and in (b) the second (1986) and (c) the sixth (1990) year after inundation. From Nummi & Kuuluvainen (2013), with kind permission from Boreal Environment Research.*

(legislation and recommendations in forestry, draining, hydroelectric power, and to some extent agriculture) rather than local land management should be supported to increase efficiency in achieving conservation goals. In this context, the tight link between breeding Teal and Beaver ponds (cf. Box 4.3) suggests that the large-scale promotion of this mammal and of similar shallow pools, including man-made flowages, should greatly benefit Teal (Sjöberg & Danell 1983, Nummi & Pöysä 1995b, Nummi *et al.* 2010, 2013). Ducks Unlimited Canada planned a specific Beaver pond-management programme that included Beaver dam removal on the least productive (older) ponds to enhance plant regeneration, and promotion of aspen or poplar trees *Populus* spp. attractive to Beaver. This management programme was discontinued because the costs were judged too high, but large areas of Beaver pond land are still protected in Ontario and produce vast numbers of young waterfowl annually (Batt 2012). Artificial damming can also create highly valuable Beaver-like Teal habitat (Figure 9.11).

To summarise, habitat-management procedures that provide safe and food-rich shallow wetlands will in most cases result in increased Teal numbers, almost anywhere along the species' flyways. In this respect, Teal are more rewarding species to work with for managers than most other ducks: they are very opportunistic birds, and respond quickly to new favourable habitat elements in the landscape as well as to improved conditions within a wetland.

As can be judged from the paragraphs above, very few Teal-specific management procedures are known, and most existing ones are only empirically derived from common sense observation of the ecological constraints faced by these birds (need for shallow water owing to small body size, and so on). A valuable perspective for future research would be to set up experimental trials of different management procedures, so as to test new options and adequately measure which ones benefit Teal the most.

We have to acknowledge, though, that we still know little concerning which time of the year habitat management is the most efficient and the most needed. A rather dense matrix of protected areas has been set up throughout western Europe to provide suitable habitat to autumn-staging and wintering waterbirds, and many of these provide safe refuges for Common Teal during periods of cold weather (Ridgill & Fox 1990). The extent to which this network would remain efficient under a changing climate leading to geographic shifts of Teal populations (see Chapter 7) needs to be assessed in future years. Teal breed at such low densities that it seems unrealistic to establish protected sites or to manage the land locally in order to promote their nesting. Rather, it may call for large-scale habitat protection across the boreal zone (see above). What seems like the priority in terms of Teal habitat management now would instead be to spend more effort towards management of spring migration habitats and stopover sites, which play a major role in the Teal annual cycle (Chapter 4) yet are poorly studied and vastly under-represented in the reserve network, both in Europe and in North America.

CHAPTER 10

The Teal's future

GENERAL NUMBERS AND TRENDS

As we write these lines, the latest report on duck-breeding population surveys for spring 2013 has just been published, again recording *c.* 3 million breeding Green-winged Teal in North America (Zimpfer *et al.* 2013), well above the 1.9 million population target (NAWMP 2012). Similarly, Common Teal numbers in most parts of Europe are at their highest since the initiation of winter counts (Delany *et al.* 2008). The two species are hence doing well over large parts of their geographic range, and there is no cause for short- or medium-term concern in areas where systematic surveys have been conducted for some time.

However, the situation may not be as favourable in other regions, especially in Asia, north-eastern Africa and the eastern Mediterranean, where Common Teal numbers have not fared as well (Delany *et al.* 2008). Admittedly, this concern is also a reflection of a greater uncertainty over the counts, largely due to a lack of regular monitoring with reliable methods in some areas, and an almost complete lack of duck counts in others. This situation is not yet alarming, but improving the coverage of the international waterbird counts in these parts of the flyways should be a priority in the near future (see also UNEP/AEWA 2013).

Notwithstanding the generally favourable situation in many flyways, unexpected changes are occurring at a more local scale. For example, although Common Teal numbers show a clear long-term increase in north-west Europe, their numbers have been decreasing over recent decades at both ends of the flyway; that is, in Doñana (Spain), a major stronghold for the species in winter (Rendón *et al.* 2008), and in Finland, the European country where breeding Teal have been monitored the most intensively (Pöysä *et al.* 2013b). Several hypotheses, ranging from widespread degradation of key habitats to global climate change, can be proposed to explain local trends in numbers. This highlights that we have been able to rather accurately demonstrate long-term population changes in Teal numbers so far, but are still largely unable to understand the processes behind these changes, let alone forecast future population trajectories.

BLACK BOXES: TEALS' BIG REMAINING SECRETS

From the very long list of literature references provided in this book, the reader may think that Teal have been such a favourite research species to so many scientists that all is known about them. This is clearly not the case, resulting in a lack of ability to provide in-depth coverage or quantified measurements for many topics treated in this book. In many others the collective knowledge is partial in either of the two species (for example, we know very little about the Common Teal's breeding ecology in Europe, and virtually nothing has been published on Green-winged Teal at staging areas), but this is not necessarily a problem, as knowledge from one continent sometimes complements that from the other. The big problem is rather the common areas of deficient information. But there is also good news: the current amassed knowledge is vast, providing a firm basis for future studies and allowing the identification of the most crucial knowledge gaps. The present monograph shows that the autumn and winter ecology of Teal is fairly well known and understood (Chapter 4), especially in western Europe and the southern USA. On the other hand, and despite the fairly detailed narrative in Chapter 6, crucial parts of Teals' lives at higher latitudes remain understudied and insufficiently understood. Below we describe these 'black boxes'; that is, some of the most problematic knowledge gaps. We finish the chapter with a 'wish list' of desired actions that would enable the research community to address these uncertainties.

Peeking into these 'black boxes' is not only about curiosity and completeness. It is rather a prerequisite if we are to understand the links between seasons, construct a useful annual life-cycle model and thus gain a flyway-wide picture of Teal ecology and management. This is urgent and tempting work, as we already know that there are seasonal carry-over effects affecting individual fitness (Arzel *et al.* 2006, Guillemain *et al.* 2008b).

Black box 1: the last weeks before settling on breeding wetlands

Spring migration is still a largely neglected period in the study of temperate-breeding dabbling ducks (Arzel *et al.* 2006), not least in Teal. Yet this phase in the annual cycle is likely to be key for breeding success in an income breeder like Teal. Most Teal have a very long migration journey to complete. Although flyways and the general timing of migration at the population level are well known, this is far from the case when it comes to the migration strategy of individuals, and how this is affected by age, sex and pairing status. As has been shown in Chapter 5, spring migration is undertaken just after or during a presumed diet shift from seeds to invertebrates, but very little is known about habitat-specific food availability and, consequently, Teal diets at this time of year. We need more studies of feeding energetics and general ecology of migrating Teal. In particular, it seems critical that we acquire a better understanding of the nutrient dynamics (mainly lipid and protein) during the last leg of migration and just prior to and during breeding. This is challenging since many birds would have to be captured and bled to measure triglycerides and isotopes, for example. Such field studies, however, seem necessary to get a better appreciation of the full suite of seasonal carry-over effects, or to determine where Teal are positioned on the income–capital breeding continuum.

Black box 2: secretive weeks on the nest and with the brood

As mentioned in Chapter 6, only a handful of active Teal nests have ever been studied, so that some vital rates (nesting success, for example) are based on very small sample sizes and hence lack robustness. It should thus be a priority to find, measure (egg numbers and weight, for example) and monitor more Teal clutches throughout the breeding range of the two species. The behaviour and fate of individual hens and broods also need more study. These are the Teal's most secretive weeks of the entire annual cycle, and finding nests and studying hens will remain a challenge. Nevertheless, the development of new cameras and telemetry systems has made this work more realistic than it was a mere decade ago.

Black box 3: when Teal almost disappear – moult ecology and strategies

In research on passerines and several other bird groups it has been obvious for a long time that knowledge about moult (timing, duration, physiological costs, pre-moult movements, and so on) is absolutely key to understanding a species' ecology as well as the behaviour of individuals. In other words, it is likely to be problematic that so little is known about moult in Teal. How many make regular pre-moult migrations, and how long are they? Where are there concentrations of moulting Teal and how well are these key wetlands being incorporated into habitat-conservation networks? How representative is the only study of moulting Teal to date (Sjöberg 1988)?

Black box 4: linking the seasons

Carry-over effects between seasons have already been documented in Teal (Chapter 7), and there is likely to be a lot more to discover in this research field. Studying carry-over effects between seasons helps us understand where bottlenecks and constraints are the most significant on an annual basis. To better understand the processes behind such population-level patterns we need to study individuals. For example, the behaviour of individual birds in winter and spring needs to be linked to their later breeding success. In practice, this requires that we somehow study individual Teal throughout the year. Capture and individual marking schemes (using nasal saddles or GPS devices) now exist in many countries, but these are rarely coordinated in a rigorous manner, and a lot of energy is spent trying to catch birds where this is difficult, on the wintering grounds. Although it may bias the sample towards males and towards females that failed breeding, it should be more productive to set up capture operations as close as possible to the breeding wetlands. An example of such places is major post-breeding moulting sites, where Teal might be caught in large numbers. Following the same individuals across the year is still a distant goal, but a first step forward could be to use a pool of marked individuals to study movements and behaviour within each season (wintering, spring migration, breeding and autumn migration).

Increased ringing efforts at moulting sites and elsewhere in late summer would also fill another big knowledge gap, namely that about movements and mortality patterns in late summer and early autumn. Mortality rates calculated thus far are biased in the sense that they are based mainly on data from birds ringed late in the autumn and already far away from their breeding or natal areas. This is a problem also in the sense that processes necessary for harvest mortality to be compensatory instead of additive (see Chapter 8) remain poorly understood. Until large-scale individual marking schemes closer to the breeding grounds can be set up, it is desirable to more intensively monitor annual breeding success at the population level. Because adults and juveniles cannot be told from each other from a distance in the field, an indirect method based on examination of wings from ducks shot by hunters would probably be the best option. Such a coordinated parts collection scheme already exists in North America and would be very valuable to implement in Europe. Unfortunately, long-term European wing surveys currently only exist in Denmark (Mitchell *et al.* 2008), hence the need to be more systematic and coordinated at the flyway level. In both Europe and North America, these data can be used to frame and formulate more detailed hypotheses about the limiting factors and environmental constraints affecting annual breeding output in Common and Green-winged Teal.

LIMITING FACTORS FOR TEAL POPULATIONS

The recent results from Finland and Doñana illustrate our ability to track population trends in Teal, but also the difficulties in determining the factors responsible for such changes; that is, population limiting factors.

The most commonly considered limiting factor, though this may not be a current threat to populations, is harvest (Chapter 7). However, there are clear signs that the number of hunters, including that of duck hunters, is gradually decreasing worldwide (Vrtiska *et al.* 2013 for North America). As explained by these authors, this ironically means less funding and other forms of support for waterfowl habitat conservation, especially in North America where duck hunters directly contribute financially to wetland (and upland) conservation initiatives. This may not necessarily be a big issue on the wintering grounds, where the most important sites have been part of a relatively dense reserve network for a long time, at least in Europe and North America. However, the shrinking funding available for wetland conservation may hamper our ability to protect and adequately manage Teal migration stopovers, and to a lesser extent breeding areas. This is particularly unfortunate given that degradation of some of these habitats may already affect Teal populations (Pöysä *et al.* 2013b).

It should also be kept in mind that the current network of wetland nature reserves was largely established at a time when a major concern was to provide these birds with safe refuges in southern areas as a safeguard in cold winters (for instance in Europe, cf. Ridgill & Fox 1990). In a scenario of general decrease in hunter numbers and global climate change, the question is whether these reserves will still match the birds' distribution and needs in the future. The decline in Teal wintering numbers in Spain and breeding numbers in Finland may actually both reflect a gradual northwards shift of the population during these two parts of the year. The response of Teal to climate change, in terms of distribution, phenology and general ecology, is another main avenue for future research (see also Guillemain *et al.* 2013b).

As explained in Chapter 7, there are other potentially limiting factors whose actual effects have hardly been quantified. One example is mortality from disease and the negative effect of parasites, which will require specific studies or monitoring. Another example is actual predation rates during the breeding season, which would provide valuable information for obtaining a circum-annual perspective on Teal demography. Similarly, it seems crucial to better quantify the magnitude of commercial harvest still practised in some regions, especially as many of these areas are also those where population monitoring schemes are the weakest.

A PROACTIVE APPROACH TO COMMON TEAL POPULATION MANAGEMENT IN EURASIA

Harvest and conservation policies for waterfowl in North America are based on a management approach that in turn relies on robust scientific data (breeding population size, regular wing and bag size surveys; see Chapter 9). No such scheme exists in other parts of the world, and in Common Teal harvest rules and general management are mostly based *a posteriori* on long-term winter counts, rather independently from one country to another. Developing an adaptive harvest management system for Common Teal populations, by which harvest and management are regularly evaluated and compared to population demographic parameters, would be a major achievement for duck scientists, wildlife managers and conservationists alike.

Integrative modelling approaches that allow the testing of different harvest and habitat management scenarios and comparing of their consequences for duck populations already exist (Mattsson *et al.* 2012 for Northern Pintail). These provide a valuable framework for similar development in Common Teal. The current situation, in which Common Teal populations fare well in western Europe, would be very appropriate for such an initiative. Indeed, this would probably lead to relatively liberal recommendations, hence a widespread consensus and support by both hunters and non-hunters. If anything, such a work would at least force us to regularly collect Teal vital rates at the flyway level, and promote international collaboration.

BOX 10.1. A wish list for Teal research, monitoring and conservation in the near future (numbers do not indicate priority)

1. Expand the international waterbird counts to poorly or irregularly covered areas of Asia, north-east Africa and the eastern Mediterranean.
2. Studies of feeding energetics and general ecology of Teal during the last leg of migration prior to settling on breeding wetlands.
3. Studies of nesting Teal. Very few Teal clutches have been measured and monitored from egg-laying to hatching. Also, the behaviour and fate of Teal broods must be ascertained (for example, by telemetry).
4. Set up capture operations on post-breeding moulting sites and elsewhere – but as close as possible to breeding areas – in order to increase the number of individually marked Teal.
5. Obtain more information about population age structure in autumn, through the collection and analysis of wings from harvested birds in Europe, as practised in North America. On both continents, analyse those data to better understand the factors affecting annual breeding success at the population level.
6. Promote agreements on general policy principles to reduce human-caused impacts in key areas, and secure funding for the conservation of Teal wetland habitats, especially at migration stopovers and in the breeding zone.
7. Study the response of Teal populations to climate change, in terms of geographic distribution, phenology and general ecology.
8. Quantify the effects of putative limiting factors such as commercial harvest, predation, parasites and diseases on Teal demographic rates and population trends.
9. Develop an adaptive harvest management scheme for Common Teal in Eurasian flyways.

Appendix 1

Winter weights and measurements of Common Teal (monthly mean ± SE, sample size in brackets). Based on data from 14 Teal ringing sites scattered throughout France, from winter 2002–2003 to 2012–2013 (ONCFS data).

Body measurements	October	November	December	January	February	March
Adult males						
Body mass (g)	329.5 ± 3.7 (70)	356.9 ± 2.5 (296)	347.0 ± 2.7 (301)	340.3 ± 2.1 (404)	329.5 ± 2.0 (502)	345.5 ± 2.1 (330)
Wing length (mm)	188.7 ± 0.6 (71)	190.1 ± 0.3 (296)	188.7 ± 0.3 (300)	189.9 ± 0.2 (402)	189.0 ± 0.2 (501)	189.0 ± 0.3 (332)
Tarsus length (mm)	30.7 ± 0.1 (69)	30.8 ± 0.1 (287)	31.2 ± 0.1 (298)	30.8 ± 0.0 (400)	30.7 ± 0.0 (496)	30.6 ± 0.1 (327)
Bill length (mm)	36.3 ± 0.2 (70)	36.2 ± 0.1 (287)	36.0 ± 0.1 (295)	36.3 ± 0.1 (401)	36.3 ± 0.1 (496)	36.6 ± 0.1 (327)
Adult females						
Body mass (g)	301.4 ± 3.3 (90)	312.2 ± 2.5 (236)	313.2 ± 2.5 (305)	311.8 ± 2.8 (219)	303.5 ± 2.5 (210)	302.0 ± 3.0 (139)
Wing length (mm)	181.0 ± 0.6 (89)	181.4 ± 0.2 (236)	180.7 ± 0.3 (305)	181.4 ± 0.3 (221)	180.8 ± 0.3 (209)	181.3 ± 0.4 (139)
Tarsus length (mm)	29.8 ± 0.2 (91)	29.7 ± 0.1 (233)	30.1 ± 0.1 (298)	29.8 ± 0.1 (204)	29.7 ± 0.3 (207)	30.0 ± 0.1 (139)
Bill length (mm)	34.8 ± 0.2 (90)	34.6 ± 0.1 (236)	34.4 ± 0.1 (295)	34.7 ± 0.1 (202)	34.4 ± 0.1 (209)	34.8 ± 0.1 (136)
Juvenile males						
Body mass (g)	319.8 ± 3.1 (136)	336.2 ± 2.0 (479)	333.0 ± 2.0 (581)	333.6 ± 2.0 (473)	326.7 ± 1.7 (545)	333.5 ± 2.2 (273)
Wing length (mm)	186.8 ± 0.3 (137)	187.7 ± 0.2 (484)	186.9 ± 0.2 (586)	187.7 ± 0.2 (482)	187.3 ± 0.2 (527)	188.0 ± 0.3 (270)
Tarsus length (mm)	30.6 ± 0.1 (134)	30.5 ± 0.1 (477)	30.7 ± 0.1 (567)	30.5 ± 0.1 (472)	30.5 ± 0.0 (542)	30.6 ± 0.1 (263)
Bill length (mm)	36.0 ± 0.1 (134)	35.9 ± 0.1 (481)	35.9 ± 0.1 (569)	36.0 ± 0.1 (469)	36.0 ± 0.1 (542)	36.3 ± 0.1 (268)
Juvenile females						
Body mass (g)	293.0 ± 3.2 (127)	305.7 ± 1.7 (537)	299.9 ± 1.8 (540)	301.7 ± 1.8 (465)	290.3 ± 1.7 (485)	298.0 ± 2.1 (319)
Wing length (mm)	179.9 ± 0.4 (127)	180.6 ± 0.2 (544)	179.7 ± 0.2 (546)	180.1 ± 0.2 (465)	179.8 ± 0.2 (483)	180.4 ± 0.2 (326)
Tarsus length (mm)	29.8 ± 0.1 (125)	29.7 ± 0.1 (529)	29.9 ± 0.1 (534)	29.6 ± 0.1 (439)	29.6 ± 0.0 (471)	29.7 ± 0.1 (300)
Bill length (mm)	34.6 ± 0.1 (127)	34.4 ± 0.1 (530)	34.4 ± 0.1 (533)	34.5 ± 0.1 (437)	34.6 ± 0.1 (471)	34.8 ± 0.1 (317)

Appendix 2

Scientific names of plants and animals cited in main text. Only those with at least genus name in main text are listed. Species only mentioned in appendices (for example, those in Teal diet) are not listed here.

Taxonomy and authorities after Flora Europaea (provided online by the Royal Botanic Garden of Edinburgh, accessed in 2011) for European plants, the PLANTS Database (provided online by the United States Department of Agriculture, Natural Resources Conservation Service, http://plants.usda.gov, accessed in 2012) for American plants, and the Integrated Taxonomic Information System (http://www.itis.gov/, accessed in 2012) for animals.

PLANTS

Bulrushes	*Scirpus* spp.
Corn	*Zea mays*
Irish Potato	*Solanum tuberosum*
Glasswort	*Salicornia* spp.
Poplars	*Populus* spp.
Rice	*Oryza sativa*
Sedges	*Carex* spp.
Smartweeds	*Polygonum* spp.
Stoneworts	*Chara* spp.
Tamarisk	*Tamarix* spp.
Wheat	*Triticum aestivum*
Willows	*Salix* spp.

INVERTEBRATES

Caddisflies	Order Trichoptera
Mayflies	Order Ephemeroptera
Midges	Family Chironomidae
Seed shrimps	Class Ostracoda

Fish

Northern Pike *Esox lucius*

Birds

American Black Duck *Anas rubripes*
American Crow *Corvus brachyrhynchos*
American Wigeon *Anas americana*
Baikal Teal *Anas formosa*
Bald Eagle *Haliaeetus leucocephalus*
Black-billed Magpie *Pica hudsonia*
Barnacle Goose *Branta leucopsis*
Blue-winged Teal *Anas discors*
Brant *Branta bernicla*
Canvasback *Aythya valisineria*
Cinnamon Teal *Anas cyanoptera*
Common Buzzard *Buteo buteo*
Common Eider *Somateria mollissima*
Common Pochard *Aythya ferina*
Common Raven *Corvus corax*
Common Teal *Anas crecca*
Eagle Owl *Bubo bubo*
Eastern Phoebe *Sayornis phoebe*
Eurasian Jackdaw *Corvus monedula*
Eurasian Wigeon *Anas penelope*
Falcated Duck *Anas falcata*
Gadwall *Anas strepera*
Garganey *Anas querquedula*
Golden Eagle *Aquila chrysaetos*
Goshawk *Accipiter gentilis*
Greater Scaup *Aythya marila*
Greater Snow Goose *Chen caerulescens atlantica*
Greater Spotted Eagle *Aquila clanga*
Green-winged Teal *Anas carolinensis*
Gyrfalcon *Falco rusticolus*
Harris Hawk *Parabuteo unicinctus*
Lesser Scaup *Aythya affinis*
Lesser Snow Goose *Chen caerulescens caerulescens*
Mallard *Anas platyrhynchos*
Marsh Harrier *Circus aeruginosus*
Moorhen *Gallinula chloropus*
Mottled Duck *Anas fulvigula*
Northern Harrier *Circus cyaneus*

Northern Pintail	*Anas acuta*
Northern Shoveler	*Anas clypeata*
Peregrine Falcon	*Falco peregrinus*
Redhead	*Aythya americana*
Ring-necked Duck	*Aythya collaris*
Ross's Goose	*Anser rossii*
Snow Goose	*Chen caerulescens*
Speckled Teal	*Anas flavirostris*
Tufted Duck	*Aythya fuligula*
Wood Duck	*Aix sponsa*
Yellow-legged Gull	*Larus michahellis*

Mammals

American Badger	*Taxidea taxus*
American Beaver	*Castor canadensis*
American Mink	*Neovison vison*
Beaver	*Castor fiber*
Brown Rat	*Rattus norvegicus*
Common Raccoon	*Procyon lotor*
Coyote	*Canis latrans*
Domestic Cat	*Felis silvestris*
Ground Squirrel	*Spermophilus* sp.
Red Fox	*Vulpes vulpes*
Striped Skunk	*Mephitis mephitis*
Wild Boar	*Sus scrofa*

Appendix 3

Most important Waterbird Conservation Regions (WCR) for Green-winged Teal in North America. As opposed to Eurasia, where importance to a bird species is generally assessed at the site level, in the Nearctic important areas are designated at the regional level instead. Only the WCRs of 'Moderately High' to 'High' importance are reported here. Data from North American Waterfowl Management Plan (2004), which should also be consulted for details and for a map of the WCRs. WCR IDs given here are based on the names of the equivalent Bird Conservation Regions (BCRs), whose boundaries sometimes differ slightly from WCRs.

Wcr Id	Wcr Number	Period of Importance	Level of Importance
Western Alaska	2	Breeding	Moderately High
Northwestern Interior Forest	4	Breeding	Moderately High
Northern Pacific Rainforest	5	Non-Breeding	Moderately High
Boreal Taiga Plains	6	Breeding	High
Boreal Taiga Plains (Northern Part)	6.1	Breeding	Moderately High
Boreal Softwood Shield (Western Half)	8.1	Breeding	Moderately High
Great Basin	9	Non-Breeding	Moderately High
Prairie Potholes	11	Breeding	High
Atlantic Northern Forests	14	Breeding	Moderately High
Oaks and Prairies	21	Non-Breeding	Moderately High
West Gulf Coastal Plain/Ouachitas	25	Non-Breeding	Moderately High
Mississippi Alluvial Valley	26	Non-Breeding	Moderately High
Southeastern Coastal Plain (Atlantic Coast)	27.1	Non-Breeding	Moderately High
Southeastern Coastal Plain (Gulf of Mexico Coast)	27.2	Non-Breeding	Moderately High
California Central Valley	32	Non-Breeding	High
Gulf Coastal Prairie	37	Non-Breeding	High
Sonora, Sinaloa & Nayarit (Mexico)*	102	Non-Breeding	Moderately High
Central Northern Mexico	103	Non-Breeding	Moderately High
Central Mexico	104	Non-Breeding	Moderately High
Gulf of Mexico Coast (Mexico)	106	Non-Breeding	Moderately High
Yucatan (Mexico)	107	Non-Breeding	Moderately High

* WCR name after Beardmore (2006)

Appendix 4

Most important sites for Common Teal within the AEWA region, compiled from the Critical Site Network Tool, developed by Wetlands International, BirdLife International and the World Conservation Monitoring Centre (UNEP-WCMC) in the framework of the Wings Over Wetlands (WOW) Project. Details about geographic coverage and criteria for site inclusion are provided in WOW (2010). Basically, these are the sites identified as critical for this species because they host a significant proportion of the relevant population (Ramsar criterion) and the sites for which Teal numbers contributed to designation as an Important Bird Area (IBA).

Country	Site Name and Period of Importance	Mean Number	Period of Reference
ALBANIA	Karavasta Lagoon (Winter)	7,480 individuals	(1996)
AZERBAIJAN	Gizilagach State Reserve (Winter)	44,000 individuals	(1996–2004)
	Lake Sarysu (Winter)	14,830 individuals	(1993–2004)
BELGIUM	Durme en Middenloop van de Schelde (Winter)	14,500 individuals	-
	IJzervallei-De Blankaart (Winter)	3,000 individuals	-
	Schorren en Polder van de Beneden-Schelde (Winter)	3,200 individuals	-
DENMARK	Ballum og Husum Enge, Kamper strandenge (Passage)	4,165 individuals	(1989)
	Nissum Fjord (Non-breeding)	4,000 individuals	(1989)
	Ringkøbing Fjord (Non-breeding)	6,900 individuals	(1994)
	Rømø (Passage)	11,000 individuals	(1983)
	Saltbæk Vig (Non-breeding)	5,500 individuals	(1983)
ESTONIA	Luitemaa (Passage)	4,120 individuals	(1999–2003)
FINLAND	Väinameri (Non-breeding)	20,000 individuals	(1995–2000)
	Oulu region wetlands (Passage)	50,000 individuals	(1997)
	Pori archipelago and wetlands (Passage)	4,000 individuals	(1997)
FRANCE	Baie du Mont Saint Michel et Ile des Landes (Winter)	850 individuals	(1991)
	Camargue (Winter)	20,741 individuals	(1997)
	Domaine d'Orx (Winter)	4,215 individuals	(1997)

Appendices 225

	Estuaire de la Loire (Passage)	20,000 individuals	(2003)
	Estuaire de la Loire (Winter)	8,674 individuals	(1997)
	Ile d'Oléron, marais de Brouage-Saint-Agnant (Winter)	7,563 individuals	(2005)
	Lac de Grand-Lieu (Winter)	5,574 individuals	(2005)
	Lac du Der-Chantecoq et étangs latéraux (Passage)	3,500 individuals	(1997)
	Lacs de la Forêt d'Orient (Winter)	7,584 individuals	(2005)
	Marais Poitevin et baie de l'Aiguillon (Winter)	8,943 individuals	(2005)
	Petite Camargue fluvio-lacustre (Non-breeding)	23,960 individuals	(1991–2009)
GEORGIA	Kolkheti (Non-breeding)	17,200 individuals	(1998)
GERMANY	Lower reaches of River Weser, unembanked area (Winter)	13,525 individuals	(1992)
	Lower Saxony Wadden Sea National Park (Passage)	7,958 individuals	(1990)
	Untersee of Lake Constance (Passage)	7,600 individuals	(1998)
GREECE	Amvrakikos gulf (Winter)	19,000 individuals	(1996)
	Evros Delta (Winter)	13,400 individuals	(1995)
	Lake Kerkini (Winter)	31,500 individuals	(1995)
	Reservoirs of former Lake Karla (Winter)	11,000 individuals	(1996)
HUNGARY	Hortobágy (Passage)	30,000 individuals	(1996)
	Lake Fertő (Passage)	10,000 individuals	(2001)
IRAQ	Haur Al Hammar (Winter)	42,784 individuals	(1973–1979)
	Haur Al Suwayqiyah (Non-breeding)	37,725 individuals	(1991–2007)
IRELAND	River Little Brosna callows: New Bridge-River Shannon (Winter)	4,000 individuals	(1996)
ISLAMIC REPUBLIC	Anzali Mordab complex (Winter)	540,000 individuals	(1970–1977)
OF IRAN	Arjan Protected Area (Winter)	45,000 individuals	(1970–1977)
	Fereidoonkenar marshes (Winter)	80,000 individuals	(1987–1992)
	Gavekhoni lake and marshes of the lower Zaindeh Rud (Winter)	24,550 individuals	(1977–1992)
	Gomishan marshes and Turkoman steppes (Winter)	30,000 individuals	(1987–1992)
	Horeh Bamdej (Winter)	15,850 individuals	(1970–1977)
	Karun river marshes (Winter)	15,300 individuals	(1970–1977)
	Kavir region (Non-breeding)	6,037 individuals	(1998)
	Kohneh Rudposht Ab-bandan (Non-breeding)	19,610 individuals	(2002–2005)
	Lake Bakhtegan, Lake Tashk and Kamjan marshes (Winter)	53,800 individuals	(1977–1992)
	Lake Maharlu (Passage)	16,500 individuals	(1977)
	Miankaleh Peninsula and Gorgan Bay (Winter)	57,000 individuals	(1970–1977)

Country	Site Name and Period of Importance	Mean Number	Period of Reference
	Paeen Rudposht Ab-bandan (Non-breeding)	50,250 individuals	(1996–2007)
	Seyed Mohalli, Zarin Kola and Larim Sara (Winter)	86,000 individuals	(1970–1977)
	Shadegan marshes and tidal mudflats of Khor-al Amaya and Khor Musa (Winter)	348,000 individuals	(1970–1977)
	Shur Gol, Yadegarlu and Dorgeh Sangi lakes (Passage)	45,100 individuals	(1977)
	South Caspian shore, from Astara to Gomishan (Winter)	171,000 individuals	(1970–1977)
ITALY	Venice lagoon (Non-breeding)	25,427 individuals	(1993–2006)
KAZAKHSTAN	Chardara Reservoir (Winter)	15,500 individuals	(2003)
	Korgalzhyn State Nature Reserve (Breeding)	27,500 individuals	(2006)
	Kulykol-Taldykol Lake System (Passage)	90,840 individuals	(2005)
	Sarykopa Lake System (Passage)	11,000 individuals	(1971)
MONTENEGRO	Lake Skadar (Winter)	20,500 individuals	(1996)
NETHERLANDS	Dollard (Non-breeding)	4,639 individuals	(1991)
	Lauwersmeer (Non-breeding)	12,783 individuals	(1991)
	Oostvaardersplassen (Non-breeding)	52,580 individuals	(1990)
	Wadden Sea (Non-breeding)	4,153 individuals	(1991)
NIGER	Dan Doutchi wetland (Winter)	470 individuals	(1994)
NORWAY	Nordre Øyeren and Sørumsneset (Passage)	7,600 individuals	(1995)
PORTUGAL	Tejo estuary (Winter)	15,150 individuals	(2002)
ROMANIA	Danube Delta (Passage)	14,500 individuals	(1990–2006)
	Lake Oltina (Winter)	12,550 individuals	(1990–2006)
RUSSIAN FEDERATION	Bulgarski (Passage)	7,000 individuals	(1989)
	Burukshunskiye limans (Passage)	47,500 individuals	(1973)
	Dvuob'ye (Breeding)	3,750 breeding pairs	(2001)
	Dvuob'ye (Non-breeding)	69,500 individuals	(1989–2001)
	Dvuob'ye (Passage)	12,000 individuals	(2001)
	Kamsko-Ikski area (Breeding)	6,000 breeding pairs	(1989)
	Kamsko-Ikski area (Passage)	16,250 Individuals	(1989)
	Koporski Bay (Breeding)	7,500 breeding pairs	(1997)
	Kuloy river (Breeding)	4,500 breeding pairs	(1999)
	Lower Ob' (Breeding)	17,500 breeding pairs	(2004)
	Lower Ob' (Passage)	45,000 individuals	(2004)
	Western Ilmen area (Non-breeding)	40,000 individuals	(1995)

SPAIN	Charca de Sierra Brava (Unknown)	20,970 individuals	(2002–2007)
	Ebro Delta (Winter)	10,150 individuals	(1996)
	Guadalquivir marshes (Winter)	115,000 individuals	(1989)
SWEDEN	Falsterbo-Bay of Foteviken (Winter)	5,000 individuals	(1992–2006)
	Getterön (Breeding)	2,600 breeding pairs	(1996)
	Lake Hornborgasjön (Passage)	6,500 individuals	(1996)
	Ledskär-Karlholm bay (Passage)	2,500 individuals	(1996)
	Umeälven Delta (Passage)	3,750 Individuals	(2001)
TURKEY	Burdur Lake (Winter)	20,000 individuals	(1987)
	Çorak Lake (Winter)	20,200 individuals	(1986)
	Gediz Delta (Winter)	11,532 individuals	(1999)
	Göksu Delta (Winter)	14,952 individuals	(1992)
	Meriç Delta (Winter)	12,400 individuals	(1999)
	Seyhan Delta (Winter)	12,800 individuals	–
	Sultansazligi (Winter)	14,721 individuals	(1999)
	Tuz Lake (Passage)	35,000 individuals	–
TURKMENISTAN	Turkmenbashy Bay (Winter)	16,100 individuals	(2001)
UKRAINE	Dnister Delta (Passage)	25,000 individuals	(1992)
UNITED KINGDOM	Abberton Reservoir (Winter)	4,550 individuals	(1995)
	Dee Estuary (Winter)	5,950 individuals	(1995)
	Hamford Water (Winter)	4,020 individuals	(1995)
	Loch Leven (Passage)	5,282 individuals	(2001–2006)
	Lower Derwent Valley (Winter)	5,800 individuals	(1995)
	Mersey Estuary (Passage)	5,187 individuals	(2001–2006)
	Mersey Estuary (Winter)	12,400 individuals	(1995)
	Mid-Essex Coast (Winter)	4,150 individuals	(1995)
	Moray Basin, Firths and Bays (Winter)	4,760 individuals	(1995)
	Ouse Washes (Winter)	5,871 individuals	(2003–2008)
	Ribble and Alt Estuaries (Winter)	7,050 individuals	(1995)
	Somerset Levels and Moors (Winter)	8,900 individuals	(1995)
UZBEKISTAN	Karnabchul Steppe (Passage)	19,166 individuals	(2006)

Appendix 5

Most important sites for Common Teal in Asia; that is, those which qualify as Important Bird Areas (IBAs) due to numbers of this species (Criteria A4i: site known or thought to hold, on a regular basis, >1% of a biogeographic population of a congregatory waterbird species). Information courtesy of BirdLife International.

Country	Site Name	Period of Reference
BANGLADESH	Aila Beel	(2004)
	Hakaluki Haor	(2004)
	Jamuna-Brahmaputra river	(2004)
	Muhuri Dam	(2004)
	Tanguar Haor and Panabeel	(2004)
CHINA (MAINLAND)	Desert and wetland from Northern Urumqi to Dabancheng	(2009)
	Oasis, Desert and Wetland at Mosuowan	(2009)
	Qinghai Hu (Koko Nor)	(2009)
	Taipo Hu Nature Reserve	(2009)
	Tumen River at Jingxin-Fangchuan *	(2009)
	Xingkai Hu Nature Reserve	(2009)
INDIA	Kanjli Lake	(2004)
	Saman Bird Sanctuary	(2004)
JAPAN	Lake Kasumigaura, Ukisima	(2004)
MONGOLIA	Khar Us Lake	(2007)
	Selengiin Tsagaan Lake	(2007)
	Teshigiin Olon Lakes	(2007)
PAKISTAN	Drigh Wildlife Sanctuary	(2004)
	Haleji Wildlife Sanctuary	(2004)
	Indus Dolphin Reserve and Kandhkot wetlands	(2004)
	Manchar Lake	(2004)
	Mangla Lake	(2004)

	Marala Game Reserve	(2004)
	Mehboob Shah Lake	(2004)
	Phoosna Wetlands Complex	(2004)
	Pugri Lake	(2004)
	Rann of Kutch Wildlife Sanctuary	(2004)
	Rasool Barrage Wildlife Sanctuary	(2004)
RUSSIA (ASIAN PART)	Bolon' lake	(2004)
	Forty islands	(2004)
	Kava valley	(2004)
	Kharchinskoye lake	(2004)
	Parapol'skiy valley	(2004)
	Semyachik lagoon	(2004)
	Zhupanovskiy lagoon	(2004)
SOUTH KOREA	Cheonsu Bay	(2004)
TAIWAN – CHINA	Kuantu	(2004)
	Taipei City Waterbird Refuge	(2004)
VIETNAM	Tram Chim	(2004)

* Also considered to hold over 20,000 Teal as a migration stopover

Appendix 6

Food items found in Common and Green-winged Teal digestive tracts. Listed are only taxa for which the seeds (oogones for algae) have been found in Teal digestive tracts, as vegetative parts of plants may be ingested only incidentally.

Taxonomy and authorities after Flora Europaea (provided online by the Royal Botanic Garden of Edinburgh, accessed in 2011) for plants eaten by Common Teal, after PLANTS Database (provided online by the United States Department of Agriculture, Natural Resources Conservation Service, http://plants.usda.gov, accessed in 2012) for plants eaten by Green-winged Teal, and after the Integrated Taxonomic Information System (http://www.itis.gov/, accessed in 2012) for animal prey eaten by both species. Most of the Common Teal data come from the Supporting Information table provided online for the paper by Brochet et al. (2012a), supplemented by more references. Numbers within brackets refer to the reference list at the end of this Appendix.

Taxon	Common Teal	Green-Winged Teal
Algae		
Characeae		
Chara sp.	[1] ; [24] ; [31] ; [35] ; [44] ; [47] ; [57] ; [79] ; [84] ; [85] ; [87] ; [89]	[8] ; [16] ; [19] ; [23] ; [28] ; [32]
Plants		
Aizoaceae		
Sesuvium maritimum		[59]
Alismataceae		
Alisma sp.	[31]	
Alisma plantago-aquatica	[1] ; [33] ; [34] ; [47] ; [89]	
Baldellia ranunculoides	[1]	
Sagittaria latifolia		[20]
Sagittaria sagittifolia	[27]	
Amaranthaceae		
Amaranthus sp.		[30] ; [46] ; [48] ; [55]
Amaranthus deflexus	[89]	

Appendices

Amaranthus graecizans		[17]
Amaranthus tuberculatus		[20]
Anacardiaceae		
Toxicodendron radicans		[48]
Asteraceae		
Pluchea sp.		[42]
Symphyotrichum divaricatum		[55]
Betulaceae	[33]	
Alnus sp.	[13] ; [24] ; [33]	
Alnus glutinosa	[33]	[48]
Alnus incana	[24] ; [34] ; [45] ; [54]	[18]
Betula sp.	[1] ; [24] ; [33]	
Betula pendula		
Carpinus caroliniana		[48]
Boraginaceae	[44]	
Myosotis scorpioides	[24]	
Brassicaceae		
Brassica sp.	[89]	[17]
Cabombaceae		
Brasenia schreberi		[28]
Callitrichaceae		
Callitriche sp.	[44] ; [89]	
Caprifoliaceae		
Sambucus nigra	[1] ; [13] ; [24] ; [33]	
Sambucus racemosa	[33]	
Caryophyllaceae		
Cerastium sp.	[1]	
Lychnis flos-cuculi	[3]	
Spergula sp.	[64]	
Spergula arvensis	[3] ; [24] ; [52] ; [79]	
Spergularia marina	[1] ; [3] ; [24] ; [52] ; [89]	
Spergularia media	[3] ; [13] ; [24] ; [52]	
Spergularia salina	[13]	

Taxon	Common Teal	Green-Winged Teal
Stellaria holostea	[84]	
Stellaria media	[13] ; [24] ; [33] ; [35]	
Ceratophyllaceae		
Ceratophyllum sp.	[1] ; [24]	
Ceratophyllum demersum	[1] ; [84]	[20]
Chenopodiaceae	[1] ; [33] ; [76]	
Arthrocnemum sp.	[78]	
Arthrocnemum glaucum	[35] ; [47] ; [89]	
Atriplex sp.	[1] ; [15] ; [24] ; [65] ; [76]	[16]
Atriplex hastata	[3] ; [24] ; [47] ; [52] ; [64]	
Atriplex littoralis	[15]	
Atriplex patula	[1] ; [24]	
Bassia sp.		[70]
Bassia hirsuta	[35]	[55]
Chenopodium sp.	[1] ; [3] ; [29] ; [44] ; [52] ; [84] ; [89]	[17] ; [21] ; [22] ; [88]
Chenopodium album	[1] ; [24] ; [33] ; [49] ; [84]	[17] ; [30]
Chenopodium glaucum	[1]	
Chenopodium rubrum	[1]	
Chenopodium vulvaria	[24] ; [33]	
Halimione portulacoides	[24] ; [35]	
Kochia hirsuta	[89]	
Salicornia sp.	[3] ; [24] ; [35] ; [52] ; [64] ; [76] ; [87] ; [89]	[59] ; [70]
Salicornia europaea	[13]	
Salicornia fruticosa	[35] ; [47]	
Salicornia perennis		[16]
Salicornia rubra		[88]
Salsola sp.	[85]	
Salsola soda	[89]	
Suaeda sp.	[2] ; [85] ; [89]	[60] ; [70]
Suaeda fruticosa	[35] ; [57]	
Suaeda maritima	[3] ; [13] ; [24] ; [47] ; [52] ; [57] ; [87]	

Appendices 233

Commelinaceae
  Aneilema sp. [25]
  Murdannia keisak [32] ; [39] ; [48]
Cornaceae
  Nyssa sylvatica [48]
Compositae [48]
  Artemisia sp. [29] ; [33] ; [44] ; [76]
  Aster tripolium [31]
  Bacchais halimifolia [13] ; [24]
  Bidens sp. [64]
  Bidens tripartita [1] ; [84]
  Chamomilla recutita [1] ; [34]
  Cirsium sp. [24]
  Cirsium palustre [24]
  Cirsium vulgare [24]
  Filaginella uliginosa [24]
  Hieracium umbellatum [1] ; [84]
  Inula sp. [89]
Corylaceae
  Carpinus betulus [1]
Cruciferae [29] ; [33] ; [44]
  Coronopus squamatus [24]
  Lepidium sp. [33]
  Lepidium perfoliatum [17]
  Rorippa sp. [1]
  Rorippa amphibia [1] ; [84]
  Rorippa microphylla [24]
Cuscutaceae
  Cuscuta sp. [1] ; [33] ; [44] ; [62]
Cyperaceae [1] ; [3] ; [6] ; [9] ; [12] ; [13] ; [15] ; [24] ; [31] ; [33] ; [19] ; [46] ; [48]
  Carex sp. [34] ; [38] ; [44] ; [45] ; [52] ; [65] ; [64] ; [76] ; [85] [8] ; [16] ; [17] ; [18] ; [21] ; [22] ; [50]
  Carex acuta [1]
  Carex aquatilis [45] [17]

Taxon	Common Teal	Green-Winged Teal
Carex canescens	[34] ; [45]	[18]
Carex crinita	[1] ; [51]	
Carex disticha	[1] ; [13]	
Carex elata	[34]	
Carex elongata	[3] ; [52]	
Carex extensa		[48]
Carex festucacea	[13]	
Carex flava	[1]	
Carex hirta	[5]	
Carex irrigua		[17]
Carex lenticularis		[48]
Carex lurida	[7]	
Carex nigra	[1] ; [24] ; [76]	[17] ; [22]
Carex obnupta	[1]	
Carex otrubae	[1]	
Carex ovalis	[3] ; [52]	
Carex pallescens	[1] ; [84]	
Carex panicea	[51]	
Carex pseudocyperus	[34] ; [45]	
Carex riparia	[13]	
Carex rostrata	[3] ; [5] ; [34] ; [52]	
Carex trinervis	[3] ; [34] ; [52] ; [64]	
Carex vesicaria	[1] ; [3] ; [52] ; [64]	[19] ; [23]
Carex vulpina	[87] ; [89]	[9] ; [60]
Cladium sp.		
Cladium mariscus		[20] ; [39] ; [48]
Cyperus sp.		[20] ; [32]
Cyperus difformis		[48]
Cyperus erythrorhizos		
Cyperus esculentus		
Cyperus flavicomus	[27]	
Cyperus michelianus		

Appendices 235

Cyperus odoratus	[20] ; [32] ; [39] ; [42] ; [48]
Cyperus strigosus	[48]
Eleocharis sp.	[1] ; [34] ; [76] ; [89] [16] ; [17] ; [23] ; [28] ; [39] ; [42] ; [46] ; [55] ; [73] ; [82]
Eleocharis acicularis	[1]
Eleocharis macrostachya	[17] ; [22] ; [30]
Eleocharis multicaulis	[1]
Eleocharis ovata	[79]
Eleocharis palustris	[1] ; [3] ; [5] ; [9] ; [13] ; [15] ; [24] ; [33] ; [34] ; [35] ; [47] ; [51] ; [52] ; [79] ; [84] ; [87] ; [89] [16] ; [18] ; [21] ; [32] ; [48] ; [90]
Eleocharis parvula	[59]
Eleocharis quadrangulata	[28] ; [48]
Eleocharis uniglumis	[1] ; [6]
Fimbristylis sp.	[42]
Schoenus nigricans	[1]
Scirpus sp.	[1] ; [3] ; [31] ; [33] ; [35] ; [38] ; [44] ; [52] ; [54] ; [57] ; [64] ; [65] ; [76] ; [79] ; [80] ; [85] [17] ; [18] ; [23] ; [46] ; [59] ; [60]
Scirpus ?acustris	[1] ; [9] ; [15] ; [24] ; [34] ; [35] ; [47] ; [52] ; [64] ; [84] ; [85]
Scirpus ?litoralis	[35] ; [47] ; [72]
Scirpus maritimus	[1] ; [2] ; [3] ; [9] ; [13] ; [24] ; [35] ; [47] ; [52] ; [64] ; [72] ; [76] ; [80] ; [84] ; [85] ; [87] ; [89] [17] ; [21] ; [55] ; [88]
Scirpus mucronatus	[35] ; [47] ; [87] ; [89]
Scirpus setaceus	[24]
Scirpus ?tabernaemontani	[3] ; [24] ; [27] ; [52]
Scirpus ?triqueter	[27]
Schoenoblectus acutus	[16] ; [17] ; [88]
Schoenoblectus americanus	[16] ; [17] ; [39]
Schoenoblectus californicus	[19]
Schoenoblectus maritimus	[22]
Schoenoblectus robustus	[32] ; [39] ; [42] ; [59]
Schoenoblectus tabernaemontani	[12] ; [89] [21] ; [22] ; [32] ; [48] ; [59]
Elatinaceae	
Elatine ?ydropiper	[34]

Taxon	Common Teal	Green-Winged Teal
Empetraceae		
Empetrum sp.	[12] ; [38] ; [45]	
Empetrum nigrum	[5] ; [7] ; [13]	
Empetrum sibiricum	[65]	
Equisetaceae		
Equisetum sp.		[9] ; [36]
Fabaceae		
Kummerowia striata	[89]	[48]
Melilotus sp.		[22]
Graminae		
Agropyron sp.	[1] ; [3] ; [13] ; [29] ; [33] ; [44] ; [52] ; [64] ; [79]	[22]
Agrostis stolonifera	[85]	
Alopecurus sp.	[3] ; [24] ; [52]	
Alopecurus geniculatus	[1] ; [3] ; [33] ; [52]	
Alopecurus myosuroides	[24] ; [64]	
Alopecurus pratensis	[24]	
Apera spica	[24]	
Arrhenatherum sp.	[13]	
Avena sp.	[33]	
Avena sativa	[15]	[17]
Bromus sp.	[1] ; [3] ; [52]	[16]
Crypsis schoenoides		[55]
Digitaria ischaemum		[48]
Digitaria sanguinalis	[89]	[32] ; [48]
Distichlis sp.		[16] ; [17]
Distichlis spicata		[22] ; [59]
Eragrostis sp.	[89]	[20]
Eragrostis hypnoides		[20]
Echinochloa sp.	[29] ; [47] ; [57] ; [89]	[19] ; [20] ; [46] ; [55] ; [60] ; [73]
Echinochloa crus-galli	[1] ; [35] ; [64] ; [79] ; [84] ; [87]	[19] ; [20] ; [30] ; [43] ; [48] ; [55]
Echinochloa walteri		[20] ; [32] ; [39] ; [42] ; [48]
Festuca arundinacea	[89]	

Festuca rubra	[24]
Glyceria sp.	[1] ; [33]
Glyceria declinata	[24]
Glyceria fluitans	[3] ; [9] ; [34] ; [51] ; [52]
Glyceria maxima	[24]
Glyceria plicata	[24]
Holcus lanatus	[24]
Hordeum sp.	[15]
Hordeum distichon	[24]
Hordeum jubatum	
Hordeum vulgare	[1] ; [49] ; [51]
Leersia sp.	
Leersia oryzoides	[1] ; [34] ; [89]
Leptochloa sp.	
Leptochloa panicea	
Lolium sp.	
Milium sp.	[1] ; [89]
Oryza sativa	[31] ; [35] ; [47] ; [56] ; [57] ; [85] ; [89]
Panicum sp.	[12] ; [89]
Panicum anceps	
Panicum dichotomiflorum	
Panicum verrucosum	
Parapholis strigosa	[24]
Paspalum sp.	
Paspalum boscianum	
Paspalem denticulatum	
Paspalum distichum	[87] ; [89]
Phalaris arundinacea	[3] ; [33] ; [51] ; [52]
Phleum pratense	[24]
Phragmites sp.	[65]
Phragmites australis	[33] ; [89]
Poa annua	[24] ; [84]
Poa trivialis	[24]
Polypogon sp.	[89]

[24]
[21]
[17]
[60]
[18] ; [20]
[55]
[59]
[17]
[19]
[48]
[48]
[32] ; [39] ; [42] ; [48]
[39]
[32]
[32]
[19]
[73]
[88]

Taxon	Common Teal	Green-Winged Teal
Puccinellia distans	[24]	
Puccinellia maritima	[24]	
Setaria italica	[33]	
Setaria magna		[39] ; [42]
Setaria pumila	[9] ; [31] ; [89]	[48]
Setaria verticillata	[89]	
Setaria viridis	[9] ; [89]	
Sorghum sp.	[89]	[46]
Sorghum bicolor	[72]	[30]
Spartina townsendii	[24]	
Tridens sp.		[39]
Triticum sp.	[49] ; [79] ; [80]	[22]
Triticum aestivum	[1] ; [24] ; [84] ; [89]	[17] ; [30] ; [43]
Urochloa platyphylla		[19]
Zea mays	[1] ; [15] ; [76] ; [84]	[20] ; [28] ; [32] ; [39] ; [42] ; [43] ; [46] ; [48]
Guttiferae		
Hypericum hirsutum	[24]	
Haemodoraceae		
Lachnanthes caroliana		[39]
Haloragaceae		
Myriophyllum sp.	[44]	[8]
Myriophyllum sibiricum	[24] ; [35] ; [62]	[21]
Myriophyllum spicatum	[1] ; [3] ; [47] ; [52] ; [84] ; [87] ; [89]	[16] ; [37] ; [48]
Myriophyllum verticillatum	[1]	
Hamamelidaceae		
Liquidambar sp.		[28]
Hippuridaceae		
Hippuris sp.	[44]	[50]
Hippuris vulgaris	[12]	[22]
	[1] ; [3] ; [5] ; [13] ; [14] ; [24] ; [33] ; [38] ; [45] ; [51] ; [52] ; [76] ; [84]	
Hydrocharitaceae		
Elodea sp.		[48]

Iridaceae
  Iris pseudacorus [1]
Juncaceae [44]
  Juncus sp. [1] ; [3] ; [6] ; [24] ; [29] ; [52] ; [57] ; [64] ; [76] ; [84] ; [87] ; [89]
  Juncus acutiflorus [24]
  Juncus arcticus [3] ; [52]
  Juncus articulatus [24]
  Juncus effusus [3] ; [52]
  Juncus gerardii [24]
  Juncus maritimus [24]
Juncaginaceae
  Triglochin sp. [24]
  Triglochin maritima [33] ; [44]
Labiatae [49]
  Galeopsis speciosa [1] ; [64] ; [84] ; [89]
  Lycopus europaeus [24]
  Mentha aquatica [3]
  Prunella vulgaris [27]
  Stachys palustris [1]
Leguminosae [33]
  Lotus sp. [1] ; [64]
  Lotus corniculatus [24]
  Lotus uliginosus [76]
  Medicago arabica [1] ; [3] ; [52] ; [64] ; [76] ; [89]
  Trifolium sp. [24]
  Trifolium campestre [24] ; [64] ; [76]
  Trifolium pratense [3] ; [24] ; [52] ; [64] ; [87]
  Trifolium repens [24]
  Trifolium squamosum [89]
  Vicia sp. [24] [20] ; [22] ; [90]
  Vicia cracca [21] ; [50]
Lemnaceae [8] ; [17]
  Lemna sp. [1] ; [29] ; [44] ; [71] ; [89] [8]
  Lemna minor [20]

Appendices 239

Taxon	Common Teal	Green-Winged Teal
Lythraceae	[44]	
Lythrum salicaria	[33]	
Malvaceae	[44]	
Menyanthaceae		
Menyanthes trifoliata	[5] ; [24] ; [34] ; [38] ; [65]	[18]
Najadaceae		
Najas sp.	[35] ; [47] ; [89]	[23]
Najas marina	[1] ; [57] ; [79] ; [80] ; [84] ; [85] ; [89]	
Najas minor	[1] ; [3] ; [52] ; [57] ; [84] ; [89]	
Nymphaeaceae		
Nymphaea sp.		[28]
Onagraceae		
Epilobium sp.	[1]	
Epilobium hirsutum	[24]	
Ludwigia leptocarpa		[32]
Ludwigia peploides	[87] ; [89]	
Papaveraceae		
Chelidonium majus	[33]	
Papaver sp.	[89]	
Papilionaceae	[3] ; [44] ; [52] ; [62]	
Parnassiaceae		
Parnassia palustris	[24]	
Pinaceae		
Pinus sp.		[23]
Plantaginaceae	[44]	[48]
Plantago sp.	[1] ; [3] ; [52]	
Plantago lanceolata	[3] ; [52]	
Plantago major	[1] ; [24] ; [33] ; [51]	
Plantago maritima	[24]	
Plantago media	[24]	
Plumbaginaceae	[44]	
Armeria maritima	[24]	
Limonium vulgare	[24]	

Polygonaceae
  Fallopia convolvulus [44]
  Polygonum sp. [89]
    [1] ; [34] ; [64] ; [65] ; [79] ; [80] ; [84] ; [85]   [8] ; [16] ; [17] ; [20] ; [21] ; [23] ; [28] ; [43] ; [46] ; [48] ; [60] ; [73]

  Polygonum amphibium [1] ; [24] ; [34] ; [51] ; [84]   [16] ; [20] ; [73]
  Polygonum aviculare [1] ; [3] ; [24] ; [52] ; [76] ; [84] ; [89]   [16]
  Polygonum convolvulus [13] ; [24] ; [34] ; [49]   [48]
  Polygonum glabrum   [16] ; [18] ; [20] ; [22] ; [48]
  Polygonum hydropiper [1] ; [9] ; [13] ; [24] ; [34]   [28] ; [32]
  Polygonum hydropiperoides   [16] ; [17] ; [20] ; [22] ; [48] ; [55]
  Polygonum lapathifolium [1] ; [3] ; [9] ; [13] ; [24] ; [51] ; [52] ; [64] ; [84] ; [87] ; [89]

  Polygonum minus [1] ; [3] ; [34] ; [52]
  Polygonum persicaria [1] ; [3] ; [9] ; [13] ; [24] ; [33] ; [49] ; [51] ; [52] ; [57] ; [64] ; [76] ; [84] ; [87] ; [89]   [20] ; [22] ; [59]

  Polygonum pensylvanicum   [18] ; [20] ; [30]
  Polygonum punctatum   [20] ; [32] ; [39] ; [42] ; [48]
  Polygonum sagittatum   [32]
  Rumex sp. [1] ; [3] ; [31] ; [34] ; [76] ; [84] ; [89]   [21] ; [55] ; [73]
  Rumex acetosa [3] ; [52]
  Rumex acetosella [13]
  Rumex aquaticus [27]
  Rumex conglomeratus [1] ; [24] ; [51]
  Rumex crispus [1] ; [24] ; [51]
  Rumex hydrolapathum [1] ; [24]
  Rumex maritimus [1]   [16]
  Rumex obtusifolius [1] ; [33]
  Rumex palustris [1] ; [52]

Pontederiaceae
  Heteranthera sp.   [43]
  Heteranthera limosa [87] ; [89]
  Heteranthera reniformis [87] ; [89]

Portulacaceae
  Calandrinia ciliata   [55]

Taxon	Common Teal	Green-Winged Teal
Potamogetonaceae	[44]	[8]
Potamogeton sp.	[1] ; [15] ; [24] ; [27] ; [29] ; [31] ; [33] ; [34] ; [35] ; [38] ; [45] ; [54] ; [57] ; [62] ; [65] ; [71] ; [76] ; [79]	[16] ; [17] ; [18] ; [20] ; [21] ; [23] ; [36] ; [46]
Potamogeton berchtoldii	[24]	
Potamogeton epihydrus		[20]
Potamogeton foliosus		[20]
Potamogeton gramineus	[1] ; [84]	
Potamogeton lucens	[27]	[16]
Potamogeton natans	[1] ; [24] ; [76]	
Potamogeton nodosus	[89]	[20]
Potamogeton obtusifolius	[1]	
Potamogeton pectinatus	[1] ; [3] ; [9] ; [24] ; [27] ; [35] ; [47] ; [52] ; [54] ; [57] ; [64] ; [72] ; [76] ; [84] ; [89]	[16] ; [20] ; [22] ; [48] ; [66]
Potamogeton perfoliatus	[5]	
Potamogeton pusillus	[1] ; [3] ; [24] ; [35] ; [45] ; [47] ; [52] ; [76] ; [84] ; [87] ; [89]	[16] ; [18] ; [20]
Stuckenia filiformis		[10]
Stuckenia pectinata		[88]
Primulaceae		
Lysimachia vulgaris	[1]	
Ranunculaceae	[44] ; [62]	
Ranunculus sp.	[1] ; [13] ; [24] ; [31] ; [33] ; [35] ; [47] ; [76] ; [84] ; [87] ; [89]	[17] ; [22] ; [32]
Ranunculus acris	[1] ; [24]	
Ranunculus baudotii	[24]	
Ranunculus bulbosus	[1]	
Ranunculus flammula	[1] ; [3] ; [13] ; [24] ; [52]	[48]
Ranunculus hispidus	[4]	
Ranunculus lingua	[1]	
Ranunculus longirostris		[21]
Ranunculus repens	[1] ; [13] ; [24] ; [34] ; [51]	
Ranunculus sardous	[1] ; [3] ; [24] ; [52] ; [64] ; [76]	

Ranunculus sceleratus	[1] ; [24]
Ranunculus trichophyllus	[1] ; [76]
Rosaceae	
Comarum palustre	[44]
Crataegus monogyna	[65]
Fragaria vesca	[24]
Potentilla sp.	[33]
Potentilla anserina	[1] ; [33]
Potentilla palustris	[3] ; [52]
Rubus sp.	[6] ; [24] ; [27] ; [34] ; [45]
Rubus fruticosus	[33] ; [54] ; [64] ; [89]
	[1] ; [3] ; [24] ; [52] ; [76] ; [84]
Rubiaceae	[44]
Cephalanthus sp.	[28]
Cephalanthus occidentalis	[20] ; [48]
Galium aparine	[24] ; [49]
Ruppiaceae	
Ruppia sp.	[2] ; [29] ; [72] ; [76]
Ruppia cirrhosa	[24] ; [85] ; [89]
Ruppia maritima	[1] ; [3] ; [15] ; [24] ; [27] ; [35] ; [47] ; [52] ; [64] ; [76] ; [85] ; [89]
	[17] ; [22] ; [39] ; [42] ; [59] ; [88]
Salicaceae	
Salix sp.	[1] ; [24]
Saxifragaceae	[44]
Scrophulariaceae	
Euphrasia odontites	[13]
Linaria arvensis	[33]
Scrophularia auriculata	[24]
Scrophularia nodosa	[33]
Verbascum sp.	[33]
Veronica sp.	[33]
Veronica anagallis-aquatica	[24]
Veronica beccabunga	[24]
Veronica catenata	[51]

Taxon	Common Teal	Green-Winged Teal
Solanaceae		
Solanum sp.	[33] ; [89]	
Solanum dulcamara	[33]	
Solanum lycopersicum	[33]	
Solanum nigrum	[33]	
Solanum tuberosum	[29]	
Sparganiaceae	[44]	
Sparganium sp.	[1] ; [5] ; [27] ; [45]	[16] ; [17] ; [18] ; [22]
Sparganium americanum		[48]
Sparganium emersum	[27] ; [34] ; [84]	
Sparganium erectum	[1] ; [24] ; [27] ; [34] ; [76] ; [84]	
Typhaceae	[44]	
Typha sp.		[88]
Umbelliferae	[44]	
Cicuta virosa	[45]	
Oenanthe sp.	[1]	
Oenanthe fistulosa	[3] ; [52]	
Torilis japonica	[24]	
Urticaceae		
Boehmeria cylindrica		[48]
Urtica dioica	[33] ; [84]	
Verbenaceae		
Verbena urticifolia		[48]
Vitaceae		
Vitis sp.		[48]
Zannichelliaceae		
Zannichellia sp.	[35] ; [47] ; [54] ; [89]	
Zannichellia palustris	[1] ; [24] ; [33] ; [57] ; [76]	[16] ; [18]
Zosteraceae		
Zostera angustifolia	[65]	
Zostera japonica	[9] ; [42]	[42]
Zostera marina	[64] ; [89]	
Zostera noltii		

Appendices

ANIMALS
Rotifera
Brachionidae
   Brachionus plicatilis [83]
Ectoprocta (statoblasts) [1] ; [69] ; [89]
Phylactolaemata
Cristatellidae (statoblasts) [84]
   Cristatella mucedo (statoblasts) [1] ; [24] ; [33]
Plumatellidae
   Pectinatella magnifica [86]
   Plumatella sp. (statoblasts) [48]
Hydrozoa [12]
Annelida [64]
Hirudinea (adults and eggs) [33] ; [53] ; [69] ; [74]   [16] ; [18] ; [70]
Erpobdellidae [16] ; [73]
   Erpobdella octoculata [24]
Glossiphoniidae [1]
   Helobdella stagnalis [24]
Oligochaeta (adults and eggs) [14] ; [29] ; [33] ; [84]
Lumbricidae [1] ; [41] ; [84]
Tubificidae [1]
   Tubifex sp. (adults and eggs) [24]   [73]
Polychaeta [64]
Nereididae [12] ; [77]
   Nereis sp. [1] ; [24] ; [77]
Arachnida
Acari [1] ; [33] ; [41] ; [45] ; [76] ; [84]
Hydrachnidae [53] ; [63] ; [69] ; [72]
   Hydracana sp. [1] ; [41] ; [45] ; [76]   [20] ; [73]
Araneae [50] ; [88]
Araneidae
Lycosidae
   Trochosa ruricola [24]

Taxon	Common Teal	Green-Winged Teal
Crustacea (adult and ephippia)	[12] ; [26] ; [33] ; [64]	[8] ; [17] ; [88]
Branchiopoda		[20]
Phyllopoda	[29]	
Anostraca	[86]	[20]
Artemiidae		[88]
Artemia franciscana		[88]
Cladocera (adult and ephippia)	[1] ; [33] ; [63] ; [69] ; [78] ; [89]	[16] ; [55] ; [61] ; [88]
Chydoridae	[24]	
Daphniidae		
Ceriodaphnia sp. (ephippia)	[86]	
Daphnia sp.	[1] ; [2] ; [75] ; [76]	
Daphnia magna (ephippia)	[24] ; [68]	
Moina brachiata	[68]	
Leptodoridae	[1]	
Macrothricidae		
Macrothrix hirsuticornis	[86]	
Malacostraca		
Amphipoda	[64] ; [65] ; [77]	[16]
Ampithoidae		
Ampithoe sectimanus		[70]
Ampithoe valida		[70]
Anisogammaridae		
Eogammarus confervicolus		[70]
Gammaridae	[1] ; [12] ; [89]	[50]
Gammarus sp.	[24] ; [86]	[20]
Gammarus limnaeus		[16]
Gammarus pulex	[24] ; [33]	
Corophiidae	[1] ; [89]	
Corophium sp.	[77]	
Corophium acherusicum		[70]
Corophium insidiosum		[70]
Corophium volutator	[24]	

Eusiridae		
Rhachotropis oculata		[70]
Hyalellidae		
Hyalella azteca		[16] ; [36]
Talitridae		
Traskorchestia georgiana	[12] ; [53] ; [63] ; [64] ; [69] ; [74] ; [77]	[70]
Isopoda	[1]	[50] ; [61]
Asellidae		
Asellus sp.	[24]	
Asellus aquaticus	[33] ; [45]	[20] ; [48]
Caecidotea sp.	[14]	
Idoteidae		
Idotea sp.		
Pentidotea resecata	[89]	[70]
Sphaeromatidae		
Lekanesphaera rugicauda	[24]	
Sphaeroma hookeri	[89]	
Sphaeroma serratum	[77]	
Decapoda		
Crangonidae		
Crangon crangon	[24]	
Pinnotheridae		
Pinnixa sp.		[67]
Maxillopoda		
Copepoda	[1] ; [45] ; [51] ; [77]	[20] ; [67]
Cyclopidae	[1] ; [2] ; [24]	
Cyclops sp.		[20]
Ostracoda	[1] ; [24] ; [45] ; [51] ; [76] ; [77] ; [89]	[20] ; [30] ; [43] ; [50] ; [55] ; [67] ; [68] ; [82] ; [88]
Podocopida		
Candonidae		[20]
Candona sp.		[67]
Candona patzcuaro		[90]
Candona simpsoni		

Taxon	Common Teal	Green-Winged Teal
Cytherideidae		
Cyprideis littoralis	[57]	
Cyprideis salebrosa		[67]
Cyprideis gelica		[67]
Cyprididae		
Cypridopsis vidua	[68] ; [86]	
Cypris sp.	[57]	
Heterocypris salina	[57]	[20]
Insecta		
Coleoptera (adult and larvae)	[26] ; [49] ; [51] ; [52] ; [64] ; [69] ; [75] ; [77] ; [85]	[8] ; [9] ; [16] ; [20] ; [30] ; [73] ; [88]
	[1] ; [2] ; [29] ; [33] ; [45] ; [53] ; [57] ; [63] ; [64] ; [72] ;	[16] ; [17] ; [18] ; [20] ; [50] ; [90]
	[74] ; [76] ; [77] ; [84] ; [89]	
Carabidae (adult and larvae)	[1] ; [24] ; [41]	[20] ; [48]
Chrysomelidae		
Diabrotica undecimpunctata		[20]
Curculionidae	[1] ; [77]	
Dytiscidae (adult and larvae)	[1] ; [29] ; [53] ; [63] ; [69] ; [74]	[20] ; [60]
Agabus disentegratus		[68]
Dytiscus sp.		[16]
Elateridae	[65]	
Elmidae		[20]
Gyrinidae	[1]	[16] ; [20]
Haliplidae (adult and larvae)	[1] ; [89]	
Haliplus sp. (adult and larvae)	[24] ; [33]	
Helophoridae		
Helophorus sp. (adult and larvae)	[24] ; [40]	
Helophorus aquaticus (larvae)	[24]	
Histeridae (larvae)	[24]	
Hydraenidae	[1]	
Hydrophilidae (adult and larvae)	[1] ; [24] ; [63] ; [69] ; [89]	[68]
Berosus ingeminatus		
Ochthebius sp.	[24]	
Cercyon analis	[24]	

Appendices

Noteridae (larvae)	[89]
Scarabaeidae	[77]
Staphylinidae (adult and nymph)	[24] ; [41]
Tenebrionidae (larvae)	[1]
Collembola	[20]
Entomobryicae	[1]
Diptera (adult, larvae and pupae)	[1] ; [2] ; [24] ; [29] ; [33] ; [38] ; [45] ; [51] ; [69] ; [76] ; [77] ; [84] ; [89]	[10] ; [20] ; [50] ; [70]
Agromyzidae	[1]
Anthomyiidae	[1]
Bibionidae	[1]
Dilopinus febrilis (larvae)	[24]
Cecidomyiidae	[1]
Ceratopogonidae (adult and larvae)	[1] ; [33] ; [63] ; [84] ; [89]	[16]
Forcipomyia sp. (larvae)	[24]
Chaoboridae	[1]
Chaoborus sp.		[16]
Chironomidae (adult, larvae and pupae)	[1] ; [12] ; [14] ; [24] ; [29] ; [33] ; [41] ; [45] ; [51] ; [53] ; [63] ; [64] ; [69] ; [72] ; [74] ; [76] ; [77] ; [84] ; [89]	[16] ; [36] ; [48] ; [55] ; [60] ; [61] ; [67] ; [88] [68]
Chironomus stigmaterus		
Culicidae (adult and larvae)	[1] ; [29] ; [33]
Culex sp. (pupae)	[24]
Dolichopodidae (adult and larvae)	[1] ; [24] ; [77]
Empididae (larvae)	[2]
Ephydridae (adult and larvae)	[1] ; [33] ; [76] ; [84] ; [89]	[70] ; [88]
Muscidae (pupae)	[72]
Psychodidae (adult and larvae)	[1] ; [24] ; [33] ; [84] ; [89]
Psychoda sp. (larvae)	[24]
Ptychopteridae	[1] ; [77]
Ptychoptera sp. (larvae)	[24]
Rhagionidae (adult and larvae)	[24] ; [89]
Sciomyzidae (larvae)	[24]
Simuliidae (larvae)	[69]

Taxon	Common Teal	Green-Winged Teal
Stratiomyidae (adult and larvae)	[1] ; [24] ; [53] ; [74] ; [77] ; [84] ; [89]	
Nemoletus sp. (larvae)	[33]	
Syrphidae	[1] ; [89]	[55] ; [60]
Eristalis sp. (larvae)	[24] ; [33]	
Tabanidae	[1] ; [89]	
Thaumaleidae	[1]	
Tipulidae (adult and larvae)	[1] ; [24] ; [41] ; [53] ; [74] ; [77] ; [89]	
Dicranota sp. (larvae)	[24]	
Ephemeroptera (adult, nymph and larvae)	[1]	
Caenidae		[20]
Caenis sp. (nymph)		[16]
Ephemeridae		[20]
Hexagenia sp. (nymph)		[88]
Hemiptera	[1] ; [29] ; [33] ; [45] ; [53] ; [74] ; [89]	
Aphididae	[1] ; [24] ; [53] ; [72] ; [74]	
Corixidae (adult and nymph)	[1] ; [24] ; [45] ; [89]	[16] ; [48] ; [88]
Corixa sp.	[29] ; [33]	[20]
Trichocorixa verticalis		[68]
Gerridae	[63] ; [69]	
Gerris sp.	[35]	
Pleidae	[1] ; [89]	
Plea sp.	[24]	
Veliidae		
Microvelia sp.	[89]	
Hymenoptera	[1] ; [45] ; [69] ; [72]	[48]
Chalcididae	[33]	
Formicidae	[33] ; [69] ; [77]	[20] ; [48]
Ichneumonidae		[20]
Myrmicinae	[33]	
Orthoptera		[48]
Acrididae	[56]	

Tettigoniidae		[20]
Lepidoptera (larvae)	[2] ; [77]	[18] ; [48]
Nymphua sp. (larvae)	[24]	
Megaloptera (larvae)	[89]	
Sialidae		
Sialis sp.	[72]	
Odonata (adult and larvae)	[1] ; [2] ; [29] ; [63] ; [69] ; [89]	[16] ; [20] ; [36] ; [50] ; [55] ; [67]
Aeshnidae		
Aeshna sp.		[20]
Cordulidae	[1]	
Coenagrionidae		[20]
Ischnura sp. (nymph)	[24]	
Lestidae	[1]	
Libellulidae	[1]	
Plecoptera (larvae)	[77]	[48]
Thysanoptera	[1]	
Trichoptera (adult, nymph and larvae)	[1] ; [24] ; [45] ; [53] ; [63] ; [65] ; [69] ; [72] ; [74] ; [77]	[20] ; [36] ; [50]
Hydropsychidae	[33]	
Hydropsyche sp. (adult and larvae)	[1] ; [51]	[20]
Hydroptilidae	[33] ; [51]	
Leptoceridae	[1]	
Limnephilicae		
Phryganeidae		[16]
Phrygarea grandis	[29]	
Psychomyiicae	[33]	
Mollusca	[26] ; [49] ; [72]	[8] ; [9] ; [18] ; [20] ; [22]
Bivalvia	[85]	[50]
Corbiculidae	[1]	
Mytilidae		
Mytilus edulis	[24]	[70]
Pisidiidae		
Pisidium sp.	[51]	
Sphaerium sp.	[33]	[20]

252   *The Teal*

Taxon	Common Teal	Green-Winged Teal
Tellinidae		
Macoma balthica		[70]
Gastropoda	[1] ; [12] ; [33] ; [51] ; [52] ; [53] ; [64] ; [69] ; [76] ; [77] ; [85] ; [89]	[16] ; [20] ; [22] ; [30] ; [50] ; [55] ; [73] ; [90]
Assimineidae		
Assiminea californica	[24]	
Assiminea grayana	[1]	
Bithyniidae		
Ellobiidae		
Phytia myosotis		[70]
Hydrobiidae	[1]	
Hydrobia sp.	[24] ; [57] ; [58] ; [64] ; [77]	
Hydrobia ulvae	[14]	
Potamopyrgus sp.	[51]	
Littorinidae		
Littorina obtusata	[24]	
Littorina saxatalis	[14]	
Lymnaeidae	[1] ; [2]	
Fossaria sonomaensis		[82]
Lymnaea sp.	[85]	
Neritidae		
Theodoxus pallasi	[29]	
Physidae	[1] ; [89]	[36]
Physa sp.	[57]	[16]
Physa fontinalis	[57]	
Physa gyrina		[20]
Planorbidae	[1] ; [33] ; [89]	[36] ; [81]
Planorbis sp	[29] ; [57] ; [64]	[16]
Armiger crista	[33]	
Gyraulus sp.	[45] ; [51]	
Gyraulus parvus		[20]
Turritellidae		
Turritella sp	[85]	

Valvatidae	[33]
Valvata sp.	[51] ; [77]
Vertiginidae	[84]
Nematoda	
Secernentea	
Ascaridae	[1]
Heterakidae	[1]
Platyhelminthes	
Turbellaria (egg)	[33] ; [89]
Protozoa	
Foraminiferida	[77] ; [89]
Vertebrata	
Unidentified fish (eggs)	[12] ; [16]
Perciformes	
Scatophagidae	[1]
Salmoniformes	
Salmonidae	
Oncorhynchus gorbuscha (eggs)	[11]

REFERENCES (see main reference list for full citation details): [1] Legagneux *et al.* unpublished data ; [2] Sanchez *et al.* unpublished data ; [3] Schricke *et al.* unpublished data ; [4] MacKay (1890) ; [5] Hesselman (1897) ; [6] Holmboe (1900) ; [7] Birger (1907) ; [8] Mabbott (1920) ; [9] Phillips (1923) and references therein ; [10] Preble & McAtee (1923) ; [11] Swarth (1924) ; [12] Cottam & Knappen (1939) ; [14] Campbell (1947) ; [15] Spärck (1947) ; [16] Munro (1949) ; [17] Yocom (1951) ; [18] Coulter (1955) ; [19] Dillon 1957 ; [20] Anderson (1959) ; [21] Keith (1961) ; [22] Yocom & Keller (1961) ; [23] Quay & Critcher (1962) ; [24] Olney (1963) ; [25] Conrad (1965) ; [26] Olney (1965) ; [27] Gaevskaya (1966) ; [28] McGilvrey (1966) ; [29] Dement'ev & Gladkov (1952) ; [30] Rollo & Bolen (1969) ; [31] Sterbertz (1969) ; [32] Kerwin & Webb (1971) ; [33] Mazzucchi (1971) ; [34] Molodovsky (1971) ; [35] Tamisier (1971) ; [36] Bartonek (1972) ; [37] Florschutz (1972) ; [38] Bengtson (1975) ; [39] Landers *et al.* (1976) ; [40] Srebrodol'skaya & Pavluk (1976) ; [41] Street (1977) ; [42] Prevost *et al.* (1978) ; [43] Sell (1979) ; [44] Brouwer (1980) ; [45] Moore (1980) ; [46] Moore (1980) ; [47] Pirot (1981) ; [48] Perry & Uhler (1981) ; [49] Thomas (1981) ; [50] Hughes & Young (1982) ; [51] Thomas (1982) ; [52] Schricke (1983) ; [53] Tubbs & Tubbs 1983) ; [54] Zuur *et al.* (1983) ; [55] Euliss & Harris (1987) ; [56] Kashkarov (1987) ; [57] Llorente *et al.* (1987) ; [58] Madsen (1988) ; [59] Swiderek *et al.* (1988) ; [60] DeRoia & Bookhout (1989) ; [61] DeRoia, Harris (1989) ; [62] Kydyraliev (1990) ; [63] Nummi (1990) ; [64] GEEA (1991) ; [65] Nechaev (1991) ; [66] Rosenberg *et al.* (1991) ; [67] Gaston (1992) ; [68] Batzer *et al.* (1993) ; [69] Nummi (1993) ; [70] Baldwin & Lovvorn (1994) ; [71] Sotnikov 1999) ; [72] Suarez-R and Urios (1999) ; [73] Anderson *et al.* (2000) ; [74] Nummi & Väänänen (2001) ; [75] Gardarsson & Einarsson (2002) ; [76] Guillemain & Fritz (2002) + Guillemain *et al.* unpublished data ; [77] Rodrigues *et al.* (2002) ; [78] Figuerola *et al.* (2003) ; [79] Curtet *et al.* (2004) ; [80] Goyon Demonteil (2004) ; [81] Wersal *et al.* (2005) ; [82] Bogiatto & Karnegis (2006) ; [83] Frisch *et al.* (2007) ; [84] Mouronval *et al.* (2007) ; [85] Karmiris *et al.* (2010) . [86] Brochet *et al.* (2010b) ; [87] Brochet *et al.* (2010a) ; [88] Vest & Conover (2011) ; [89] Brochet *et al.* (2012b) ; [90] Green *et al.* (2013).

Appendix 7

Summary table of Avian Influenza Virus strains recorded in Teal. The period of sampling is given when available. References are either scientific papers presenting Avian Influenza studies, or the name of the authors who deposited the given virus genomic sequences. See main reference list for full citation details.

Virus strain	Geographic area	Period of year	Reference
H1N1	Italy	Winter	Terregino et al. (2007)
	Mazandaran, Iran	February	Fereidouni et al. (2010)
	Spain	December	Busquets et al. (2010)
	The Netherlands	November	Fouchier et al.*
	USA – Louisiana	September	Kim et al.*, Sreevatsan et al.*
	USA – Minnesota	September	Sreevatsan et al.*
	USA – Ohio	October	Spiro et al.*
	USA – Wisconsin	October	Spiro et al.*
H1N2	Norway		Germundsson et al. (2010)
	USA – California	October	Spiro et al.*
	USA – Illinois	October	Spiro et al.*
	USA – Oklahoma	April	Kocan et al. 1980
	USA – Texas	November–January	Ferro et al. 2008
H1N3	France		Le Gall-Recule et al. *
	Italy	Winter	Terregino et al. (2007)
	USA – California	January	Spiro et al.*
H2N1	USA – Ohio	November	Spiro et al.*
H2N3	The Netherlands		Wentworth et al.*
	USA – Illinois	October	Spiro et al.*
H2N9	USA – California	December	Spiro et al.*
	USA – Wisconsin	November	Spiro et al.*
H3N2	Norway		Germundsson et al. (2010)
	USA – Minnesota	September	Sreevatsan et al.*
	USA – Wisconsin	October	Spiro et al.*
H3N3	Sweden	August	Fouchier et al.*
H3N8	Canada – New Brunswick	September	Wentworth et al.*
	Kazakhstan	October	Karamendin et al.*
	Russia	October	Sayfutdinova et al.*
	Russia – Altai	August	Sivay et al.*
	Russia – Chany	August	Sivay et al.*

	Russia – Primorye Territory		Shchelkanov et al. (2007)
	Russia – Siberia	August	Sayfutdinova et al.*
	Russia – Tagarskoe lake	September	Sayfutdinova et al.*
	USA – Alaska	August, September, October	Runstadler et al. (2007), Spiro et al.*
	USA – Louisiana	September	Sreevatsan et al.*, Spiro et al.*
	USA – Minnesota	September	Sreevatsan et al.*
	USA – Mississippi	January	Spiro et al.*
	USA – Ohio		Spiro et al.*
	USA – Oregon	August, September, October	Spiro et al.*
	USA – Wisonsin	October	Spiro et al.*
H3N8		August	Obenauer et al. (2006)
	Italy	Winter	Terregino et al. (2007)
H4N2	Norway		Germundsson et al. (2010)
	USA – Ohio	September, October	Spiro et al.*
H4N3	Astrakhan		Lomakina & Yamnikova*
H4N6	Canada – Alberta	July	Spiro et al.*
	Canada – Manitoba	August	Spiro et al.*
	Canada – New Brunswick	August, September	Wentworth et al.*
	Canada – Nova Scotia	September	Spiro et al.*
	Italy	Winter	Terregino et al. (2007)
	Japan		Fujimoto et al. (2010)
	Poland		Smietanka et al. (2012)
	Russia – Chany	August	Sivay et al.*
	Spain		Busquets et al. (2010)
	USA		Spackman et al. (2005)
	USA – California	December	Spiro et al.*
	USA – Ohio		Spiro et al.*
	USA – Texas	November–January	Ferro et al. (2008)
H4N8	Norway		Germundsson et al. (2010)
	USA		Spackman et al. (2005)
H5N1	China		Li et al. (2004)
	Egypt	December	Saad et al. (2007)
	Germany	November	Bogs et al. (2010)
	Hong Kong		World Health Organization Global Influenza Program Surveillance Network (2005)
	Vietnam		Cao et al.*
H5N2	China Xianghai	October	Zhou et al.*
	Egypt	October	Saad et al. (2007)
	Italy	Winter	Spiro et al.*, Terregino et al. (2007)
	Switzerland	October	Baumer et al. (2010)
	Ukraine		Muzyka et al. *

	USA – California	November	Cardona et al.*, Kim et al.*
	USA – Delaware		Kim et al.*
	USA – Iowa		Kim et al.*
	USA – Ohio		Kim et al.*
H5N3	Iran	February	Fereidouni et al. (2010)
	Italy	Winter	Terregino et al. (2007)
	Japan	January	Fujimoto et al. (2010)
H6N1	Canada – Nova Scotia	September	Spiro et al. *
	Hong Kong		Hoffmann et al. (2000)
	USA – California	October, November	Spiro et al. *
H6N2	Iran	February	Fereidouni et al. (2010)
	Norway		Germundsson et al. (2010)
	USA – California	October	Spiro et al. *
	USA – Illinois	October	Spiro et al. *
	USA – Minnesota	September	Spiro et al.*, Sreevatsan et al.*
	USA – Ohio	October	Spiro et al.*, Spackman et al. (2005)
	USA – Texas	November–January	Ferro et al. (2008)
	USA – Wisconsin	October	Spiro et al.*
H6N8	The Netherlands		Spiro et al.*
	USA – Minnesota	September	Lebarbenchon et al. (2012)
	USA – Ohio	October	Spiro et al.*
H7N1	Taiwan		Banks et al. (2000)
H7N3	USA – California	December, January, February	Carona et al.*, Spiro et al.*
	USA – Minnesota	September	Sreevatsan et al.*
	USA – Ohio		Kim et al.*
	USA – Texas	February	Hanson et al. (2005)
H7N6	USA – California	January	Spiro et al.*
H7N7	Germany		Fereidouni et al.*
	Iceland		Gorman et al. (1991)
	Italy	Winter	Terregino et al. (2007)
	USA – Minnesota	September	Sreevatsan et al.*
	USA – Mississippi	January	Spiro et al.*
H7N9	Spain	January	Busquets et al. (2010)
H8N3	USA – California	January	Spiro et al.*
H8N4	Spain	January	Busquets et al. (2010)
	USA		Spackman et al. (2005)
	The Netherlands		Spiro et al.*
	USA – California	January	Spiro et al.*
	USA – Texas	February	Hanson et al. (2005), Kim et al.*
H8N8	Russia – Chany	August	Sivay et al.*
H9N1	Northern Ireland		Slomka et al.*

H9N2	China – Shanghai Norway Russia – Primorie Switzerland USA – Texas	October, December October October	Zhou & Yao*, Zhou et al.* Jonassen et al.* Lomakina & Yamnikova* Baumer et al. (2010) Ferro et al. *
H10N2	USA – Louisiana		Kim et al.*
H10N4	Japan Spain		Fujimoto et al. (2010) Busquets et al. (2010)
H10N6	Russia – Tynda	September	Sayfutdinova et al.*
H10N7	Egypt Italy USA – California USA – Illinois	 Winter December, January October	Lin et al. (2009) Terregino et al. (2007) Spiro et al.* Spiro et al.*
H11N1	Iran	February	Fereidouni et al. (2010)
H11N2	USA – Ohio		Spiro et al.*
H11N5	USA – California	October	Cardona et al. *, Spiro et al.*
H11N9	Iran The Netherlands USA – California USA – Minnesota USA – Mississippi USA – Ohio	February November November, January September January	Fereidouni et al. (2010) Fouchier et al.* Spiro et al.* Sreevatsan et al.* Spiro et al.* Spiro et al.*
H12N2	Norway		Germundsson et al. (2010)
H12N5		August	Obenauer et al. (2006)
H15N4	Russia – Chany	August	Sivay et al.*
H16N3	Russia – Volga area		Yamnikova & Lomakina*

* not a scientific paper but a genomic sequence whose details are available at: http://www.fludb.org

Appendix 8

Parasites found in Common or Green-winged Teal. The table is not intended to be exhaustive. Some taxonomic names may have changed since description in the reference indicated. Table 7.2 provides prevalence values (% birds infested by a given parasite) for the limited number of parasite species in which this information is available. See main reference list for full citation details.

Parasite	Geographic area	Reference
Parasites in digestive tract		
Trematodes		
Amphimerus anatis	Eurasia	McDonald (1969)
Apatemon gracilis	Eurasia, Africa, America	McDonald (1969)
	Nova Scotia, New Brunswick, Canada	Turner & Threlfall (1975)
	Oklahoma, USA	Shaw & Kocan (1980)
A. skrjabini	Asia	McDonald (1969)
Australapatemon minor	Poland	Sulgostowska (2007)
Catatropis verrucosa	Eurasia, Africa, N America	McDonald (1969)
Cotylurus brevis	Mexico	Martínez-Haro *et al.* (2012)
C. cornutus	Eurasia, Africa, America	McDonald (1969)
	Nova Scotia, New Brunswick, Canada	Turner & Threlfall (1975)
	Poland	Sulgostowska (2007)
C. erraticus	Eurasia, N America	McDonald (1969)
C. flabelliformis	Eurasia, N America	McDonald (1969)
C. japonicus	Asia	McDonald (1969)
C. orientalis	Asia	McDonald (1969)
C. platycephalus	Nova Scotia, New Brunswick, Canada	Turner & Threlfall (1975)
Cryptocotyle concavum	Eurasia, N America	McDonald (1969)
Cyathocotyle bushiensis	Québec, Canada	Hoeve & Scott (1988)
C. prussica	Eurasia, USA	McDonald (1969)
Dendritobilharzia pulverulenta	Texas, USA	Canaris *et al.* (1981)
Diplostomum spathaceum	Eurasia, Africa	McDonald (1969)
Echinochasmus dietzevi	Eurasia	McDonald (1969)
Echinoparyphium aconiatum	Eurasia, N America	McDonald (1969)
E. baculus	Eurasia, N America	McDonald (1969)
E. recurvatum	Worldwide	McDonald (1969)
	Nova Scotia, New Brunswick, Canada	Turner & Threlfall (1975)
	Poland	Kavetska *et al.* (2008)
	Mexico	Martínez-Haro *et al.* (2012)
Echinostoma chloropodis	Eurasia, N America	McDonald (1969)

Parasite	Geographic area	Reference
E. crecci	Asia	McDonald (1969)
E. grandis	Eurasia	McDonald (1969)
E. miyagawai	Eurasia	McDonald (1969)
E. paraulum	Eurasia	McDonald (1969)
E. revolutum	Europe	Naumann (1897)
	Worldwide	McDonald (1969)
	Nova Scotia, New Brunswick, Canada	Turner & Threlfall (1975)
	Texas, USA	Canaris *et al.* (1981)
E. robustum	Eurasia	McDonald (1969)
E. sarcinum	Eurasia	McDonald (1969)
Hypoderaeum conoideum	Eurasia, Africa, N America	McDonald (1969)
	Nova Scotia, New Brunswick, Canada	Turner & Threlfall (1975)
	Texas, USA	Canaris *et al.* (1981)
H. gnedini	Eurasia	McDonald (1969)
Levinseniella belopolskoi	Asia	McDonald (1969)
Maritrema subdolum	Eurasia	McDonald (1969)
Metorchis nettioni	India	McDonald (1969)
M. xanthosomus	Eurasia	McDonald (1969)
M. zacharovi	Europe	McDonald (1969)
Microphallus primas	Nova Scotia, New Brunswick, Canada	Turner & Threlfall (1975)
Notocotylus attenuatus	Eurasia, N America, Australia	McDonald (1969)
	Nova Scotia, New Brunswick, Canada	Turner & Threlfall (1975)
N. dafilae	Asia, N America	McDonald (1969)
N. ephemera	Europe	McDonald (1969)
N. imbricatus	Eurasia, N America	McDonald (1969)
N. parviovatus	Eurasia	McDonald (1969)
N. seineti	Europe, N America	McDonald (1969)
N. stagnicolae	Texas, USA	Canaris *et al.* (1981)
Opisthorchis longissimus	Eurasia	McDonald (1969)
Parechinostomum cinctum	Eurasia	McDonald (1969)
P. alveatum	Nova Scotia, New Brunswick, Canada	Turner & Threlfall (1975)
Paramonostomum harwoodi	India	McDonald (1969)
P. nettioni	India	McDonald (1969)
Petasiger skrjabini	Eurasia	McDonald (1969)
Plagiorchis laricola	Eurasia	McDonald (1969)
Prosthogonimus cuneatus	Nova Scotia, New Brunswick, USA	Turner & Threlfall (1975)
Psilochasmus oxyurus	Eurasia, Africa, America	McDonald (1969)
	Nova Scotia, New Brunswick, Canada	Turner & Threlfall (1975)
Psilorchis ajgainis	India	McDonald (1969)
Psilostomum sp.	Nova Scotia, New Brunswick, Canada	Turner & Threlfall (1975)
Psilotrema oligoon	Eurasia	McDonald (1969)
Sphaeridiotrema globulus	Eurasia, N America	McDonald (1969)
	Québec, Canada	Hoeve & Scott (1988)
Strigea falconis	Worldwide	McDonald (1969)
Tracheophilus cymbius	Oklahoma, USA	Shaw & Kocan (1980)
Typhlocoelum cucumerinum	Texas, USA	Canaris *et al.* (1981)
Zygocotyle lunata	Asia, Africa, N America	McDonald (1969)
	Nova Scotia, New Brunswick, Canada	Turner & Threlfall (1975)
	Oklahoma, USA	Shaw & Kocan (1980)
	Texas, USA	Canaris *et al.* (1981)

Parasite	Geographic area	Reference
Cestodes		
Anatinella spinulosa	Asia	McDonald (1969)
Anomotaenia ciliata	Eurasia, N America	McDonald (1969)
Aploparaksis fuligulosa	Europe	McDonald (1969)
A. furcigera	Eurasia, N America	McDonald (1969)
Cloacotaenia megalops	Europe	Naumann (1897)
	Worldwide	McDonald (1969)
	Texas, USA	Canaris *et al.* (1981)
	France	Green *et al.* (2011)
	Poland	Królaczyk *et al.* (2011)
	Mexico	Martínez-Haro *et al.* (2012)
Diorchis acuminata	Alaska, USA	Schiller (1953)
	Eurasia, N America	McDonald (1969)
D. bulbodes	Eurasia, N America	McDonald (1969)
D. formosensis	Eurasia	McDonald (1969)
D. longicirrosa	Europe, Egypt	McDonald (1969)
D. inflata	Eurasia, Africa, N America	McDonald (1969)
D. jacobii	Europe	McDonald (1969)
D. longiovum	Alaska, USA	Schiller (1953)
	USA	McDonald (1969)
D. nyrocae	Eurasia, N America	McDonald (1969)
D. nyrocoides	Asia	McDonald (1969)
D. ransomi	Eurasia, N America	McDonald (1969)
D. spinata	Eurasia, N America	McDonald (1969)
D. stefanskii	Eurasia	McDonald (1969)
D. tuvensis	Asia	McDonald (1969)
D. vigisi	Asia	McDonald (1969)
D. wigginsi	Europe, N America	McDonald (1969)
Diploposthe laevis	Europe	Naumann (1897)
	Wordwide	McDonald (1969)
Echinocotyle clerci	Eurasia	McDonald (1969)
E. skrjabini	Asia	McDonald (1969)
Fimbriaria fasciolaris	Alaska, USA	Schiller (1953),
	Wordwide	McDonald (1969)
Gastrotaenia cygni	Texas, USA	Canaris *et al.* (1981)
G. dogieli	Eurasia	McDonald (1969)
Hymenolepis abortiva	Eurasia, Egypt	McDonald (1969)
H. aequabilis	Eurasia, N America	McDonald (1969)
H. anatina	Eurasia, Africa, N America	McDonald (1969)
H. andrejewoi	Eurasia	McDonald (1969)
H. arcuata	Eurasia, N America	McDonald (1969)
H. bisaccata	Eurasia, S America	McDonald (1969)
H. brachycephala	Eurasia, Egypt	McDonald (1969)
H. collaris	Alaska, USA	Schiller (1953)
	Eurasia, Africa, N America, Australia	McDonald (1969)
H. compressa	Eurasia, N America	McDonald (1969)
H. coronula	Eurasia, Africa, N America	McDonald (1969)
H. crecca	Asia	McDonald (1969)
H. dafilae	Asia, N America	McDonald (1969)
H. echinocotyle	Eurasia, N America	McDonald (1969)
H. fausti	Eurasia, N America	McDonald (1969)
H. fragilis	Asia	McDonald (1969)

H. giranensis	Eurasia	McDonald (1969)
H. gladium	Asia	McDonald (1969)
H. gracilis	Eurasia, Africa, N America	McDonald (1969)
	Texas, USA	Canaris *et al.* (1981)
H. inflatocirrosa	Asia	McDonald (1969)
H. jaegerskioeldi	Eurasia, N America	McDonald (1969)
H. krabbeella	Eurasia	McDonald (1969)
H. longicirrosa	Eurasia	McDonald (1969)
H. longivaginata	Eurasia, N America	McDonald (1969)
H. macrocephala	Eurasia, N America	McDonald (1969)
H. meggitti	Eurasia	McDonald (1969)
H. microsoma	Eurasia, N America	McDonald (1969)
H. octacantha	Eurasia, Africa	McDonald (1969)
H. oweni	Eurasia	McDonald (1969)
H. paramicrosoma	Eurasia, N America	McDonald (1969)
H. parvula	Eurasia, N America	McDonald (1969)
H. przewalskii	Eurasia	McDonald (1969)
H. pseudosetigera	Iceland	McDonald (1969)
H. querquedula	Asia, N America, Antarctica	McDonald (1969)
H. retracta	Asia	McDonald (1969)
H. rosseteri	Eurasia, Africa, N America	McDonald (1969)
H. setigera	Eurasia, N America	McDonald (1969)
H. simplex	Eurasia, N America	McDonald (1969)
H. spiralicirrata	Kazakhstan	McDonald (1969)
H. teresoides	Eurasia, N America	McDonald (1969)
H. trichoryncha	Eurasia	McDonald (1969)
Lallum magniparuterina	India	Johri (1960)
Lateriporus skrjabini	Eurasia	McDonald (1969)
L. teres	Eurasia, N America	McDonald (1969)
Sobolevicanthus krabbeella	Texas, USA	Canaris *et al.* (1981)
	Mexico	Martínez-Haro *et al.* (2012)
Taenia fragilis	Europe	Naumann (1897)

Acanthocephalans

Corynosoma constrictum	N America	McDonald (1969)
	Nova Scotia, New Brunswick, Canada	Turner & Threlfall (1975)
	Oklahoma, USA	Shaw & Kocan (1980)
	Texas, USA	Canaris *et al.* (1981)
Filicollis anatis	Europe	Naumann (1897)
	Eurasia	McDonald (1969)
Polymorphus actuganensis	Asia	McDonald (1969)
P. acutis	Eurasia, N America	McDonald (1969)
P. diploinflatus	Eurasia	McDonald (1969)
P. kostylewi	Asia	McDonald (1969)
P. magnus	Eurasia	McDonald (1969)
P. minutus	Eurasia, N America	McDonald (1969)
	Texas, USA	Canaris *et al.* (1981)
P. trochus	Asia, N America	McDonald (1969)
Pseudocorynosoma constrictum	Mexico	Martínez-Haro *et al.* (2012)

Parasite	Geographic area	Reference
Nematodes		
Amidostomum acutum	Eurasia, Africa, N America, Australia	McDonald (1969)
	Nova Scotia, New Brunswick, Canada	Turner & Threlfall (1975)
	Texas, USA	Canaris *et al.* (1981)
	The Netherlands	Borgsteede *et al.* (2006)
	Poland	Kavetska *et al.* (2011)
A. anseris	Eurasia, Africa, N America	McDonald (1969)
	Texas, USA	Canaris *et al.* (1981)
A. fulicae	Eurasia, Africa, N America	McDonald (1969)
A. henryi	Eurasia	McDonald (1969)
A. monodon	Eurasia	McDonald (1969)
Capillaria anatis	Eurasia, America	McDonald (1969)
	Nova Scotia, New Brunswick, Canada	Turner & Threlfall (1975)
C. contorta	Worldwide	McDonald (1969)
	Nova Scotia, New Brunswick, Canada	Turner & Threlfall (1975)
Capillaria sp.	Mexico	Martínez-Haro *et al.* (2012)
Contracaecum microcephalum	Eurasia, Africa, N America	McDonald (1969)
Cosmocephalus obvelatus magnus	Eurasia, Africa, N America	McDonald (1969)
Echinuria uncinata	Eurasia, Africa, N America	McDonald (1969)
	Texas, USA	Canaris *et al.* (1981)
Epomidiostomum orispinum	Eurasia, Africa	McDonald (1969)
E. querquedulae	Asia, Africa	McDonald (1969)
E. uncinatum	Eurasia, Africa, N America	McDonald (1969)
	Nova Scotia, New Brunswick, Canada	Turner & Threlfall (1975)
Eustrongylides mergorum	Eurasia, N America	McDonald (1969)
Gnathostoma spinigerum	Asia, N America, Australia	McDonald (1969)
Hystrichis neglectus	Europe	McDonald (1969)
Hystrichis sp.	Europe	Naumann (1897)
Porrocaecum crassum	Eurasia	McDonald (1969)
P. ensicaudatum	Eurasia, Africa, N America	McDonald (1969)
P. heteroura	Eurasia	McDonald (1969)
P. semiteres	Europe, Egypt	McDonald (1969)
Sarconema pseudolabiata	Eurasia	McDonald (1969)
Streptocara crassicauda	Eurasia, N America	McDonald (1969)
Strongylus sp.	Europe	Naumann (1897)
Tetrameres crami	Eurasia, N America	McDonald (1969)
	Oklahoma, USA	Shaw & Kocan (1980)
	Texas, USA	Canaris *et al.* (1981)
T. fissispina	Worldwide	McDonald (1969)
T. ryjikovi	Asia	McDonald (1969)
	Nova Scotia, New Brunswick, Canada	Turner & Threlfall (1975)
Parasites of the blood system		
Trematodes		
Bilharziella indica	Asia	McDonald (1969)
B. lali	Asia	McDonald (1969)
B. polonica	Eurasia, Africa, N America	McDonald (1969)
Cercaria mieensis	Japan	McDonald (1969)
Dendritobilharzia asiaticus	India	McDonald (1969)

Parasite	Geographic area	Reference
D. pulverulenta	Eurasia, Africa, N America	McDonald (1969)
Trichobilharzia indica	India	McDonald (1969)
T. kowalewskii	Eurasia	McDonald (1969)
T. ocellata	Eurasia, N America	McDonald (1969)
T. querquedulae	Nova Scotia, New Brunswick, Canada	Turner & Threlfall (1975)
T. yokogawai	Asia	McDonald (1969)
Nematodes		
Filaria sp.	Massachusetts, USA	Bennett et al. (1974)
	New Brunswick, Nova Scotia, Prince Albert Island, Canada	Bennett et al. (1975)
	Labrador, Newfoundland, Canada	Bennett et al. (1991)
	Oklahoma, USA	Shaw & Kocan (1980)
Apicomplexa		
Haemoproteus nettionis	Massachusetts, USA	Bennett et al. (1974)
	New Brunswick, Nova Scotia, Prince Albert Island, Canada	Bennett et al. (1975)
	Labrador, Newfoundland, Canada	Bennett et al. (1991)
	North America	Herman (1965)
Leucocytozoon simondi	Massachusetts, USA	Bennett et al. (1974)
	New Brunswick, Nova Scotia, Prince Albert Island, Canada	Bennett et al. (1975)
	Labrador, Newfoundland, Canada	Bennett et al. (1991)
	North America	Herman (1965)
Plasmodium circumflexum	Massachusetts, USA	Bennett et al. (1974)
	New Brunswick, Nova Scotia, Prince Albert Island, Canada	Bennett et al. (1975)

Parasites of the respiratory system		
Trematodes		
Cyclocoelum mutabile	Eurasia, N America	McDonald (1969)
Hyptiasmus arcuatus	Eurasia, N America	McDonald (1969)
Typhlocoelum cucumerinum	Eurasia, Africa, N America, Australia	McDonald (1969)
T. sisovi	Worldwide	McDonald (1969)
Orchipedum tracheicola	Eurasia, N America	McDonald (1969)

Parasites in reproductive tract		
Trematodes		
Prosthogonimus cuneatus	Eurasia, Africa, America	McDonald (1969)
P. macrorchis	Eurasia, N America	McDonald (1969)
P. ovatus	Worldwide	McDonald (1969)
P. pellucidus	Eurasia, America	McDonald (1969)
Schistogonimus rarus	Eurasia	McDonald (1969)

Parasites in muscles		
Coccidia		
Sarcocystis sp.	North Dakota, USA	Hoppe (1976)

Parasite	Geographic area	Reference
Ectoparasites		
Mallophaga- bird lice		
Anaticola buccinator	Europe	Naumann (1897)
A. crassicornis	Europe	Naumann (1897)
	Nova Scotia, New Brunswick, Canada	Turner & Threlfall (1975)
	Texas, USA	Canaris *et al.* (1981)
	Spain	Castresana *et al.* (1999)
Anatoecus dentatus	Nova Scotia, New Brunswick, Canada	Turner & Threlfall (1975)
	Spain	Castresana *et al.* (1999)
A. icterodes	Europe	Naumann (1897)
	Texas, USA	Canaris *et al.* (1981)
	Spain	Castresana *et al.* (1999)
Holomenopon clypeilargum	Thailand, USA, Sweden	Price (1971)
H. leucoxanthum	Thailand	Price (1971)
	Spain	Castresana *et al.* (1999)
H. setigerum	USA ?	Price (1971)
	Texas, USA	Canaris *et al.* (1981)
Trinoton querquedulae	Nova Scotia, New Brunswick, Canada	Turner & Threlfall (1975)
	Texas, USA	Canaris *et al.* (1981)
	Spain	Castresana *et al.* (1999)
Acarina		
Epidermoptes sp.	Texas, USA	Canaris *et al.* (1981)
Hirudinea		
Theromyzon tessulatum	Europe ?	Sage (1958)
	Eurasia, Africa, America	McDonald (1969)
T. rude	Canada	Bartonek & Trauger (1975)
Undetermined leeches	Alberta, Canada	McKinney & Derrickson (1979)

References

ACKERMAN, J. T. 2002. Of mice and mallards: positive indirect effects of coexisting prey on waterfowl nest success. *Oikos* 99: 469–480.

ADLER, M. 2010. Sexual conflict in waterfowl: why do females resist extrapair copulations? *Behavioral Ecology* 21: 182–192.

AEWA. 2008. *L'Accord sur la Conservation des Oiseaux Migrateurs d'Afrique-Eurasie (AEWA) –Tableau 1. Etat des populations d'Oiseaux d'Eau Migrateurs (version adoptée par la MOP4)*. AEWA, Bonn, Germany.

AFTON, A. D. 1978. Incubation rhythms and egg temperatures of an American Green-winged Teal and a renesting Pintail. *Prairie Naturalist* 10: 115–119.

AFTON, A. D. & PAULUS, S. L. 1992. Incubation and brood care. Pp. 62–108 in Batt B. D. J., Afton, A. D. Anderson, M. G., Ankney C. D., Johnson, D. H., Kadlec J. A. & Krapu G. L. (eds), *Ecology and Management of Breeding Waterfowl*. University of Minnesota Press, Minneapolis, USA.

ALBRECHT, T. & KLVANA, P. 2004. Nest crypsis, reproductive value of a clutch and escape decisions in incubating female Mallards *Anas platyrhynchos*. *Ethology* 110: 603–613.

ALERSTAM, T. & HEDENSTRÖM, A. 1998. The development of bird migration theory. *Journal of Avian Biology* 29: 343–369.

ALERSTAM, T., ROSÉN, M., BÄCKMAN, J., ERICSON, P. G. P. & HELLGREN, O. 2007. Flight speeds among bird species: allometric and phylogenetic effects. *PLoS Biology* 5: e197.

ALHAINEN, M., VÄÄNÄNEN, V. M., PÖYSÄ, H. & ERMALA, A. 2010. Duck hunting bag in Finland – what do wing samples tell us about the species composition and age structure in a bag? *Suomen Riista* 56: 40–47. [In Finnish with English summary]

ALI, S. & RIPLEY, S. D. 1968. *Handbook of the Birds of India and Pakistan, together with those of Nepal, Sikkim, Bhutan and Ceylon. Volume 1. Divers to Hawks*. Oxford University Press, Bombay, India.

ALISAUSKAS, R. T. & ANKNEY, C. D. 1992. The cost of egg laying and its relationship to nutrient reserves in waterfowl. Pp. 30–61 in Batt, B. D. J., Afton, A. D., Anderson, M. G., Ankney, C. D., Johnson, J. A., Kadlec, D. H. & Krapu, G. L. (eds), *Ecology and Management of Breeding Waterfowl*. University of Minnesota Press, Minneapolis, USA.

ALLEVA, E., FRANCIA, N., PANDOLFI, M., DE MARINIS, A. M., CHIAROTTI, F. & SANTUCCI, D. 2006. Organochlorine and heavy metal contaminants in wild mammals and birds of Urbino-Pesaro Province, Italy: an analytic overview for potential bioindicators. *Archives of environmental contamination and ecotoxicology* 51: 123–134.

AMAT, J. A. 1990. Age-related pair bonding by male Eurasian Wigeons in relation to courtship activity. *The Auk* 107: 197–198.

AMBEDKAR, V. C. & DANIEL, J. C. 1990. A study of the migration of common teal *Anas crecca crecca* Linnaeus based on ring recoveries in India and U.S.S.R. *Journal of the Bombay Natural History Society* 87: 405–420.

AMERICAN ORNITHOLOGISTS' UNION. 1886. *The Code of Nomenclature and Check-list of North American Birds adopted by the American Ornithologists' Union being the Report of the Committee of the Union on Classification and Nomenclature.* American Ornithologists' Union, New York, USA.

AMERICAN ORNITHOLOGISTS' UNION. 1899. Ninth supplement to the American Ornithologists' Union check-list of North American birds. *The Auk* 16: 97–133.

AMERICAN ORNITHOLOGISTS' UNION. 1944. Nineteenth supplement to the American Ornithologists' Union check-list of North American birds. *The Auk* 61: 441–464.

AMERICAN ORNITHOLOGISTS' UNION. 1973. Thirty-second supplement to the American Ornithologists' Union check-list of North American birds. *The Auk* 90: 411–419.

AMERICAN ORNITHOLOGISTS' UNION. 1998. *Check-list of North American Birds. Seventh Edition.* American Ornithologists' Union, Washington, D.C, USA.

AMUNDSON, C. L. & ARNOLD, T. W. 2011. The role of predator removal, density-dependence, and environmental factors on Mallard duckling survival in North Dakota. *Journal of Wildlife Management* 75: 1330–1339.

ANDERSON, A. 1963. Patagial tags for waterfowl. *Journal of Wildlife Management* 27: 284–288.

ANDERSON, D. R. & BURNHAM, K. P. 1976. *Population Ecology of the Mallard VI. The effect of exploitation on survival.* Fish and Wildlife Service Resource Publication 128. U.S. Department of the Interior, Washington, D.C., USA.

ANDERSON, H. G. 1959. Food habits of migratory ducks in Illinois. *Illinois Natural History Survey Bulletin* 27: 289–344.

ANDERSON, J. T. & SMITH, L. M. 1999. Carrying capacity and diel use of managed playa wetlands by nonbreeding waterbirds. *Wildlife Society Bulletin* 27: 281–291.

ANDERSON, J. T., SMITH, L. M. & HAUKOS, D. A. 2000. Food selection and feather molt by nonbreeding American Green-winged Teal in Texas Playas. *Journal of Wildlife Management* 64: 222–230.

ANDERSON, M. G., RHYMER, J. M. & ROHWER, F. C. 1992. Philopatry, dispersal, and the genetic structure of waterfowl populations. Pp. 365–395 in Batt, B. D. J., Afton, A. D., Anderson, M. G., Ankney, C. D., Johnson, D. H., Kadlec J. A. & Krapu G. L. (eds), *Ecology and Management of Breeding Waterfowl.* University of Minnesota Press, Minneapolis, USA.

ANKNEY, C. D. & ALISAUSKAS, R. T. 1991. The use of nutrient reserves by breeding waterfowl. *Acta XX Congressus Internationalis Ornithologici*: 2170–2176.

ANKNEY, C. D., AFTON, A. D. & ALISAUSKAS, R. T. 1991. The role of nutrient reserves in limiting waterfowl reproduction. *The Condor* 93: 1029–1032.

ARNOTT, W. G. 2007. *Birds in the Ancient World from A to Z.* Routledge, Oxon, UK & New York, USA.

ARTEMYEV, Y. T. & POPOV, V. A. 1977. *Birds of Volga-Kama Territory – Non passerines.* Nauka, Moscow, Russia. [In Russian]

ARZEL, C. & ELMBERG, J. 2004. Time use, foraging behavior and microhabitat use in a temporary guild of spring-staging dabbling ducks (*Anas* spp.). *Ornis Fennica* 81: 157–168.

ARZEL, C., ELMBERG, J. & GUILLEMAIN, M. 2006. Ecology of spring-migrating Anatidae: a review. *Journal of Ornithology* 147: 167–184.

ARZEL, C., ELMBERG, J. & GUILLEMAIN, M. 2007b. A flyway perspective of foraging activity in Eurasian Green-winged Teal, *Anas crecca crecca*. *Canadian Journal of Zoology* 85: 81–91.

ARZEL, C., DESSBORN, L., PÖYSÄ, H., ELMBERG, J., NUMMI, P. & SJÖBERG, K. 2014. Early springs and breeding performance in two sympatric bird species with different migration strategies. *Ibis*, 156: 288–298.

ARZEL, C., ELMBERG, J., GUILLEMAIN, M., LEPLEY, M., BOSCA, F., LEGAGNEUX, P. & NOGUES, J. B. 2009. A flyway perspective on food resource abundance in a long-distance migrant, the Eurasian teal (*Anas crecca*). *Journal of Ornithology* 150: 61–73.

ARZEL, C., ELMBERG, J., GUILLEMAIN, M., LEGAGNEUX, P., BOSCA, F., CHAMBOULEYRON, M., LEPLEY, M., PIN, C., ARNAUD, A. & SCHRICKE, V. 2007a. Average mass of seeds encountered by foraging dabbling ducks in Western Europe. *Wildlife Biology* 13: 328–336.

ASFERG, T. 2012. *Vildtudbyttestatistik for jagtsæsonen 2011/12*. Aarhus Universitet, Denmark. [In Danish]

ASHCROFT, R. E. 1976. A function of the pairbond in the common eider. *Wildfowl* 27: 101–105.

ATKINSON, P. W., ROBINSON, R. A., CLARK, J. A., MIYAR, T., DOWNIE, I. S., DU FEU, C. R., FIEDLER, W., FRANSSON, T., GRANTHAM, M. J., GSCHWENG, M., SPINA, F. & CRICK, H. Q. P. 2007. *Migratory Movements of Waterfowl: a Web-based Mapping Tool*. EURING report to the EU Commission. (http://blx1.bto.org/ai-eu/)

ATKINSON, P. W., CLARK, J. A., DELANY, S., DIAGANA, C. H., DU FEU, C., FIEDLER, W., FRANSSON, T., GAUTHIER-CLERC, M., GRANTHAM, M., GSCHWENG, M., HAGEMEIJER, W., HELMINK, T., JOHNSON, A., KHOMENKO, S., MARTAKIS, G., OVERDIJK, O., ROBINSON, R. A., SOLOKHA, A., SPINA, F., SYLLA, S. I., VEEN, J. & VISSER, D. 2006. *Urgent Preliminary Assessment of Ornithological Data Relevant to the Spread of Avian Influenza in Europe*. Wetlands International / EURING report to the European Commission, Wetlands International, Wageningen, The Netherlands.

AUDUBON, J. J. 1860a. *The Birds of America, from drawings made in the United States and their territories. Volume I*. Audubon, Lockwood & Son, New York, USA.

AUDUBON, J. J. 1860b. *The Birds of America, from drawings made in the United States and their territories. Volume VI*. Audubon, Lockwood & Son, New York, USA.

AVERY, D. & WATSON, R. T. 2009. Regulation of lead-based ammunition around the world. Pp. 161–168 in Watson, R. T., Fuller, M., Pokras M. & Hunt W. G. (eds), *Ingestion of Lead from Spent Ammunition: implications for wildlife and humans*. The Peregrine Fund, Boise, Idaho, USA.

BABENKO, V. G. 2000. *Birds of lower Cis-Amur River area*. Prometei Press, Moscow, Russia. [In Russian]

BAKER, K. 1993. *Identification Guide to European Non-passerines*. The British Trust for Ornithology (BTO Guide 24), Thetford, UK.

BALDASSARRE, G. A. 1982. Field-feeding ecology of waterfowl wintering on the Southern High Plains of Texas. PhD Thesis, Texas Tech University, Lubbock, USA.

BALDASSARRE, G. A. & BOLEN, E. G. 1984. Field-feeding ecology of waterfowl wintering on the Southern High Plains of Texas. *Journal of Wildlife Management* 48: 63–71.

BALDASSARRE, G. A. & BOLEN, E. G. 1986. Body weight and aspects of pairing chronology of Green-winged Teal and Northern Pintails wintering on the Southern High Plains of Texas. *The Southwestern Naturalist* 31: 361–366.

BALDASSARRE, G. A. & BOLEN, E. G. 1994. *Waterfowl Ecology and Management. First Edition.* John Wiley & Sons, Inc. New York, USA.

BALDASSARRE, G. A. & BOLEN, E. G. 2006. *Waterfowl Ecology and Management. Second Edition.* Krieger Publishing Company, Malabar, Florida.

BALDASSARRE, G. A., QUINLAN, E. E. & BOLEN, E. G. 1988a. Mobility and site fidelity of Green-winged Teal wintering on the Southern High Plains of Texas. Pp. 483–493 in Weller, M. W. (ed), *Waterfowl in Winter.* University of Minnesota Press, Minneapolis, USA.

BALDASSARRE, G. A., WHYTE, R. J. & BOLEN, E. G. 1986. Body weight and carcass composition of nonbreeding Green-winged Teal on the Southern High Plains of Texas. *Journal of Wildlife Management* 50: 420–426.

BALDASSARRE, G. A., PAULUS, S. L., TAMISIER, A. & TITMAN, R. D. 1988b. Workshop summary: techniques for timing activity of wintering waterfowl. Pp. 181–188 in Weller, M. W. (ed), *Waterfowl in Winter.* University of Minnesota Press, Minneapolis, USA.

BALDWIN, J. R. & LOVVORN, J. R. 1994. Habitats and tidal accessibility of the marine foods of dabbling ducks and brant in Boundary Bay, British Columbia. *Marine Biology* 120: 627–638.

BALMAKI, B. & BARATI, A. 2006. Harvesting status of migratory waterfowl in northern Iran: a case study from Gilan Province. Pp. 868–869 in Boere, G. C., Galbraith, C. A. & Stroud, D. A. (eds), *Waterbirds around the World.* The Stationery Office, Edinburgh, UK.

BALSER, D. S., DILL, H. H. & NELSON, H. K. 1968. Effect of predator reduction on waterfowl nesting success. *Journal of Wildlife Management* 32: 669–682.

BALUJA, G., HERNANDEZ, L. M., GONZALEZ, J. & CLAVERO, R. 1983. Mercury distribution in an ecosystem of the "Parque Nacional de Doñana", Spain. *Bulletin of Environmental Contamination and Toxicology* 30: 544–551.

BANKS, J., SPEIDEL, E. C., MCCAULEY, J. W. & ALEXANDER, D. J. 2000. Phylogenetic analysis of H7 haemagglutinin subtype influenza A viruses. *Archives of Virology* 145: 1047–1058.

BANKS, R. C., CICERO, C., DUNN, J. L., KRATTER, A. W., RASMUSSEN, P. C., REMSEN, J. V. JR., RISING, J. D. & STOTZ, D. F. 2002. Forty-third supplement to the American Ornithologists' Union Check-list of North American Birds. *The Auk* 119: 897–906.

BARTONEK, J. C. 1972. Summer foods of American Widgeon, Mallards, and a Green-winged Teal near Great Slave Lake, N.W.T. *The Canadian Field-Naturalist* 86: 373–376.

BARTONEK, J. C. & DANE, C. W. 1964. Numbered nasal discs for waterfowl. *Journal of Wildlife Management* 28: 688–692.

BARTONEK, J. C. & TRAUGER, D. L. 1975. Leech (*Hirudinea*) infestations among waterfowl near Yellowknife, Northwest Territories. *Canadian Field Naturalist* 89: 234–243.

BATT, B. 2012. *The Marsh Keepers Journey: the Story of Ducks Unlimited Canada.* Ducks Unlimited Canada.

BATTEN, L. A. 1977. Sailing on reservoirs and its effect on water birds. *Biological Conservation* 11: 49–58.

BATZER, D. P., McGEE, M. & RESH, V. H. 1993. Characteristics of invertebrates consumed by Mallards and prey response to wetland flooding schedules. *Wetlands* 13: 41–49.

BAUMER, A., FELDMANN, J., RENZULLO, S., MULLER, M., THUR, B. & HOFMANN, M. A. 2010. Epidemiology of avian influenza virus in wild birds in Switzerland between 2006 and 2009. *Avian diseases* 54: 875–884.

BEARD, E.B. 1964. Duck brood behavior at the Seney National Wildlife Refuge. *Journal of Wildlife Management* 28: 492–521.

BEARDMORE, C. J. (ed). 2006. *Sonoran Joint Venture: Bird Conservation Plan, Version 1.0.* Sonoran Joint Venture, Tucson, USA.

BEHROOZ, R. D., ESMAILI-SARI, A., GHASEMPOURI, S. M., BAHRAMIFAR, N. & HOSSEINI, S. M. 2009. Organochlorine pesticide and polychlorinated biphenyl in feathers of resident and migratory birds of South-West Iran. *Archives of Environmental Contamination and Toxicology* 56 : 803–810.

BELL, D. V. & AUSTIN, L. W. 1985. The game-fishing season and its effects on overwintering wildfowl. *Biological Conservation* 33: 65–80.

BELLROSE, F. C. 1951. Effects of ingested lead shot upon waterfowl populations. *Transactions of the North American Wildlife Conference* 16:125–135.

BELLROSE, F. C. 1953. A preliminary evaluation of cripple losses in waterfowl. *Transactions of the North American Wildlife Conference* 18: 337–360.

BELLROSE, F. C. 1959. Lead poisoning as a mortality factor in waterfowl populations. *Illinois Natural History Survey Bulletin* 27: 236–288.

BELLROSE, F. C. 1976. *Ducks, Geese and Swans of North America.* Stackpole Books, Harrisburg, USA.

BELLROSE, F. C., SCOTT, T. G., HAWKINS, A. S. & LOW, J. B. 1961. Sex ratios and age ratios in North American ducks. *Illinois Natural History Survey Bulletin* 27: 391–474.

BELON, P. 1555. *L'histoire de la nature des oiseaux, avec leurs descriptions, et naïfs portraicts retirez du naturel, écrite en sept livres.* Gilles Corrozet, Paris, France.

BENGTSON, S. A. 1971. Habitat selection of duck broods in Lake Mývatn area, North-East Iceland. *Ornis Scandinavica* 2: 17–26.

BENGTSON, S. A. 1972. Reproduction and fluctuations in the size of duck populations at Lake Mývatn, Iceland. *Oikos* 23: 35–58.

BENGTSON, S. A. 1975. Food of ducklings of surface feeding ducks at Lake Mývatn, Iceland. *Ornis Fennica* 52: 1–4.

BENNETT, G. F., STOTTS, V. D. & BATEMAN, M. C. 1991. Blood parasites of black ducks and other anatids from Labrador and insular Newfoundland. *Canadian Journal of Zoology* 69: 1405–1407.

BENNETT, G. F., BLANDIN, W., HEUSMANN, H. W. & CAMPBELL, A. G. 1974. Hematozoa of the Anatidae of the Atlantic flyway. 1. Massachusetts. *Journal of Wildlife Diseases* 10: 442–451.

BENNETT, G. F., SMITH, A. D., WHITMAN, W. & CAMERON, M. 1975. Hematozoa of the Anatidae of the Atlantic flyway. II. The Maritime provinces of Canada. *Journal of Wildlife Diseases* 11: 280–289.

BENNETT, J. W. & BOLEN, E. G. 1978. Stress response in wintering Green-winged Teal. *Journal of Wildlife Management* 42: 81–86.

BENT, A. C. 1962. *Life histories of North American Wild Fowl.* Dover Publications, New York, USA.

BEREZOVIKOV, N. 2001. Age structure of populations of river and diving ducks on the lake Markakol. In *Problems of Study and Protection of Waterfowl in Eastern Europe and North Asia, 16–17.* Working Group meeting on geese and swans of Eastern Europe and North Asia (January 25–27, 2001).

BERKHOUDT, H. 1980. The morphology and distribution of cutaneous mechanoreceptors (Herbst and Grandry corpuscules) in bill and tongue of the Mallard (*Anas platyrhynchos* L.). *Netherlands Journal of Zoology* 30: 1–34.

BETHKE, R. W. & NUDDS, T. D. 1993. Variation in the diversity of ducks along a gradient of environmental variability. *Oecologia* 93: 242–250.

BETHKE, R. W. & NUDDS, T. D. 1995. Effects of climate change and land use on duck abundance in Canadian prairie-parklands. *Ecological Applications* 5: 588–600.

BEYER, W. N. & HEINZ, G. H. 2000. Implications of regulating environmental contaminants on the basis of wildlife populations and communities. *Environmental Toxicology and Chemistry* 19: 1703–1704.

BEYER, W. N., SPANN, J. & DAY, D. 1999. Metal and sediment ingestion by dabbling ducks. *The Science of the Total Environment* 231: 235–239.

BEYER, W. N., DALGARN, J., DUDDING, S., FRENCH, J. B., MATEO, R., MIESNER, J., SILEO, L. & SPANN, J. 2004. Zinc and lead poisoning in wild birds in the Tri-State Mining District (Oklahoma, Kansas, and Missouri). *Archives of Environmental Contamination and Toxicology* 48: 108–117.

BEZZEL, E. 1959. Beiträge zur Biologie der Geschechter bei Entenvögel. *Anzeiger der Ornithologischen Gesellschaft in Bayern* 5: 269–356.

BIOLOGISCHE STATION 'RIESELFELDER MÜNSTER'. 1982. Control of vegetation at Rieselfelder Münster, Federal Republic of Germany. Pp. 38–43 in Scott, D. A. (ed), *Managing Wetlands and their Birds – A Manual of Wetland and Waterfowl Management*. IWRB, Slimbridge, UK.

BIRDLIFE INTERNATIONAL. 2004a. *Birds in Europe: population estimates, trends and conservation status*. BirdLife International, Cambridge, UK.

BIRDLIFE INTERNATIONAL. 2004b. *Birds in the European Union: A status assessment*. BirdLife International, Wageningen, The Netherlands.

BIRGER, S. 1907. Über endozoische Samenverbreitung durch Vögel. *Svensk Botanisk Tidskrift* 1: 1–31.

BLACK, J. M. 1996. Introduction: pair bonds and partnerships. Pp. 3–20 in Black, J. M. (ed), *Partnerships in Birds – The Study of Monogamy*. Oxford University Press, New York.

BLANC, R., GUILLEMAIN, M., MOURONVAL, J. B., DESMONTS, D. & FRITZ, H. 2006. Effects of non-consumptive leisure disturbance to wildlife. *Revue d'Ecologie (Terre et Vie)* 61: 117–133.

BLAND, T. C. 1950. Punt-guns. Pp. 80–100 in Andrews, J., Bland, T. C., Blockey, R., Farmiloe, R., Harrison, R., Inge, J., Lee-Elliott, C., McArthur, A., Service, D. & Shephard, M. (eds), *Wildfowling*. Seeley, Service & Co. Limited, London, UK.

BLAIZE, C., BRETAGNOLLE, V. & SCHRICKE, V. 2003. Premiers résultats sur l'hivernage des Anatidés dans l'estuaire de la Seine. *Le Cormoran* 13: 9–12.

BLINOVA, T. K. & BLINOV, V. N. 1999. *The Birds of Southern Transuralia: Forest-Steppe and Steppe*. Nauka, Sierian Branch, Novosibirsk, Russia. [In Russian]

BLUHM, C. K. 1985. Mate preferences and mating patterns of Canvasbacks (*Aythya valisineria*). Pp. 45–56 in Gowaty, P. A. & Mock, D. W. (eds), *Avian Monogamy*. The American Ornithologists' Union (Ornithological Monographs 37), Washington, D.C., USA.

BLUHM, C. K. 1988. Temporal patterns of pair formation and reproduction in annual cycles and associated endocrinology in waterfowl. In Johnston R. F. (ed), *Current Ornithology* Vol. 5: 123–185. Plenum Press, New York, USA.

BLUHM, C. K. & GOWATY, P. A. 2004. Social constraints on female mate preferences in mallards, *Anas platyrhynchos*, decrease offspring viability and mother productivity. *Animal Behaviour* 68: 977–983.

BLUMS, P. & MEDNIS, A. 1996. Secondary sex ratios in Anatinae. *The Auk* 113: 505–511.

BLUMS, P., CLARK, R. G. & MEDNIS, A. 2002. Patterns of reproductive effort and success in birds: path analyses of long-term data from European ducks. *Journal of Animal Ecology* 71: 280–295.

BLUMS, P., MEDNIS, A., BAUGA, I., NICHOLS, J. D. & HINES, J. E. 1996. Age-specific survival and philopatry in three species of European ducks: a long-term study. *The Condor* 98: 61–74.

BLUS, L. J. 1996. DDT, DDD, and DDE in birds. Pp. 49–71 in Beyer, W. N., Heinz, G. H. & Redmon-Norwood, A. W. (eds) *Environmental Contaminants in Wildlife – Interpreting Tissue Concentrations.* SETAC Special Publications Series, Lewis Publishers, Boca Raton, Florida, USA.

BOGIATTO, R. J. & KARNEGIS, J. D. 2006. The use of Eastern Sacramento Valley vernal pools by ducks. *California Fish and Game* 92: 125–141.

BOGS, J., VEITS, J., GOHRBANDT, S., HUNDT, J., STECH, O., BREITHAUPT, A., TEIFKE, J. P., METTENLEITER, T. C. & STECH, J. 2010. Highly Pathogenic H5N1 Influenza Viruses carry virulence determinants beyond the polybasic hemagglutinin cleavage site. *PLoS ONE* 5(7): e11826.

BOHONAK, A. J. & WHITEMAN, H. H. 1999. Dispersal of the fairy shrimp *Branchinecta coloradensis* (Anostraca): effects of hydroperiod and salamanders. *Limnology and Oceanography* 44: 487–493.

BOLLINGER, T. K., EVELSIZER, D. D. & ZIMMER, M. 2011. Efficacy of carcass clean-up during avian botulism outbreaks. Pp. 16–35 in Bollinger, T. K., Evelsizer, D. D., Dufour, K. W., Soos, C., Clark, R. G., Wobeser, G., Kehoe, F. P., Guyn, K. L. & Pybus, M. J. (eds), *Ecology and Management of Avian Botulism on the Canadian Prairies.* Report to the Prairie Habitat Joint Venture, Canada.

BØNLØKK, J., MADSEN, J., THORUP, K., PEDERSEN, K. T., BJERRUM, M. & RAHBEK, C. 2006. *Dansk Traekfugleatlas – The Danish bird migration atlas.* Forlaget Rhodos A/S & Zoologisk Museum, Københarns Universitet, Denmark.

BOOS, J. D., NUDDS, T. D. & SJÖBERG, K. 1989. Posthatch brood amalgamation by Mallards. *Wilson Bulletin* 101: 503–505.

BOOS, M., ZIMMER, C., CARRIERE, A., ROBIN, J. P. & PETIT, O. 2007. Post-hatching parental care behaviour and hormonal status in a precocial bird. *Behavioural Processes* 76: 206–214.

BORGSTEEDE, F. H. M., KAVETSKA, K. M. & ZOUN, P. E. F. 2006. Species of the nematode genus *Amidostomum* Railliet and Henry, 1909 in aquatic birds in the Netherlands. *Helminthologia* 43: 98–102.

BOYD, H. 1957. Mortality and kill amongst British-ringed Teal *Anas crecca. Ibis* 99: 157–177.

BOYD, H. 1961. Reported casualties to ringed ducks in the spring and summer. *Wildfowl Trust Annual Report* 12: 144–146.

BOYD, H. 1962. Population dynamics and the exploitation of ducks and geese. Pp. 85–95 in Le Cren, E. D. & Holdgate, M. W (eds), *The Exploitation of Natural Animal Populations. A Symposium of the British Ecological Society, Durham 28th–31st March 1960.* John Wiley & Sons, Inc. New York, USA.

BOYD, H. 1964. Wildfowl and other water-birds found dead in England and Wales in January-March 1963. *Wildfowl Trust Annual Report* 15: 20–22.

BOYLE, S. A. & SAMSON, F. B. 1985. Effects of nonconsumptive recreation on wildlife: a review. *Wildlife Society Bulletin* 13: 110–116.

BOZORGMEHRI-FARD, M. H. & KEYVANFAR, H. 1979. Isolation of Newcastle Disease Virus from Teals (*Anas crecca*) in Iran. *Journal of Wildlife Diseases* 15: 335–337.

BØNLØKK, J., MADSEN, J., THORUP, K., PEDERSEN, K. T., BJERRUM, M. & RAHBEK, C. 2006. *The Danish bird migration atlas*. Forlaget Rhodos A/S & Zoologisk Museum, Københarns Universitet, Copenhagen, Denmark. [In Danish]

BRAUNE, B. M. & MALONE, B. J. 2006. Mercury and selenium in livers of waterfowl harvested in Northern Canada. *Archives of Environmental Contamination and Toxicology* 50: 284–289.

BREGNBALLE, T., MADSEN, J. & RASMUSSEN, P. A. F. 2004. Effects of temporal and spatial hunting control in waterbird reserves. *Biological Conservation* 119: 93–104.

BREGNBALLE, T., SPEICH, C., HORSTEN, A. & FOX, A. D. 2009b. An experimental study of numerical and behavioural responses of spring staging dabbling ducks to human pedestrian disturbance. *Wildfowl*, Special Issue 2: 131–142.

BREGNBALLE, T., ANDERSEN, U. D., CLAUSEN, P., KJAER, P. A. & FOX, A. D. 2009a. Habitat use and home range size of autumn staging radio-marked Teal *Anas crecca* at Ulvshale-Nyord, Denmark. *Wildfowl*, Special Issue 2: 100–114.

BRENNAN, P. L. R. & PRUM, R. O. 2012. the limits of sexual conflict in the narrow sense: new insights from waterfowl biology. *Philosophical Transactions of the Royal Society B* 367: 2324–2338.

BRENNAN, P. L., PRUM, R. O., MCKRAKEN, K. G., SORENSON, M. D., WILSON, R. E. & BIRKHEAD, T. R. 2007. Coevolution of male and female genital morphology in waterfowl. *PLoS ONE* 2(5): e418.

BROCHET, A. L., GUILLEMAIN, M., FRITZ, H., GAUTHIER-CLERC, M. & GREEN, A. J. 2010a. Plant dispersal by teal (*Anas crecca*) in the Camargue: duck guts are more important than their feet. *Freshwater Biology* 55: 1262–1273.

BROCHET, A. L., GAUTHIER-CLERC, M., MATHEVET, R., BECHET, A., MONDAIN-MONVAL, J. Y. & TAMISIER, A. 2009. Marsh management, reserve creation, hunting periods and carrying capacity for wintering ducks and coots. *Biodiversity and Conservation* 18: 1879–1894.

BROCHET, A. L., DESSBORN, L., LEGAGNEUX, P., ELMBERG, J., GAUTHIER-CLERC, M. FRITZ, H. & GUILLEMAIN, M. 2012a. Is diet segregation between dabbling ducks due to food partitioning? A review of seasonal patterns in the Western Palearctic. *Journal of Zoology* 286: 171–178.

BROCHET, A. L., GAUTHIER-CLERC, M., GUILLEMAIN, M., FRITZ, H., WETARKEYN, A., BALTANÁS, Á. & GREEN, A. J. 2010b. Field evidence of dispersal of branchiopods, ostracods and bryozoans by teal (*Anas crecca*) in the Camargue (southern France). *Hydrobiologia* 637: 255–261.

BROCHET, A. L., MOURONVAL, J. B., AUBRY, P., GAUTHIER-CLERC, M., GREEN, A. J., FRITZ, H. & GUILLEMAIN, M. 2012b. Diet and feeding habitats of Camargue dabbling ducks: what has changed since the 1960s? *Waterbirds* 35: 555–576.

BROCK, H. H. 1907. Recent occurrence of the European Teal and the Marbled Godwit near Portland, Maine. *The Auk* 24: 94.

BRODSKY, L. M. & WEATHERHEAD, P. J. 1985. Time and energy constraints on courtship in wintering American Black Ducks. *The Condor* 87: 33–36.

BRODSKY, L. M., ANKNEY, C. D. & DENNIS, D. G. 1988. The influence of male dominance on social interactions in black ducks and mallards. *Animal Behaviour* 36: 1371–1378.

BROGDEN, K. A. & RHOADES, K. R. 1983. Prevalence of serologic types of *Pasteurella multocida* from 57 species of birds and mallards in the United States. *Journal of Wildlife Diseases* 19: 315–320.

BROUWER, C. 1980. Contribution à l'étude du régime alimentaire des sarcelles d'hiver *(Anas crecca) en Bretagne Méridionale*. Anatidés de Bretagne Meridionale, rapport d'activité 1979/1980. Station Biologique de Paimpont, Paimpont, France.

BROWN, D. J., HUBERT, W. A. & ANDERSON, S. H. 1996. Beaver ponds create wetland habitat for birds in mountains of Southeastern Wyoming. *Wetlands* 16: 127–133.

BROWN, L. H., URBAN, E. K. & NEWMAN, K. 1982. *The Birds of Africa Volume 1*. Academic Press, London, UK.

BROWNIE, C., ANDERSON, D. R., BURNHAM, K. P. & ROBSON, D. S. 1985. *Statistical Inference from Band Recovery Data – a Handbook*. Fish and Wildlife Service Resource Publication No. 156. U.S. Department of the Interior, Washington, D.C., USA.

BRUA, R. B. 1998. Negative effects of patagial tags on Ruddy Ducks. *Journal of Field Ornithology* 69: 530–535.

BRUDERER, B. & BOLDT, A. 2001. Flight characteristics of birds: I. Radar measurements of speeds. *Ibis* 143: 17–204.

BRUNO, J. F., STACHOWICZ, J. J. & BERTNESS, M. D. 2003. Inclusion of facilitation into ecological theory. *Trends in Ecology and Evolution* 18: 119–125.

BRUINZEEL, L. W., VAN EERDEN, M. R., DRENT, R. H. & VULINK, J. T. 1997. Scaling metabolisable energy intake and daily energy expenditure in relation to the size of herbivorous waterfowl: limits set by available foraging time and digestive performance. Pp. 111–132 in Van Eerden, M. R. (ed), Patchwork – Patch use, habitat exploitation and carrying capacity for water birds in Dutch freshwater wetlands. PhD Thesis, University of Gröningen, Gröningen, The Netherlands.

BUB, H. 1991. *Bird Trapping and Bird Banding. A handbook for trapping methods all over the world*. Cornell University Press, Ithaca, NY, USA.

BUFFON (LECLERC, G. L., COMTE DE -). 1853. *Œuvres complètes de Buffon, Tome sixième, Oiseaux II*. Furne & Cie, Paris, France.

BURTON, N. H. K., REHFISH, M. M. & CLARK, N. A. 2002. Impacts of disturbance from construction work on the densities and feeding behavior of waterbirds using the intertidal mudflats of Cardiff Bay, UK. *Environmental Management* 30: 865–871.

BUSQUETS, N., ALBA, A., NAPP, S., SÁNCHEZ, A., SERRANO, E., RIVAS, R., NÚÑEZ, J. I. & MAJÓ, N. 2010. Influenza A virus subtypes in wild birds in North-Eastern Spain (Catalonia). *Virus Research* 149: 10–18.

BUSTNES, J. O. & ERIKSTAD, K. E. 1990. Effects of patagial tags on laying date and egg size in Common Eider. *Journal of Wildlife Management* 54: 216–218.

CALLENGE, C., GUILLEMAIN, M., GAUTHIER-CLERC, M. & SIMON, G. 2010. A new exploratory approach to the study of the spatio-temporal distribution of ring recoveries: the example of Teal (*Anas crecca*) ringed in Camargue, Southern France. *Journal of Ornithology* 151: 945–950.

CALLICUTT, J. T., HAGY, H. M. & SCHUMMER, M. L. 2011. The food preference paradigm: a review of autumn–winter food use by North American dabbling ducks (1900–2009). *Journal of Fish and Wildlife Management* 2: 29–40.

CALVO, B. & FURNESS, R. W. 1992. A review of the use and the effects of marks and devices on birds. *Ringing & Migration* 13: 129–151.

CAMPBELL, J. W. 1947. The food of some British wildfowl. *Ibis* 89: 429–432.

CAMPREDON, P. 1983. Sexe et âge ratios chez le canard siffleur *Anas penelope* L. en période hivernale en Europe de l'Ouest. *Revue d'Ecologie (Terre Vie)* 37: 117–128.

CAMPREDON, S., CAMPREDON, P., PIROT, J. Y. & TAMISIER, A. 1982. *Manuel d'analyse des contenus stomacaux de canards & de foulques*. CNRS-ONC, Paris, France.

CANARIS, A. G., MENA, A. C. & BRISTOL, J. R. 1981. Parasites of waterfowl from southwest Texas: III. The Green-winged Teal, *Anas crecca*. *Journal of Wildlife Diseases* 17: 57–64.

CAO, L., BARTER, M. & MEI, G. 2008. New Anatidae population estimates for eastern China: implications for current flyway estimates. *Biological Conservation* 141: 2301–2309.

CAPPERS, R. T. J., BEKKER, R. M. & JANS, J. E. A. 2006. *Digital Seed Atlas of the Netherlands*. Barkhuis Publishing, Eelde, The Netherlands.

CARBONE, C. & OWEN, M. 1995. Differential migration of the sexes of Pochard *Aythya ferina*: results from an international survey. *Wildfowl* 46: 99–108.

CARNEY, S. M. 1992. *Species, Age and Sex Identification of Ducks using Wing Plumage*. U.S. Department of the Interior / U.S. Fish and Wildlife Service, Washington, D.C., USA.

CARPENE, E., SERRA, R. & ISANI, G. 1995. Heavy metals in some species of waterfowl of Northern Italy. *Journal of Wildlife Diseases* 31: 49–56.

CASTRESANA, L., NOTARIO, A. & MARTÍN MATEO, P. 1999. Estudio de los malófagos ectoparásitos de anátidas (*Insecta, Mallophaga*) en la Península Ibérica. Identificación, características biométricas y aspectos biológicos. *Zoologica baetica* 10: 63–86.

CASWELL, H. 2001. *Matrix Population Models: Construction, Analysis, and Interpretation*. Sinauer Associates, Sunderland, Massachusetts.

CHAMPAGNON, J., GUILLEMAIN, M., ELMBERG, J., FOLKESSON, K. & GAUTHIER-CLERC, M. 2010. Changes in Mallard *Anas platyrhynchos* bill morphology after 30 years of supplemental stocking. *Bird Study* 57: 344–351.

CHANSIGAUD, V. 2009. *The History of Ornithology*. New Holland Publishers, London, UK.

CHODACHEK, K. D. & CHAMBERLAIN, M. J. 2006. Effects of predator removal on upland nesting ducks in North Dakota grassland fragments. *The Prairie Naturalist* 38: 25–37.

CHU, D. S., NICHOLS, J. D., HESTBECK, J. B. & HINES, J. E. 1995. Banding reference areas and survival rates of Green-winged Teal, 1950–89. *Journal of Wildlife Management* 59: 487–498.

CITES. 2011. *Appendices I, II and III valid from 27 April 2011 – Interpretation*. CITES Appendices: http://www.cites.org/eng/app/index.php

CLAUSEN, P., NOLET, B. A., FOX, A.D. & KLAASSEN, M. 2002. Long-distance endozoochorous dispersal of submerged macrophyte seeds by migratory waterbirds in northern Europe – a critical review of possibilities and limitations. *Acta Oecologica* 23: 191–203.

CMS. 2009. *Appendices I and II of the Convention on the Conservation of Migratory Species of Wild Animals (CMS) as amended by the Conference of the Parties in 1985, 1988, 1991, 1994, 1997, 1999, 2002, 2005 and 2008*. CMS: http://www.cms.int/documents/appendix/cms_app1_2.htm

COKER, C. R., McKINNEY, F., HAYS, H., BRIGS, S. V. & CHENG, K. M. 2002. Intromittent organ morphology and testis size in relation to mating system in waterfowl. *The Auk* 119: 403–413.

CONOMY, J. T., COLLAZO, J. A., DUBOVSKY, J. A. & FLEMING, W. J. 1998. Dabbling duck behaviour and aircraft activity in coastal North Carolina. *Journal of Wildlife Management* 62: 1127–1134.

CONRAD, W. B. JR. 1965. A food habits study of ducks wintering on the Lower Pee Dee and Waccamaw rivers, Georgetown, South Carolina. *Proceedings of the Annual Conference of the Southeastern Association of Fish and Wildlife Agencies* 19: 93–98.

CONVERSE, K. A. 2007. Aspergillosis. Pp. 360–374 in Thomas, N. J., Hunter, D. B. & Atkinson, C. T. (eds), *Infectious Diseases of Wild Birds*. Blackwell Publishing, Oxford, UK.

CONVERSE, K. A. & KIDD, G. A. 2001. Duck plague epizootics in the United States, 1967–1995. *Journal of Wildlife Diseases* 37: 347–357.

COOKE, A. 1975. The effects of fishing on waterfowl on Grafham Water. *Cambridge Bird Club Report* 48: 40–46.

COOPER, B. A. & RITCHIE, R. J. 1995. The altitude of bird migration in east-central Alaska: a radar and visual study. *Journal of Field Ornothology* 66: 590–608.

COTTAM, C. & KNAPPEN, P. 1939. Food of some uncommon North American birds. *The Auk* 56: 138–169.

COULTER, M. W. 1955. Spring food habits of surface-feeding ducks in Maine. *Journal of Wildlife Management* 19: 263–267.

COX, R. R. JR., HANSON, M. A., ROY, C. C., EULISS, N. H. JR., JOHNSON, D. H. & BUTLER, M. G. 1998. Mallard duckling growth and survival in relation to aquatic invertebrates. *Journal of Wildlife Management* 62: 124–133.

CRAMP, S. & SIMMONS, K. E. L. 1977. *The Birds of the Western Palearctic, Vol. 1*. Oxford University Press, Oxford, UK.

CRISTOL, D. A., BAKER, M. B. & CARBONE, C. 1999. Differential migration revisited. Latitudinal segregation by age and sex class. In Nolan V. Jr, Ketterson, E. D. & Tompson, C. F. (eds), *Current Ornithology* vol 15: 33–88. Kluwer, New York, USA.

CROCHET, P. A. & JOYNT, G. 2011. *AERC TAC list of Western Palearctic birds. March 2011 version*. Available at http://www.aerc.eu/tac.html

CRUICKSHANK, A. D. 1936. Some observations of the European Teal. *The Auk* 53: 321–322.

CUNNINGHAM, E. J. A. 2003. Female mate preferences and subsequent resistance to copulation in the mallard. *Behavioural Ecology* 14: 326–333.

CURTET, L., HÉRAULT, L., HUGUET, L., FOURNIER, J. Y. & BROYER, J. 2004. Etangs piscicoles et alimentation des anatidés en période internuptiale: principaux faciès utilisés. *Faune Sauvage* 262: 4–11.

DALBY, L. 2013. Waterfowl, duck distributions and a changing climate. PhD Thesis, Institute for Bioscience, Aarhus University, Denmark.

DALBY, L., FOX, A. D., PETERSEN, I. K., DELANY, S. & SVENNING, J. C. 2013. Temperature does not dictate the wintering distributions of European dabbling duck species. *Ibis* 155: 80–88.

DANELL, K. & SJÖBERG, K. 1977. Seasonal emergence of chironomids in relation to egglaying and hatching of ducks in a restored lake (northern Sweden). *Wildfowl* 28: 129–135.

DANELL, K. & SJÖBERG, K. 1980. Foods of wigeon, teal, mallard and pintail during the summer in a northern swedish lake. *Viltrevy* 11: 141–167.

DANELL, K. & SJÖBERG, K. 1982. Seasonal and diel changes in the feeding behaviour of some dabbling duck species on a breeding lake in northern Sweden. *Ornis Scandinavica* 13: 129–134.

DARWIN, C. R. 1859. *On the Origin of Species by Means of Natural Selection, or the Preservation of Favoured Races in the Struggle for Life.* First Edition. John Murray, London, UK.

DARWIN, C. R. 1872. *The Origin of Species by Means of Natural Selection, or the Preservation of Favoured Races in the Struggle for Life.* Sixth edition; with additions and corrections. John Murray, London, UK.

DASSOW, J. A., EICHHOLZ, M. W., STAFFORD, J. D. & WEATHERHEAD, P. J. 2012. Increased nest defence of upland-nesting ducks in response to experimentally reduced risk of nest predation. *Journal of Avian Biology* 43: 61–67.

DAVIS, E. S. 2002a. Male reproductive tactics in the mallard, *Anas platyrhynchos*: social and hormonal mechanisms. *Behavioral Ecology and Sociobiology* 52: 224–231.

DAVIS, E. S. 2002b. Female choice and the benefits of mate guarding by male mallards. *Animal Behaviour* 64: 619–628.

DAWSON, R. D. & CLARK, R. G. 2000. Effect of hatching date and egg size on growth, recruitment and adult size of lesser scaup. *The Condor* 102: 930–935.

DAY, A. M. 1949. *North American Waterfowl.* Stackpole Co., New York, USA.

DECARIE, R., MORNEAU, F., LAMBERT, D., CARRIERE, S. & SAVARD, J. P. L. 1995. Habitat use by brood-rearing waterfowl in subarctic Québec. *Arctic* 48: 383–390.

DECEUNINCK, B. 1997. *Synthèse des dénombrements de canards et foulques hivernant en France 1967–1995.* Ligue Pour la Protection des Oiseaux / Wetlands International, Rochefort sur Mer, France.

DE JUANA, E. & GARCIA, E. F. J. 2010. Vagrancy or migration: why do American Teals cross the Atlantic? *Ardeola* 57: 417–430.

DEKKER, D. 1980. Hunting success rates, foraging habits, and prey selection of Peregrine falcons migrating through Central Alberta. *The Canadian Field Naturalist* 94: 371–382.

DELACOUR, J. 1956. *The Waterfowl of the World. Volume Two – The Dabbling Ducks.* Country Life Limited, London, UK.

DELANY, S., DODMAN, T., SCOTT, D., BUTCHART, S., MARTAKIS, G. & HELMINK, T. 2008. *Report on the Conservation Status of Migratory Waterbirds in the Agreement Area. Fourth Edition.* AEWA/MOP 4.8. Wetlands International, Wageningen, The Netherlands.

DEMENT'EV, G. P. & GLADKOV, N. A. (eds). 1952. *Birds of the Soviet Union – Volume IV.* "Soviet Science" State publisher, Moscow, Russia. [In Russian, English translation and publication by Israel Program for Scientific Translations, Jerusalem, Israel, 1967]

DENNIS, J. V. 1994. Transatlantic migration by ringed birds from North America. *Dutch Birding* 16: 235–237.

DEROIA, D. M. 1989. Spring and autumn feeding ecology of Blue-winged and Green-winged Teals on the Lake Erie marshes. M.S. Thesis, Ohio State University, Columbus, USA.

DEROIA, D. M. & BOOKHOUT, T. A. 1989. Spring feeding ecology of teal on the Lake Erie marshes. *Ohio Journal of Science* 89: 3.

DERVIEUX, A. & TAMISIER, A. 1979. Quelques aspects éthologiques et physiologiques de l'activité sexuelle de Sarcelles d'hiver en captivité. *L'oiseau et la Revue Française d'Ornithologie* 49: 299–322.

DESSBORN, L., ELMBERG, J. & ENGLUND, G. 2011a. Pike predation affects breeding success and habitat selection of ducks. *Freshwater Biology* 56: 579–589.

DESSBORN, L., ENGLUND, G., ELMBERG, J. & ARZEL, C. 2012. Innate responses of mallard duckling towards aerial, aquatic and terrestrial predators. *Behaviour* 159: 1299–1317.

DESSBORN, L., ELMBERG, J., NUMMI, P., PÖYSÄ, H. & SJÖBERG, K. 2009. Hatching in dabbling ducks and emergence in chironomids: a case of predator–prey synchrony? *Hydrobiologia* 636: 319–329.

DESSBORN, L., BROCHET, A. L., ELMBERG, J., LEGAGNEUX, P., GAUTHIER-CLERC, M. & GUILLEMAIN, M. 2011b. Geographical and temporal patterns in the diet of pintail *Anas acuta*, wigeon *Anas penelope*, mallard *Anas platyrhynchos* and teal *Anas crecca* in the Western Palearctic. *European Journal of Wildlife Research* 57: 1119–1129.

DEVINEAU, O. 2007. Dynamique et gestion des populations exploitées: l'exemple de la sarcelle d'hiver. PhD Thesis, University Montpellier II, France.

DEVINEAU, O., GUILLEMAIN, M., JOHNSON, A. R. & LEBRETON, J. D. 2010. A comparison of green-winged teal *Anas crecca* survival and harvest between Europe and North America. *Wildlife Biology* 16: 12–24.

DEVINK, J. M., CLARK, R. G., SLATTERY, S. M. & WAYLAND, M. 2008. Is selenium affecting body condition and reproduction in boreal breeding scaup, scoters, and ring-necked ducks? *Environmental Pollution* 152: 116–122.

DE VRIES, V. 1939. Contribution to the food biology of four species of ducks, following material from Vlieland and Terschelling. *Limosa* 12: 87–98. [In Dutch]

DILLON, O. W. JR. 1957. Food habits of wild ducks in the rice-marsh transition area of Louisiana. *Proceedings of the Annual Conference of the Association of Fish and Wildlife Agencies* 12: 114–119.

DOBINSON, H. M. & RICHARDS, A. J. 1964. The effects of the severe winter of 1962/63 on birds in Britain. *British Birds* 57: 373–433.

DOMBROVSKI, V. 2010. The diet of the greater spotted eagle (*Aquila clanga*) in Belarusian Polesis. *Slovak Raptor Journal* 4: 23–36.

DRENT, R. H. & DAAN, S. 1980. The prudent parent: energetic adjustments in avian breeding? *Ardea* 68: 225–252.

DREVER, M. C., WINS-PURDY, A., NUDDS, T. D. & CLARK, R. G. 2004. Decline of duck nest success revisited: relationships with predators and wetlands in dynamic prairie environments. *The Auk* 121: 497–508.

DUBOWY, P. J. 1988. Waterfowl communities and seasonal environments: temporal variability in interspecific competition. *Ecology* 69: 1439–1453.

DUEBBERT, H. F. & KANTRUD, H. A. 1974. Upland duck nesting related to land use and predator reduction. *Journal of Wildlife Management* 38: 257–265.

DUEBBERT, H. F. & LOKEMOEN, J. T. 1980. High duck nest success in a predator-reduced environment. *Journal of Wildlife Management* 44: 428–437.

DU FEU, C. R., JOYS, A. C., CLARK, J. A., FIEDLER, W., DOWNIE, I. S., VAN NOORDWIJK, A. J., SPINA, F., WASSENAAR, R. & BAILLIE, S. R. 2009. EURING Data Bank geographical index 2009. (http://www.euring.org/edb)

DUFOUR, K. W. & BOLLINGER, T. K. 2011. Late-summer survival of mallards exposed to avian botulism: an investigation using band recovery analyses. Pp. 60–76 in Bollinger, T. K., Evelsizer, D. D., Dufour, K. W., Soos, C., Clark, R. G., Wobeser, G., Kehoe, F. P., Guyn, K. L. & Pybus, M. J. (eds), *Ecology and Management of Avian Botulism on the Canadian Prairies*. Report to the Prairie Habitat Joint Venture, Canada.

DUNCAN, D. C. 1986. Survival of dabbling duck broods on prairie impoundments in Southeastern Alberta. *The Canadian Field-Naturalist* 100: 110–113.

DUNCAN, P., HOFFMAN L., LAMBERT, R. & WALMSLEY, J. G. 1982. Management of a day roost for wintering waterfowl, especially Teal *Anas crecca* and Gadwall *Anas strepera*, in the Camargue, France. Pp. In 73–82 Scott, D. A. (ed), *Managing Wetlands and their Birds – A Manual of Wetland and Waterfowl Management*. IWRB, Slimbridge, UK.

DUNCAN, P., HEWISON, A. J. M., HOUTE, S., ROSOUX, R., TOURNEBIZE, T., DUBS, F., BUREL, F. & BRETAGNOLLE, V. 1999. Long-term changes in agricultural practices and wildfowling in an internationally important wetland, and their effects on the guild of wintering ducks. *Journal of Applied Ecology* 36: 11–23.

DUNN, P. O., AFTON, A. D., GLOUTNEY, M. L. & ALISAUSKAS, R. T. 1999. Forced copulation results in few extrapair fertilizations in Ross's and lesser snow goose. *Animal Behaviour* 57: 1071–1081.

DURANEL, A. 1999. Effets de l'ingestion de plombs de chasse sur le comportement alimentaire et la condition corporelle du canard colvert (*Anas platyrhynchos*). Thèse pour le diplôme d'état de Docteur Vétérinaire, Ecole Nationale Vétérinaire de Nantes, France.

DZUS, E. H. & CLARK, R. G. 1997. Brood size manipulation in mallard ducks: effects on duckling survival and brooding efficiency. *Ecoscience* 4: 437–445.

DZUS, E. H. & CLARK, R. G. 1998. Brood survival and recruitment of mallards *Anas platyrhynchos* in relation to wetland density and hatch date. *The Auk* 115: 311–318.

EADIE, J. McA., KEHOE, F. P. & NUDDS, T. D. 1988. Pre-hatch and post-hatch brood amalgamation in North American Anatidae: a review of hypotheses. *Canadian Journal of Zoology* 66: 1709–1721.

EICHHOLZ, M. W. & TOWERY, B. N. 2010. Potential influence of egg location on synchrony of hatching of precocial birds. *The Condor* 112: 696–700.

ELDRIDGE, J. L. & KRAPU, G. L. 1988. The influence of diet quality on clutch size and laying pattern in Mallards. *The Auk* 105: 102–110.

ELIASON, C. A. & SHAWKEY, M. D. 2011. Decreased hydrophobicity of iridescent feathers: a potential cost of shiny plumage. *The Journal of Experimental Biology* 214: 2157–2163.

ELMBERG, J., DESSBORN, L. & ENGLUND, G. 2010. Presence of fish affects lake use and breeding success in ducks. *Hydrobiologia* 641: 215–223.

ELMBERG, J., FOLKESSON, K., GUILLEMAIN, M. & GUNNARSSON, G. 2009. Putting density dependence in perspective: nest density, nesting phenology, and biome, all matter to survival of simulated mallard *Anas platyrhynchos* nests. *Journal of Avian Biology* 40: 317–326.

ELMBERG, J., NUMMI, P., PÖYSÄ, H. & SJÖBERG, K. 1993. Factors affecting species number and density of dabbing duck guilds in North Europe. *Ecography* 16: 251–260.

ELMBERG, J., NUMMI, P., PÖYSÄ, H. & SJÖBERG, K. 2003. Breeding success of sympatric dabbling ducks in relation to population density and food resources. *Oikos* 100: 333–341.

ELMBERG, J., PÖYSÄ, H., SJÖBERG, K. & NUMMI, P. 1997. Interspecific interactions and co-existence in dabbling ducks: observations and an experiment. *Oecologia* 111: 129–136.

ELMBERG, J., SJÖBERG, K., PÖYSÄ, H. & NUMMI, P. 2000. Abundance–distribution relationships on interacting trophic levels: the case of lake-nesting waterfowl and dysticid water beetles. *Journal of Biogeography* 27: 821–827.

ELMBERG, J., NUMMI, P., PÖYSÄ, H., GUNNARSSON, G. & SJÖBERG, K. 2005. Early breeding teal *Anas crecca* use the best lakes and have the highest reproductive success. *Annales Zoologici Fennici* 42: 37–43.

ENGEL, S., KLAASSEN, R. H. G., KLAASSEN, M. & BIEBACH, H. 2006. Exhaled air temperature as a function of ambient temperature in flying and resting ducks. *Journal of Comparative Physiology B* 176: 527–534.

ERICSON, P. G. P. & TYRBERG, T. 2004. *The Early History of the Swedish Avifauna. A Review of the Subfossil Record and Early Written Sources.* Kungl. Vitterhets historie och antikvitets akademien. Stockholm, Sweden.

EULISS, N. H. & HARRIS, S. W. 1987. Feeding ecology of Northern Pintails and Green-winged Teal wintering in California. *Journal of Wildlife Management* 51: 724–732.

EVANS, C. D. & BLACK, K. E. 1956. *Duck Production Studies on the Prairie Potholes of South Dakota.* U.S. Fish and Wildlife Service Special Scientific Report – Wildlife 32. Washington, D.C., USA.

EVELSIZER, D. D., CLARK, R. G. & BOLLINGER, T. K. 2010. Relationships between local carcass density and risk of mortality in molting mallards during avian botulism outbreaks. *Journal of Wildlife Diseases* 46: 507–513.

FEDYNICH, A. M. 1987. Interlake movements of wintering waterfowl on the Southern High Plains. M.Sc. Thesis, Texas Tech University, Lubbock, USA.

FEDYNICH, A. M. & GODFREY, R. D. JR. 1989. Gadwall pair recaptured in successive winters on the Southern High Plains of Texas. *Journal of Field Ornithology* 60: 168–170.

FELDHEIM, C. L. 1997. The length of incubation in relation to nest initiation date and clutch size in dabbling ducks. *The Condor* 99: 997–1001.

FEREIDOUNI, S. R., WERNER, O., STARICK, E., BEER, M., HARDER, T. C., AGHAKHAN, M., MODIRROUSTA, H., AMINI, H., MOGHADDAM, M. K., BOZORGHMEHRIFARD, M. H., AKHAVIZADEGAN, M. A., GAIDET, N., NEWMAN, S. H., HAMMOUMI, S., CATTOLI, G., GLOBIG, A., HOFFMANN, B., SEHATI, M. E., MASOODI, S., DODMAN, T., HAGEMEIJER, W., MOUSAKHANI, S. & METTENLEITER, T. C. 2010. Avian influenza virus monitoring in wintering waterbirds in Iran, 2003–2007. *Virology Journal* 7: 43.

FERRO, P. J., EL-ATTRACHE, J., FANG, X., ROLLO, S. N., JESTER, A., MERENDINO, T., PETERSON, M. J. & LUPIANI, B. 2008. Avian Influenza Surveillance in hunter-harvested Waterfowl from the Gulf Coast of Texas (November 2005–January 2006). *Journal of Wildlife Diseases* 44: 434–439.

FIGUEROLA, J. & GREEN, A. J. 2000. Haematozoan parasites and migratory behaviour in waterfowl. *Evolutionary Ecology* 14: 143–153.

FIGUEROLA, J. & GREEN, A. J. 2002. Dispersal of aquatic organisms by waterbirds: a review of past research and priorities for future studies. *Freshwater Biology* 47: 483–494.

FIGUEROLA, J. & GREEN, A. J. 2006. A comparative study of egg mass and clutch size in the Anseriformes. *Journal of Ornithology* 147: 57–68.

FIGUEROLA, J., GREEN, A. J. & SANTAMARÍA, L. 2003. Passive internal transport of aquatic organisms by waterfowl in Doñana, south-west Spain. *Global Ecology and Biogeography* 12: 427–436.

FIGUEROLA, J., MATEO, R., GREEN, A. J., MONDAIN-MONVAL, J. Y., LEFRANC, H. & MENTABERRE, G. 2005. Grit selection in waterfowl and how it determines exposure to ingested lead shot in Mediterranean wetlands. *Environmental Conservation* 32: 1–9.

FINNISH GAME AND FISHERIES RESEARCH INSTITUTE. 2012. *Official Statistics of Finland – Hunting 2011*. Finnish Game and Fisheries Research Institute, Helsinki. [In Finnish]

FISHER, R. A. 1930. *The Genetical Theory of Natural Selection*. Clarendon Press, Oxford, UK.

FITZSIMMONS, O. N., BALLARD, B. M., MERENDINO, M. T., BALDASSARRE, G. A. & HARTKE, K. M. 2012. Implications of coastal wetland management to nonbreeding waterbirds in Texas. *Wetlands* 32: 1057–1066.

FJELDSÅ, J. 1977. *Guide to the Young of European Precocial Birds*. Skarv Nature Publications, Tisvildeleje, Denmark.

FLEISCHLI, M. A., FRANSON, J. C., THOMAS, N. J., FINLEY, D. L. & RILEY, W. JR. 2004. Avian mortality events in the United States caused by anticholinesterase pesticides: a retrospective summary of National Wildlife Health Center Records from 1980 to 2000. *Archives of Environmental Contamination and Toxicology* 45: 542–550.

FLEMING, K. & HOWELL, D. (compilers). 2013. An assessment of the harvest potential of North American Teal. Unpublished report of the joint USFWS-Flyway-CWS Teal Harvest Potential Working Group.

FLINT, V. Y. & KRIVENKO, V. G. 1990. The recent status and trends of waterfowl in the USSR. pp. 23–26 in Matthews, G. V. T. (ed), *Managing Waterfowl Populations. Proceedings of an IWRB symposium, Astrakhan, USSR, 2–5 October 1989*. IWRB Special Publication 12, IWRB, Slimbridge, UK.

FLORSCHUTZ, O. JR. 1972. The importance of Eurasian Milfoil (*Myriophyllum spicatum*) as a waterfowl food. *Proceedings of the Annual Conference of the Southeastern Association of Fish and Wildlife Agencies* 26: 189–194.

FOUQUE, C., GUILLEMAIN, M. & SCHRICKE, V. 2009. Trends in the numbers of Coot *Fulica atra* and wildfowl *Anatidae* wintering in France, and their relationships with hunting activity at wetland sites. *Wildfowl*, Special Issue 2: 42–59.

FOX, A. D. 1986. The breeding Teal (*Anas crecca*) of a coastal raised mire in central West Wales. *Bird Study* 33: 18–23.

FOX, A. D. 2005a. Population dynamics. Pp. 132–151 in Kear, J. (ed), *Ducks, Geese and Swans*. Oxford University Press, Oxford, UK.

FOX, A. D. 2005b. Eurasian and American Green-winged Teal *Anas crecca/carolinensis*. Pp. 609–613 in Kear, J. (ed), *Ducks, Geese and Swans*. Oxford University Press, Oxford, UK.

FOX, A. D. & MADSEN, J. 1997. Behavioural and distributional effects of hunting disturbance on waterbirds in Europe: implications for refuge design. *Journal of Applied Ecology* 34: 1–13.

FOX, A. D., KING, R. & WATKIN, J. 1992. Seasonal variation in weight, body measurements and condition of free-living Teal. *Bird Study* 39: 53–62.

FRANSSON, T. & PETTERSSON, J. 2001. *Swedish bird ringing atlas. Vol. 1. Divers-raptors*. Naturhistoriska riksmuseet & Sveriges Ornitologiska Förening, Stockholm, Sweden. [In Swedish]

FRISCH, D., GREEN, A. J. & FIGUEROLA, J. 2007. High dispersal capacity of a broad spectrum of aquatic invertebrates via waterbirds. *Aquatic Science* 69: 568–574.

FRIEDMANN, H. 1948. The Green-winged Teal of the Aleutian Islands. *Proceedings of the Biological Society of Washington* 61: 157–158.

FRIEND, M. & FRANSON, J. C. 1999. *Field Manual of Wildlife Diseases – General field procedures and diseases of birds*. Information and Technology Report 1999–001. U.S. Department of the Interior, U.S. Geological Survey, Washington, D.C., USA.

FRITZ, H., GUILLEMAIN, M. & DURANT, D. 2002. The cost of vigilance for intake rate in the mallard (*Anas platyrhynchos*): an approach through foraging experiments. *Ethology, Ecology & Evolution* 14: 91–97.

FRITZ, H., GUILLEMAIN, M., & GUÉRIN, S. 2000. Changes in the frequency of prospecting fly-overs by Marsh harriers *Circus aeruginosus* in relation to short-term fluctuations in dabbling duck abundance. *Ardea* 88: 9–16.

FUJIMOTO, Y., ITO, H., SHIVAKOTI, S., NAKAMORI, J., TSUNEKUNI, R., OTSUKI, K. & ITO, T. 2010. Avian influenza virus and paramyxovirus isolation from migratory waterfowl and shorebirds in San-in district of Western Japan from 2001 to 2008. *Journal of Veterinary Medical Science* 72: 963–967.

FURNESS, R. W. 1996. Cadmium in birds. Pp. 389–404 in Beyer, W., Heinz, G. H. & Redmon-Norwood, A. W. (eds), *Environmental Contaminants in Wildlife – Interpreting Tissue Concentrations*. SETAC Special Publications Series, Lewis Publishers, Boca Raton, Florida, USA.

FURNISS, O. C. 1935. The sex ratio in ducks. *The Wilson Bulletin* 47: 277–278.

GAEVSKAYA, N. S. 1966. *The role of higher aquatic plants in the nutrition of the animals of freshwater basins*. Nauka, Moscow, Russia. [In Russian]

GAIDET, N., CATTOLI, G., HAMMOUMI, S., NEWMAN, S. H., HAGEMEIJER, W., TAKEKAWA, J. Y., CAPPELLE, J., DODMAN, T., JOANNIS, T., GIL, P., MONNE, I., FUSARO, A., CAPUA, I., MANU, S., MICHELONI, P., OTTOSSON, U., MSHELBWALA, J. H., LUBROTH, J., DOMENECH, J. & MONICAT, F. 2008. Evidence of infection by H5N2 highly pathogenic avian influenza viruses in healthy wild waterfowl. *PLoS Pathogens* 4: e1000127.

GAILLARD, J. M., PONTIER, D., ALLAINÉ, D., LEBRETON, J. D., TROUVILLIEZ, J. & CLOBERT, J. 1989. An analysis of demographic tactics in birds and mammals. *Oikos* 56: 59–76.

GARDARSSON, A. 1979. Waterfowl populations of Lake Mývatn and recent changes in numbers and food habits. *Oikos* 32: 250–270.

GARDARSSON, A. & EINARSSON, A. 2002. The food relations of the waterbirds of Lake Mývatn, Iceland. *Verhandlungen des Internationalen Verein Limnologie* 28: 754–763.

GARGIULO, A., SENSALE, M., MARZOCCO, L., FIORETTI, A., MENNA, L. F. & DIPINETO, L. 2011. *Campylobacter jejuni, Campylobacter coli*, and cytolethal distending toxin (CDT) genes in common teals (*Anas crecca*). *Veterinary microbiology* 150: 401–404.

GARRETTSON, P. R. & ROHWER, F. C. 2001. Effects of mammalian predator removal on production of upland-nesting ducks in North Dakota. *Journal of Wildlife Management* 65: 398–405.

GARRISON, D. A., FEDYNICH, A. M., SMITH, A. J., FERRO, P. J., BUTLER, D. A., PETERSON, M. J. & LUPIANI, B. 2011. Ingestion of lead and nontoxic shot by Green-winged Teal (*Anas crecca*) and Northern Shovelers (*Anas clypeata*) from the Mid-Gulf Coast of Texas, USA. *Journal of Wildlife Diseases* 47: 784–786.

GASTON, G. R. 1992. Green-winged Teal ingest epibenthic meiofauna. *Estuaries* 15: 227–229.

GASTON, G. R. & NASCI, J. C. 1994. Behavioral ecology of two teal species (Blue-winged Teal, *Anas discors*, and Green-winged Teal, *Anas crecca*) overwintering in marshes of coastal Louisiana, USA. *Gulf Research Reports* 9: 39–48.

GAUTHIER-CLERC, M., LEBARBENCHON, C. & THOMAS, F. 2007. Recent expansion of highly pathogenic Avian Influenza H5N1: a critical review. *Ibis* 149: 202–214.

GAUTHREAUX, S. A. JR. 1978. The ecological significance of behavioral dominance. In Bateson, P. P. G. & Klopfer, P. H. (eds), *Perspectives in Ethology*. Vol. 3. Social Behaviour: 17–54. Plenum Press, New York, USA & London, UK.

GENDRON, M. H. & SMITH, A. C. 2012. National Harvest Survey website. Bird Populations Monitoring, National Wildlife Research Centre, Canadian Wildlife Service, Ottawa, Ontario. Accessed 11 April 2013.

GERASIMOV, Y. N. & GERASIMOV, N. N. 1990. Anseriformes in hunters' bags in Kamchatka. Pp. 118–119 in Matthews, G. V. T. (ed), *Managing Waterfowl Populations. Proceedings of an IWRB symposium, Astrakhan, USSR, 2–5 October 1989*. IWRB Special Publication 12, IWRB, Slimbridge, UK.

GERASIMOV, Y. N., SAL'NIKOV, G. M. & BUSLAEV, S. V. 2000. *Birds of the Ivanovo Region*. Kamchatka Institute of Ecology and Land Management, Moscow, Russia. [In Russian]

GERMUNDSSON, A., MADSLIEN, K. I., HJORTAAS, M. J., HANDELAND, K. & JONASSEN, C. M. 2010. Prevalence and subtypes of Influenza A viruses in wild waterfowl in Norway 2006–2007. *Acta Veterinaria Scandinavica* 52: 28.

GERSTENBERGER, S. L. 2004. Mercury concentrations in migratory waterfowl harvested from Southern Nevada Wildlife Management Areas, USA. *Environmental Toxicology* 19: 35–44.

GESNER, C. 1555. *Historiae animalium: Liber III qui est de avium natura*. Aput Christoph. Froschoverum.

GILBURN, A. S. & KIRBY, J. S. 1992. *Winter Status, Distribution and Habitat Use by Teal in the United Kingdom*. Report to the Joint Nature Conservation Committee, Wildfowl and Wetlands Trust, Slimbridge, UK.

GILG, O. & YOCCOZ, N. G. 2010. Explaining bird migration. *Science* 327: 276–277.

GILL, F. & DONSKER, D. (eds). 2012. *IOC World Bird Names (v 3.1)*. Accessed on 25 September 2012 at: http://www.worldbirdnames.org

GIMÉNEZ-ANAYA, A., HERRERO, J., ROSELL, C., COUTO, S. & GARCÍA-SERRANO, A. 2008. Food habits of wild boars (*Sus scrofa*) in a Mediterranean coastal wetland. *Wetlands* 28: 197–203.

GIRAUDEAU, M., DUVAL, C., GUILLON, N., BRETAGNOLLE, V., GUTTIERREZ, C. & HEEB, P. 2010a. Effects of access to preen gland secretions on mallard plumage. *Naturwissenschaften* 97: 577–581.

GIRAUDEAU, M., DUVAL, C., CZIRJÁK, G. Á., BRETAGNOLLE, V., ERAUD, C., McGRAW, K. J. & HEEB, P. 2010b. Maternal investment of female mallards is influenced by male carotenoid-based coloration. *Proceedings of the Royal Society B* 278: 781–788.

GITAY, H., FOX, A. D. & RIDGILL, S. C. 1990. Survival estimates of Teal (*Anas crecca*) ringed at three stations in Britain. *Ring*, 13: 45–58.

GLASGOW, L. L. & BARDWELL, J. L. 1962. Pintail and teal foods in south Louisiana. *Proceedings of the Annual Conference of the Southeast Association of Fish and Wildlife Agencies* 16: 175–184.

GLOUTNFY, M. L. & CLARK, R. G. 1991. The significance of body mass to female dabbling ducks during late incubation. *The Condor* 93: 811–816.

GLOUTNEY, M. L., CLARK, R. G., AFTON, A. L. & HUFF, G. J. 1993. Timing of nest searches for upland nesting waterfowl. *Journal of Wildlife Management* 57: 597–601.

GOLLOP, J. B. & MARSHALL, W. H. 1954. *A Guide for Aging Duck Breeds in the Field*. Mississippi Flyway Council Technical Section, USA.

GÓMEZ, G., BAOS, R., GÓMARA, B., JIMÉNEZ, B., BENITO, V., MONTORO, R., HIRALDO, F. & GONZÁLEZ, M. J. 2004. Influence of a mine tailing accident near Doñana National Park (Spain) on heavy metals and arsenic accumulation in 14 species of waterfowl (1998 to 2000). *Archives of Environmental Contamination and Toxicology* 47: 521–529.

GONZALES, J., DÜTTMANN, H. & WINK, M. 2009. Phylogenetic relationships based on two mitochondrial genes and hybridization patterns in Anatidae. *Journal of Zoology* 279: 310–318.

GOODBURN, S. F. 1984. Mate guarding in the Mallard *Anas platyrhynchos*. *Ornis Scandinavica* 15: 261–265.

GOODMAN, S. M., MEININGER, P. L., BAHA EL DIN, S. M., HOBBS, J. J. & MULLIÉ, W. C. 1989. *The Birds of Egypt*. Oxford University Press, Oxford, UK.

GORMAN, O. T., BEAN, W. J., KAWAOKA, Y., DONATELLI, I., GUO, Y. J. & WEBSTER, R. G. 1991. Evolution of influenza-A virus nucleoprotein genes – Implications for the origins of H1N1 human and classical swine viruses. *Journal of Virology* 65: 3704–3714.

GOYON DEMONTEIL, M. C. 2004. Examen du contenu stomacal des canards sauvages de la Dombes: conséquences pour la gestion floristique des étangs. Veterinary Thesis, Ecole Nationale Vétérinaire de Lyon, University Claude Bernard, Lyon, France.

GREEN, A. J. & FIGUEROLA, J. 2005. Recent advances in the study of long-distance dispersal of aquatic invertebrates via birds. *Diversity and Distributions* 11: 149–156.

GREEN, A. J., FIGUEROLA, J. & SÁNCHEZ, M. I. 2002. Implications of waterbird ecology for the dispersal of aquatic organisms. *Acta Oecologica* 23: 177–189.

GREEN, A. J., FRISCH, D., MICHOT, T. C., ALLAIN, L. K. & BARROW, W. C. 2013. Endozoochory of seeds and invertebrates by migratory waterbirds in Oklahoma, USA. *Limnetica* 32: 39–46.

GREEN, A. J., GEORGIEV, B. B., BROCHET, A. L., GAUTHIER-CLERC, M., FRITZ, H. & GUILLEMAIN, M. 2011. Determinants of the prevalence of the cloacal cestode *Cloacotaenia megalops* in teal wintering in the French Camargue. *European Journal of Wildlife Research* 57: 275–281.

GREENWOOD, R. J. & SOVADA, M. A. 1996. Prairie duck populations and predation management. *Transactions of the North American Wildlife and Natural Resources Conference* 61: 31–42.

GREENWOOD, R. J., SARGEANT, A. B., JOHNSON, D. H., COWARDIN, L. M. & SHAFFER, T. L. 1995. Factors associated with duck nest success in the prairie pothole region of Canada. *Wildlife Monographs* 128: 1–57.

GRENQUIST, P. 1965. Changes in abundance of some duck and sea-bird populations off the coast of Finland 1949–1963. *Finnish Game Research* 27: 114pp.

GRINNELL, J., BRYANT, H. C. & STORER, T. I. 1918. *The Game Birds of California*. University of California Press, Berkeley, USA.

GROUPE D'ETUDES EN ECOLOGIE APPLIQUÉE 1991. *Etude des conditions de l'hivernage des Sarcelles d'hiver sur le Bassin d'Arcachon*. Rapport de synthèse, Fédération Départementale des Chasseurs de la Gironde et Association de Chasse Maritime du Bassin d'Arcachon, Ludon Medoc, France.

GUILLEMAIN, M. 2011. Determinants of bird ring return: a questionnaire to duck hunters. *Wildlife Biology* 16: 440–444.

GUILLEMAIN, M. & FRITZ, H. 2002. Temporal variation in feeding tactics: exploring the role of competition and predators in wintering dabbling ducks. *Wildlife Biology* 8: 81–90.

GUILLEMAIN, M., DUNCAN, P. & FRITZ, H. 2001. Switching to a feeding method that obstructs vision increases head-up vigilance in dabbling ducks. *Journal of Avian Biology* 32: 345–350.

GUILLEMAIN, M., FOUQUE, C. & FIGUEROLA, J. 2012a. Consistent opposition between eyelid and iris brightness supports a role for vigilance signalling in ducks. *Ibis* 154: 461–467.

GUILLEMAIN, M., FRITZ, H. & DUNCAN, P. 2002a. The importance of protected areas as nocturnal feeding grounds for dabbling ducks wintering in western France. *Biological Conservation* 103: 183–198.

GUILLEMAIN, M., FRITZ, H. & DUNCAN, P. 2002b. Foraging strategies of granivorous dabbling ducks wintering in protected areas of the French Atlantic coast. *Biodiversity and Conservation* 11: 1721–1732.

GUILLEMAIN, M., HOUTE, S. & FRITZ, H. 2000a. Activities and food resources of wintering Teal (*Anas crecca*) in a diurnal feeding site: a case study in Western France. *Revue d'Ecologie (Terre et Vie)* 55: 1–11.

GUILLEMAIN, M., FRITZ, H. & GUILLON, N. 2000b. Foraging behavior and habitat choice of wintering Northern Shoveler in a major wintering quarter in France. *Waterbirds* 23: 355–364.

GUILLEMAIN, M., MARTIN, G. R. & FRITZ, H. 2002c. Feeding methods, visual fields and vigilance in dabbling ducks (Anatidae). *Functional Ecology* 16: 522–529.

GUILLEMAIN, M., PRADEL, R. & GAUTHIER-CLERC, M. 2012b. *Can demographic heterogeneity explain Teal population growth rates?* Poster, 3rd Pan-European Duck Symposium, Jindřichův Hradec, Czech Republic, April 2012.

GUILLEMAIN, M., SADOUL, N. & SIMON, G. 2005a. European flyway permeability and abmigration in Teal *Anas crecca*, an analysis based on ringing recoveries. *Ibis* 147: 688–696.

GUILLEMAIN, M., BLANC, R., LUCAS, C. & LEPLEY, M. 2007e. Ecotourism disturbance to wildfowl in protected areas: historical, empirical and experimental approaches in the Camargue, Southern France. *Biodiversity and Conservation* 16: 3633–3651.

GUILLEMAIN, M., DEHORTER, O., JOHNSON, A. R. & SIMON, G. 2005b. A test of the wintering strategy hypothesis with teal (*Anas crecca*) ringed in the Camargue, southern France. *Journal of Ornithology* 146: 184–187.

GUILLEMAIN, M., FRITZ, H., GUILLON, N. & SIMON, G. 2002d. Ecomorphology and coexistence in dabbling ducks: the role of lamellar density and body length in winter. *Oikos* 98: 547–551.

GUILLEMAIN, M., FRITZ, H., JOHNSON, A. R. & SIMON, G. 2007f. What type of lean ducks do hunters kill? Weakest local ones rather than migrants. *Wildlife Biology* 13: 102–107.

GUILLEMAIN, M., LEPLEY, M., FRITZ, H. & HECKER, N. 2007d. Marsh Harriers *Circus aeruginosus* target Teals *Anas crecca* at roosts according to potential availability of vulnerable prey not total availability. *Bird Study* 54: 268–270.

GUILLEMAIN, M., MONDAIN-MONVAL, J. Y., JOHNSON, A. R. & SIMON, G. 2005c. Long-term climatic trend and body size variation in teal *Anas crecca*. *Wildlife Biology* 11: 81–88.

GUILLEMAIN, M., VAN WILGENBURG, S. L., LEGAGNEUX, P. & HOBSON, K. A. 2014. Assessing geographic origins of Teal through analysis of stable-hydrogen (δ^2H) isotopes and ring-recoveries. *Journal of Ornithology* 155: 165–172.

GUILLEMAIN, M., ELMBERG, J., ARZEL, C., JOHNSON, A. R. & SIMON, G. 2008b. The income-capital breeding dichotomy revisited: late winter body condition is related to breeding success in an income breeder. *Ibis* 150: 172–176.

GUILLEMAIN, M., FUSTER, J., LEPLEY, M., MOURONVAL, J. B. & MASSEZ, G. 2009c. Winter site fidelity is higher than expected for Eurasian Teal *Anas crecca* in the Camargue, France. *Bird Study* 56: 272–275.

GUILLEMAIN, M., MONDAIN-MONVAL, J. Y., WEISSENBACHER, E., BROCHET, A. L. & OLIVIER. A. 2008a. Hunting bag and distance from nearest day-roost in Camargue ducks. *Wildlife Biology* 14: 379–385.

GUILLEMAIN, M., ARZEL, C., MONDAIN-MONVAL, J. Y., SCHRICKE, V., JOHNSON, A. R. & SIMON, G. 2006. Spring migration dates of teal *Anas crecca* ringed in the Camargue, southern France. *Wildlife Biology* 12: 163–169.

GUILLEMAIN, M., DEVINEAU, O., LEBRETON, J. D., MONDAIN-MONVAL, J. Y., JOHNSON, A. R. & SIMON, G. 2007c. Lead shot and teal (*Anas crecca*) in the Camargue, Southern France: effects of embedded and ingested pellets on survival. *Biological Conservation* 137: 567–576.

GUILLEMAIN, M., HEARN, R., KING, R., GAUTHIER-CLERC, M., SIMON, G. & CAIZERGUES, A. 2009a. Comparing migration of Teal from two main wintering areas of Western Europe: a long term study from Essex, England, and Camargue, France. *Ringing and Migration* 24: 273–276.

GUILLEMAIN, M., HEARN, R., KING, R., GAUTHIER-CLERC, M., SIMON, G. & CAIZERGUES, A. 2009b. Differential migration of the sexes cannot be explained by the body size hypothesis in Teal. *Journal of Ornithology* 150: 685–689.

GUILLEMAIN, M., DEVINEAU, O., BROCHET, A. L., FUSTER, J., FRITZ, H., GREEN, A. J. & GAUTHIER-CLERC, M. 2010c. What is the spatial unit for a wintering teal *Anas crecca*? Weekly day roost fidelity inferred from nasal saddles in the Camargue, Southern France. *Wildlife Biology* 16: 215–220.

GUILLEMAIN, M., DEVINEAU, O., GAUTHIER-CLERC, M., HEARN, R., KING, R., SIMON, G. & GRANTHAM, M. 2011. Changes in ring recovery rates over the last 50 years: shall we continue to ring ducks? *Journal of Ornithology* 152: 55–61.

GUILLEMAIN, M., ELMBERG, J., GAUTHIER-CLERC, M., MASSEZ, G., HEARN, R., CHAMPAGNON, J. & SIMON, G. 2010a. Wintering French Mallard and Teal are heavier and in better body condition than 30 years ago: effects of a changing environment? *Ambio* 39: 170–180.

GUILLEMAIN, M., BERTOUT, J. M., CHRISTENSEN, T. K., PÖYSÄ, H., VÄÄNÄNEN, V. M., TRIPLET, P., SCHRICKE, V. & FOX, A. D. 2010b. How many juvenile Teal *Anas crecca* reach the wintering grounds? Flyway-scale survival rate inferred from wing age-ratios. *Journal of Ornithology* 151: 51–60.

GUILLEMAIN, M., FOX, A. D., PÖYSÄ, H., VÄÄNÄNEN, V. M., CHRISTENSEN, T. K., TRIPLET, P., SCHRICKE, V. & KORNER-NIEVERGELT, F. 2013a. Autumn survival inferred from wing age ratios: Wigeon juvenile survival half that of adults at best? *Journal of Ornithology* 154: 351–358.

GUILLEMAIN, M., ARZEL, C., LEGAGNEUX, P., ELMBERG, J., FRITZ, H., LEPLEY, M., PIN, C., ARNAUD, A. & MASSEZ, G. 2007b. Predation risk constrains the plasticity of foraging behaviour in teals, *Anas crecca*: a flyway-level circumannual approach. *Animal Behaviour* 73: 845–854.

GUILLEMAIN, M., POISBLEAU, M., DENONFOUX, L., LEPLEY, M., MOREAU, C., MASSEZ, G., LERAY, G., CAIZERGUES, A., ARZEL, C., RODRIGUES, D. & FRITZ, H. 2007a. Multiple tests of the effect of nasal saddles on dabbling ducks: combining field and aviary approaches. *Bird Study* 54: 35–45.

GUILLEMAIN, M., PÖYSÄ, H., FOX, A. D., ARZEL, C., DESSBORN, L., EKROOS, J., GUNNARSSON, G., HOLM, T. E., CHRISTENSEN, T. K., LEHIKOINEN, A., MITCHELL, C., RINTALA, J. & MØLLER, A. P. 2013b. Effects of climate change on European ducks: what do we know and what do we need to know? *Wildlife Biology*, 19: 404–419.

GUNNARSSON, G. & ELMBERG, J. 2008. Density-dependent nest predation – an experiment with simulated Mallard nests in contrasting landscapes. *Ibis* 150: 259–269.

GUNNARSSON, G., ELMBERG, J. & WALDENSTRÖM, J. 2011. Trends in body mass of ducks over time: the hypotheses in Guillemain *et al.* revisited. *Ambio* 40: 338–340.

GUNNARSSON, G., ELMBERG, J., SJÖBERG, K., PÖYSÄ, H. & NUMMI, P. 2004. Why are there so many empty lakes? Food limits survival of mallard ducklings. *Canadian Journal of Zoology* 82: 1698–1703.

GUNNARSSON, G. ELMBERG, E., PÖYSÄ, H., NUMMI, P., SJÖBERG, K., DESSBORN, L. & ARZEL, C. 2013. Density dependence in ducks: a review of the evidence. *European Journal of Wildlife Research* 59: 305–321.

GURD, D. B. 2005. The ecology of adaptive radiation of dabbling ducks (*Anas* spp.). PhD Thesis, Simon Fraser University, British Columbia, Canada.

GURD, D. B. 2007. Predicting resource partitioning and community organization of filter-feeding dabbling ducks from functional morphology. *The American Naturalist* 169: 334–343.

HAAPANEN, A. & NILSSON, L. 1979. Breeding waterfowl populations in northern Fennoscandia. *Ornis Scandinavica* 10: 145–219.

HANSON, B. A., SWAYNE, D. E., SENNE, D. A., LOBPRIES, D. S., HURST, J. & STALLKNECHT, D. E. 2005. Avian Influenza Viruses and Paramyxoviruses in Wintering and Resident Ducks in Texas. *Journal of Wildlife Diseases* 41: 624–628.

HARIO, M., HOLLMÉN, T. E., MORELLI, T. L. & SCRIBNER, K. T. 2002. Effects of mate removal on the fecundity of common eider *Somateria mollissima* females. *Wildlife Biology* 8: 161–168.

HARRISON, C. 1975. *A Field Guide to the Nests, Eggs and Nestlings of European Birds with North Africa and the Middle East.* W. Collins Sons & Co. Ltd., Glasgow, UK.

HARRISON, C. 1978. *A Field Guide to the Nests, Eggs and Nestlings of North American Birds.* W. Collins Sons & Co. Ltd., Glasgow, UK.

HARRISON, J. M. & HARRISON, J. G. 1971. Notes on a further Pintail x Teal hybrid. *Bulletin of the British Ornithologists' Club* 91: 28–32.

HARRISON, J. M. & HUDSON, M. 1964. Some effects of severe weather on wildfowl in Kent in 1962–63. *Wildfowl Trust Annual Report* 15: 26–32.

HEATON, A. 2001. *Duck decoys.* Shire Publications Ltd, Princes Risborough, UK.

HEGSETH, M. N., CAMUS, L., HELGASON, L. B., BOCCHETTI, R., GABRIELSEN, G. W. & REGOLI, F. 2011. Hepatic antioxidant responses related to levels of PCBs and metal in chicks of three Arctic seabird species. *Comparative Biochemistry and Physiology, Part C* 154: 28–35.

HEINROTH, O. & HEINROTH, M. 1923. *Die Vögel Mitteleuropas in allen Lebens- und Entwicklungsstufen photographisch aufgenommen und in ihrem Seelenleben bei der Aufzucht vom Ei ab beobachtet. Band III.* H. Bermühler, Berlin, Germany.

HEINZ, G. H. 1996. Selenium in birds. Pp. 447–458 in Beyer, W. N., Heinz, G. H. & Redmon-Norwood, A. W. (eds), *Environmental Contaminants in Wildlife – Interpreting Tissue Concentrations.* SETAC Special Publications Series, Lewis Publishers, Boca Raton, Florida, USA.

HEITMEYER, M. E. 1995. Influences of age, body condition, and structural size on mate selection by dabbling ducks. *Canadian Journal of Zoology* 73: 2251–2258.

HEITMEYER, M. E. & VOHS, P. A. JR. 1984. Characteristics of wetlands used by migrant dabbling ducks in Oklahoma, USA. *Wildfowl* 35: 61–70.

HEITMEYER, M. E., FREDRICKSON, L. H. & HUMBURG, D. D. 1993. Further evidence of biases associated with hunter-killed Mallards. *Journal of Wildlife Management* 57: 733–740.

HÉMERY, G., HOUTSA, F., NICOLAU-GUILLAUMET, P. & ROUX, F. 1979. Distribution géographique, importance et evolution numériques des effectifs d'Anatidés et de Foulques hivernant en France (janvier 1967 à 1976). *Bulletin Mensuel de l'Office National de la Chasse*, Numéro Spécial scientifique et technique Mai 79 : 5–91.

HÉNAUX, V. & SAMUEL, M. D. 2011. Avian influenza shedding patterns in waterfowl: implications for surveillance, environmental transmission, and disease spread. *Journal of Wildlife Diseases* 47: 566–578.

HENNY, C. J., HALLOCK, R. J. & HILL, E. F. 1994. Cyanide and migratory birds at gold mines in Nevada, USA. *Ecotoxicology* 3: 45–58.

HEPP, G. R. 1985. Effects of environmental parameters on the foraging behaviour of three species of wintering dabbling ducks (Anatini). *Canadian Journal of Zoology* 63: 289–294.

HEPP, G. R. 1986. Effects of body weight and age on the time of pairing of American Black Ducks. *The Auk* 103: 477–484.

HEPP, G. R. 1989. Benefits, costs, and determinants of dominance in American Black Ducks. *Behaviour* 109: 222–234.

HEPP, G. R. & HAIR, J. D. 1983. Reproductive behaviour and pairing chronology in wintering dabbling ducks. *Wilson Bulletin* 95: 675–682.

HEPP, G. R. & HAIR, J. D. 1984. Dominance in wintering waterfowl (Anatini): effects on distribution of sexes. *The Condor* 86: 251–257.

HERMAN, C. M. 1965. The occurrence of Protozoan blood parasites in Anatidae. Pp. 341–349 in Blank, T. H. (ed), *International Union of Game Biologists – Transactions of the VIth congress*. The Nature Conservancy, London, UK.

HESSELMAN, H. L. 1897. Some observations upon the spread of plant. *Botaniska Notiser* 97: 97–112. [In Swedish]

HEUSMANN, H. W. 1999. Let's get rid of the midwinter waterfowl inventory in the Atlantic Flyway. *Wildlife Society Bulletin* 27: 552–558.

HICKEY, J. J. 1952. *Survival Studies of Banded Birds*. U.S. Fish and Wildlife Service Special Scientific Report – Wildlife 15. Washington, D.C., USA.

HIGGINS, K. F., KIRSCH, L. M., KLETT, A. T. & MILLER, H. W. 1992. *Waterfowl Production on the Woodworth Station in South-central North Dakota, 1965–1981*. U.S. Fish and Wildlife Service, Resource Publication 180.

HIGGINSON, A. D., McNAMARA, J. M. & HOUSTON, A. I. 2012. The starvation-predation trade-off predicts trends in body size, muscularity, and adiposity between and within taxa. *The American Naturalist* 179: 338–350.

HILDÉN, O. & SAUROLA, P. 1982. Speed of autumn migration of birds ringed in Finland. *Ornis Fennica* 59: 140–143.

HILL, D., WRIGHT, R. & STREET, M. 1987. Survival of Mallard ducklings *Anas platyrhynchos* and competition with fish for invertebrates on a flooded gravel quarry in England. *Ibis* 129: 159–167.

HIRONS, G. & THOMAS, G. 1993. Disturbance on estuaries: RSPB nature reserve experience. In Davidson, N. & Rothwell, P. (eds), *Disturbance to Waterfowl on Estuaries. Wader Study Group Bulletin* 68 Special Issue: 72–78.

HIRSCHFELD, A. & HEYD, A. 2005. Mortality of migratory birds caused by hunting in Europe: bag statistics and proposals for the conservation of birds and animal welfare. *Berichte zum Vogelschutz* 42: 47–74.

HOBSON, K. A. & CLARK, R. G. 1993. Turnover of ^{13}C in cellular and plasma fractions of blood: implications for nondestructive sampling in avian dietary studies. *The Auk* 110: 638–641.

HOBSON, K. A., PIATT, J. F. & PITOCCHELLI, J. 1994. Using stable isotopes to determine seabird trophic relationships. *Journal of Animal Ecology* 63: 786–798.

HOCHBAUM, H. A. 1955. *Travels and Traditions of Waterfowl.* University of Minnesota Press, Minneapolis, USA.

HOEVE, J. & SCOTT, M. E. 1988. Ecological studies on *Cyathocotyle bushiensis* (*Digenea*) and *Sphaeridiotrema globulus* (*Digenea*), possible pathogens of dabbling ducks in Southern Québec. *Journal of Wildlife Diseases* 24: 407–421.

HOFFMANN, D. J., RICE, C. P. & KUBIAK, T. J. 1996. PCBs and Dioxins in birds. Pp. 165–207 in Beyer, W. N., Heinz, G. H. & Redmon-Norwood, A. W. (eds), *Environmental Contaminants in Wildlife – Interpreting Tissue Concentrations.* SETAC Special Publications Series, Lewis Publishers, Boca Raton, Florida, USA.

HOFFMANN, L. 1955. Premier compte-rendu: 1950–54 – Appendice II. Calendrier ornithologique de la Tour du Valat. *Comptes-rendus de la Tour du Valat* 1 : 37–71.

HOFFMANN, L. 1960. Le saturnisme, fléau de la sauvagine en Camargue. *Revue d'Ecologie (Terre et Vie)* 107: 120–131.

HOFFMANN, L. & PENOT, J. 1955. Premier recensement des canards hivernant en Camargue. *Revue d'Ecologie (Terre et Vie)* 9: 315–320.

HOFFMANN, E., STECH, J., LENEVA, I., KRAUSS, S., SCHOLTISSEK, C., CHIN, P. S., PEIRIS, M., SHORTRIDGE, K. F. & WEBSTER, R. G. 2000. Characterization of the influenza A virus gene pool in avian species in southern China: Was H6N1 a derivative or a precursor of H5N1? *Journal of Virology* 74: 6309–6315.

HOHMAN, W. L., ANKNEY, C. D. & GORDON, D. H. 1992. Ecology and management of postbreeding waterfowl. Pp. 128–189 in Batt, B. D., Afton, A. D., Anderson, M. G., Ankney, C. D., Johnson, D. H., Kadlec, J. A. & Krapu, G. L. (eds), *Ecology and Management of Breeding Waterfowl.* University of Minnesota Press, Minneapolis, USA.

HOHMAN, W. L., PRITCHERT, R. D., PACE, R. M., WOOLINGTON, D. W. & HELM, R. 1990. Influence of ingested lead on body mass of wintering Canvasbacks. *Journal of Wildlife Management* 54: 211–215.

HOLMBERG, K., EDSMAN, L. & KLINT, T. 1989. Female mate preferences and male attributes in mallard duck *Anas platyrhynchos. Animal Behaviour* 38: 1–7.

HOLMBOE, J. 1900. Notizen über die endozoische Samenverbreitung der Vögel. *Nyt Magazin für Naturvidenskaberne* 38: 303–320.

HOMES, R. C. 1942. Sex ratios in winter duck flocks. *British Birds* 36: 42–50.

HOPPE, D. M. 1976. Prevalence of macroscopically detectable *Sarcocystis* in North Dakota ducks. *Journal of Wildlife Diseases* 12: 27–29.

HOUHAMDI, M. & SAMRAOUI, B. 2001. Diurnal time budget of wintering Teal *Anas crecca* at Lac des Oiseaux, northeast Algeria. *Wildfowl* 52: 87–96.

HOVETTE, C. 1972. Le saturnisme des Anatidés en Camargue. *Alauda* 150: 1–17.

HOVETTE, C. 1973. Le saturnisme des Anatidés en Camargue. Thèse de Doctorat de l'Université de Provence, Marseille, France.

HOWELL, T. R. 1959. A hybrid of the Pintail and Green-winged Teal. *The Condor* 61: 226–227.

HU, H. & CUI, Y. 1990. The effect of habitat destruction on the waterfowl of lakes in the Yangtze and the Han River basins. Pp. 189–193 in Matthews, G. V. T. (ed), *Managing Waterfowl Populations. Proceedings of an IWRB symposium, Astrakhan, USSR, 2–5 October 1989.* IWRB Secial Publication 12, IWRB, Slimbridge, UK.

HUBALEK, Z. 2000. Keratinophilic fungi associated with free-living mammals and birds. Pp. 93–103 in Kushwaha, R. K. S. & Guarro, J. (eds), *Biology of Dermatophytes and other Keratinophilic Fungi*. Revista Iberoamericana de Micología, Bilbao, Spain.

HUGHES, J. H. & YOUNG, E. L. JR. 1982. Autumn foods of dabbling ducks in Southeastern Alaska. *Journal of Wildlife Management* 46: 259–263.

HUME, A. O. & MARSHALL, C. H. T. 1881. *The Game Birds of India, Burmah and Ceylon*. Calcutta, India.

HUNTLEY, B., GREEN, R. E., COLLINGHAM, Y. C. & WILLIS, S. G. 2007. *A Climatic Atlas of European Breeding Birds*. Lynx Edicions, Barcelona, Spain.

IMPEKOVEN, M. 1965. Verbreitung und Fluchtmigration von Krickenten in den kalten Wintern 1956 und 1962–63. Eine Vergleichende Analyse von Rückmeldungen in La Tour du Valat (Südfrankreich) beringter Tiere. Pp. 393–298 in Blank, T. H. (ed), *International Union of Game Biologists – Transactions of the VIth Congress*. The Nature Conservancy, London, UK.

ISOLA, C. R., COLWELL, M. A., TAFT, O. W. & SAFRAN, R. J. 2000. Interspecific differences in habitat use of shorebirds and waterfowl foraging in managed wetlands of California's San Joaquin Valley. *Waterbirds* 23: 196–203.

JACOB, J. & ZISWEILER, V. 1982. The uropygial gland. In Farner, D. S., King, J. R. & Parkes, K. C. (eds), *Avian Biology* – Volume 4: 199–324. Academic Press, New York, USA.

JALLAGEAS, M., TAMISIER, A. & ASSENMACHER, I. 1978. Comparative study of the annual cycles in sexual and thyroid function in male Peking Ducks (*Anas platyrhynchos*) and Teal (*Anas crecca*). *General and Comparative Endocrinology* 36: 201–210.

JARVIS, R. L. 1966. Occurrence of European or Aleutian Green-winged Teal in western North America, with a recent record. *The Murrelet* 47: 15–18.

JOHNSEN, A., RINDAL, E., ERICSON, P. G. P., ZUCCON, D., KERR, K. C. R., STOECKLE, M. Y. & LIFJELD, J. T. 2010. DNA barcoding of Scandinavian birds reveals divergent lineages in trans-Atlantic species. *Journal of Ornithology* 151: 565–578.

JOHNSGARD, P. A. 1960. Pair-formation mechanisms in *Anas* (Anatidae) and related genera. *Ibis* 102: 616–618.

JOHNSGARD, P. A. 1965. *Handbook of Waterfowl Behaviour*. Cornell University Press, Ithaca, New York, USA.

JOHNSGARD, P. A. & BUSS, I. O. 1956. Waterfowl sex ratios during spring in Washington state and their interpretation. *Journal of Wildlife Management* 20: 384–388.

JOHNSON, D. H., NICHOLS, J. D. & SCHWARTZ, M. D. 1992. Population dynamics of breeding waterfowl. Pp. 446–185 in Batt, B. D. J., Afton, A. D., Anderson, M. G., Ankney, C. D., Johnson, D. H., Kadlec, J. A. & Krapu, G. L. (eds), *Ecology and Management of Breeding Waterfowl*. University of Minnesota Press, Minneapolis, USA.

JOHNSON, K. 1995. Green-winged Teal. In Poole, A. & Gill, F. (eds) *The Birds of North America, No. 193*. The Academy of Natural Sciences, Philadelphia, and The American Ornithologists' Union, Washington, D.C., USA.

JOHNSON, K. P. 2000. The evolution of courtship display repertoire size in the dabbling ducks (Anatini). *Journal of Evolutionary Biology* 13: 634–644.

JOHNSON, K. P. & SORENSON, M. D. 1999. Phylogeny and biogeography of dabbling ducks (Genus: *Anas*): a comparison of molecular and morphological evidence. *The Auk* 116: 792–805.

JOHNSON, K. P., McKINNEY, F., WILSON, R. & SORENSON, M. D. 2000. The evolution of postcopulatory displays in dabbling ducks (Anatini): a phylogenetic perspective. *Animal Behaviour* 59: 953–963.

JOHNSON, L. G. & MORRIS, R. L. 1971. Pesticide and mercury levels in migrating duck populations. *Bulletin of Environmental Contamination and Toxicology* 6: 513–516.

JOHNSON, W. P. & ROHWER, F. C. 1996. Disturbance of wintering Green-winged Teal and Mallards by raptors. *The Southwestern Naturalist* 41: 331–334.

JOHNSON, W. P. & ROHWER, F. C. 1998. Pairing chronology and agonistic behaviors of wintering Green-winged Teal and Mallards. *Wilson Bulletin* 110: 311–315.

JOHNSON, W. P. & ROHWER, F. C. 2000. Foraging behavior of Green-winged Teal and Mallards on tidal mudflats in Louisiana. *Wetlands* 20: 184–188.

JOHRI, G. N. 1960. A new paruterinid cestode, *Lallum magniparuterina* gen. et sp. nov. from the intestine of a common teal, *Nettion crecca* Linn. *Parasitology* 50: 269–272.

JONSSON, J. E. & GARDARSSON, A. 2001. Pair formation in relation to climate: Mallard, Eurasian Wigeon and Eurasian Teal wintering in Iceland. *Wildfowl* 52: 55–68.

JORDE, D. G., KRAPU, G. L. & CRAWFORD, R. D. 1983. Feeding ecology of Mallards wintering in Nebraska. *Journal of Wildlife Management* 47: 1044–1053.

KALETA, E. F. & TADAY, E. M. A. 2003. Avian host range of *Chlamydophila* spp. based on isolation, antigen detection and serology. *Avian Pathology* 32: 435–462.

KALMBACH, E. R. 1937. Crow-waterfowl relationships based on preliminary studies on Canadian breeding grounds. *United States Department of Agriculture Circular* No. 433. Washington, D.C., USA.

KARELSE, D. 1994. Duck decoys in The Netherlands. *Wildfowl* 45: 260–266.

KARMIRIS, I., KAZANTZIDIS, S. & PAPACHRISTOU, T. G. 2010. Variation in diet composition of wintering waterfowl among Greek wetlands. *Avocetta* 34: 21–28.

KASHKAROV, D. Y. 1987. *Birds of Uzbekistan*. Fan Press, Tashkent, Uzbekistan. [In Russian]

KAVETSKA, K. M., RZAD, I. & SITKO, J. 2008. Taxonomic structure of Digenea in wild ducks (Anatinae) from west Pomerania. *Wiadomosci Parazytologiczne* 54: 131–136.

KAVETSKA, K. M., KRÓLACZYK, K., STAPF, A., GRZESIAK, W., KALISINSKA, E. & PILARCZYK, B. 2011. Revision of the species complex *Amidostomum acutum* (Lundahl, 1848) (Nematoda: Amidostomatidae). *Parasitological Research* 109: 105–117.

KEAR, J. 1965. The internal food reserves of hatching Mallard ducklings. *Journal of Wildlife Management* 29: 523–528.

KEAR, J. 1990. *Man and Wildfowl.* A. & C. Black Publishers Limited, London, UK.

KEAR, J. (ed). 2005. *Ducks, Geese and Swans.* Oxford University Press, Oxford, UK.

KEITH, L. B. 1961. A study of waterfowl ecology on small impoundments in Southeastern Alberta. *Wildlife Monographs* 6: 88pp.

KELLER, V. 1996. Effects and management of disturbance of waterbirds by human recreational activities: a review. *Gibier Faune Sauvage / Game & Wildlife* 13: 1039–1047.

KENNEY, D. 1993. Common Buzzard taking Common Teal in flight. *British Birds* 86: 625.

KENYON, K. W. 1961. Birds of Amchitka Island, Alaska. *The Auk* 78: 305–326.

KETTERSON, E. D. & NOLAN, V. JR. 1976. Geographic variation and its climatic correlates in the sex ratio of eastern-wintering Dark-eyed Juncos (*Junco hyemalis hyemalis*). *Ecology* 57: 679–693.

KERWIN, J. A. & WEBB, L. G. 1971. Foods of ducks wintering in coastal South Carolina, 1965–1967. *Proceedings of the Annual Conference of the Southeastern Association of Fish and Wildlife Agencies* 25: 223–245.

KIM, J. K., NEGOVETICH, N. J., FORREST, H. L. & WEBSTER, R. G. 2009. Ducks: the "Trojan Horses" of H5N1 Influenza. *Influenza and other Respiratory Viruses* 3: 121–128.

KIRBY, J. S., SALMON, D. G., ATKINSON-WILLES, G. L. & CRANSWICK, P. A. 1995. Index numbers for waterbird populations. III. Long-term trends in the abundance of wintering wildfowl in Great Britain, 1966/67–1991/92. *Journal of Applied Ecology* 32: 536–551.

KRIVONOSOV, G. A. & RUSANOV, G. M. 1990. Wintering waterfowl in the Northern Caspian. Pp. 27–31 in Matthews, G. V. T. (ed), *Managing Waterfowl Populations. Proceedings of an IWRB symposium, Astrakhan, USSR, 2–5 October 1989*. IWRB Special Publication 12, IWRB, Slimbridge, UK.

KLAASSEN, M. 2002. Relationships between migration and breeding strategies in arctic breeding birds. Pp. 237–249 in Berthold, P., Gwinner, E. & Sonnenschein, E. (eds), *Avian Migration*. Springer-Verlag, Berlin, Germany.

KLAASSEN, R. H. G., STRANDBERG, R., HAKE, M., OLOFSSON, P., TØTTRUP, A. P. & ALERSTAM, T. 2010. Loop migration in adult marsh harriers *Circus aeruginosus*, as revealed by satellite telemetry. *Journal of Avian Biology* 41: 200–207.

KLEIN, M. L., HUMPHREY, S. R. & PERCIVAL, H. F. 1995. Effects of ecotourism on distribution of waterbirds in a wildlife refuge. *Conservation Biology* 9: 1454–1465.

KOCAN, A. A., HINSHAW, V. S. & DAUBNEY, G. A. 1980. Influenza A viruses isolated from migrating ducks in Oklahoma. *Journal of Wildlife Diseases* 16: 281–285.

KOCAN, A. A., SHAW, M. G. & MORGAN, P. M. 1979. Some parasitic and infectious diseases in waterfowl in Oklahoma. *Journal of Wildlife Diseases* 15: 137–141.

KOKKO, H., GUNNARSSON, T. G., MORREL, L. J. & GILL, J. A. 2006. Why do female migratory birds arrive later than males? *Journal of Animal Ecology* 75: 1293–1303.

KOOLOOS, J. G. M. & ZWEERS, G. A. 1991. Integration of pecking, filter feeding and drinking mechanisms in waterfowl. *Acta Biotheoretica* 39:107–140.

KORTEGAARD, L. 1974. An ecological outline of a moulting area of Teal, Vejkerne, Denmark. *Wildfowl* 25: 134–142.

KORTRIGHT, F. H. 1943. *The Ducks, Geese and Swans of North America – A vade mecum for the Naturalist and the Sportsman*. The American Wildlife Institute, Washington, D.C.

KOSKIMIES, J. & LAHTI, L. 1964. Cold-hardiness of the newly hatched young in relation to ecology and distribution in ten species of European ducks. *The Auk* 81: 281–307.

KRAMER, G. W., CARRERA, E. & ZAVELETA, D. 1995. Waterfowl harvest and hunter activity in Mexico. *Transactions of the North American Wildlife and Natural Resources Conference* 60: 243–250.

KRAPU, G. L. & REINECKE, K. J. 1992. Foraging ecology and nutrition. Pp. 1–29 in Batt, B. D. J., Afton, A. D., Anderson, M. G., Ankney, C. D., Johnson, D. H., Kadlec, J. A. & Krapu, G. L. (eds), *Ecology and Management of Breeding Waterfowl*. University of Minnesota Press, Minneapolis, USA.

KRAPU, G. L., PIETZ, P. J., BRANDT, D. A. & COX, R. R. 2000. Factors limiting Mallard brood survival in prairie pothole landscapes. *Journal of Wildlife Management* 64: 553–561.

KRAUS, R. H. S., VAN HOOFT, P., MEGENS, H-J. TSVEY, A., FOKIN, S. Y., YDENBERG, R. C. & PRINS, H. H. T. 2013. Global lack of flyway structure in a cosmopolitan bird revealed by a genome wide survey of single nucleotide polymorphisms. *Molecular Ecology* 22: 41–55.

KRAUS, R. H. S., ZEDDEMAN, A., VAN HOOFT, P., SARTAKOV, D., SOLOVIEV, S. A., YDENBERG, R. C. & PRINS, H. H. T. 2011. Evolution and connectivity in the world-wide migration system of the mallard: Inferences from mitochondrial DNA. *BMC Genetics* 12: 99.

KRÓLACZYK, K., KAVETSKA, K. M., KALISINSKA, E. & NOWAK, M. R. 2011. *Cloacotaenia megalops* (Nitzsh in Creplin, 1829) (Cestoda, Hymenolepididae) in wild ducks in Western Pomerania, Poland. *Wiadomosci Parazytologiczne* 57: 123–126.

KVIST, A., KLAASSEN, M. & LINDSTRÖM, Å. 1998. Energy expenditure in relation to flight speed: what is the power of mass loss rate estimates? *Journal of Avian Biology* 29: 485–498.

KYDYRALIEV, A. K. 1990. *Birds of lakes and montane rivers of Kirghizia*. Ilim Press, Frunze, Russia. [In Russian]

LACK, D. 1968. *Ecological Adaptations for Breeding in Birds*. Methuen & Co. Ltd., London, UK.

LACK, D. 1974. *Evolution Illustrated by Waterfowl*. Blackwell Scientific Publications, Oxford, UK.

LAKE, B. C., WALKER, J. & LINDBERG, M. S. 2006. Survival of ducks banded in the boreal forest of Alaska. *Journal of Wildlife Management* 70: 443–449.

LAMPIO, T. 1974. Bag limits, sale, export, import and recolonization of waterfowl. *Finnish Game Research* 34: 51–56.

LAMPIO, T. 1982. National and local requirements for regulation of waterfowl shooting pressure. Pp. 293–301 in Scott, D. A. (ed), *Managing Wetlands and their Birds – A Manual of Wetland and Waterfowl Management*. IWRB, Slimbridge, UK.

LANDERS, J. L., JOHNSON, A. S., MORGAN, P. H. & BALDWIN, W. P. 1976. Duck foods in managed tidal impoundments in South Carolina. *Journal of Wildlife Management* 40: 721–728.

LASIEWSKI, R. C. & DAWSON, W. R. 1967. A re-examination of the relation between standard metabolic rate and body weight in birds. *The Condor* 69: 13–23.

LAURIE-AHLBERG, C.C. & McKINNEY, F. 1979. The nod-swim display of male Green-winged Teal (*Anas crecca*). *Animal Behaviour* 27: 165–172.

LEBARBENCHON, C., SREEVATSAN, S., LEFEVRE, T., YANG, M., RAMAKRISHNAN, M. A., BROWN, J. D. & STALLKNECHT, D. E. 2012. Reassortant influenza A viruses in wild duck populations: effects on viral shedding and persistence in water. *Proceedings of the Royal Society B – Biological Sciences* 279: 3967–3975.

LEBARBENCHON, C., ALBESPY, F., BROCHET, A-L., GRANDHOMME, V., RENAUD, F., THOMAS, F., VAN DER WERF, S., AUBRY, P., GUILLEMAIN, M. & GAUTHIER-CLERC, M. 2009. Spread of avian influenza viruses by common teal (*Anas crecca*) in Europe. *PLoS ONE* 4(10): e7289.

LEBRET, T. 1947. The migration of the Teal, *Anas crecca crecca* L., in Western Europe. *Ardea* 35: 79–131.

LEBRET, T. 1958. The "Jump-flight" of the Mallard, *Anas platyrhynchos* L., the Teal, *Anas crecca* L. and the Shoveler, *Spatula clypeata* L. *Ardea* 46: 68–72.

LEBRET, T. 1961. The pair formation in the annual cycle of the Mallard, *Anas platyrhynchos* L. *Ardea* 49: 97–158.

LEBRETON, J. D. 1973. Etude des déplacements saisonniers des sarcelles d'hiver, *Anas c. crecca* L., hivernant en Camargue à l'aide de l'analyse factorielle des correspondances. *Comptes-rendus de l'Académie des Sciences de Paris* 277 : 2417–2420.

LEBRETON, J. D. 2005. Dynamical and statistical models for exploited populations. *Australian & New Zealand Journal of Statistics* 47: 49–63.

LEGAGNEUX P., DUHART M. & SCHRICKE V. 2007. Seeds consumed by waterfowl in winter: a review of methods and a new web-based photographic atlas for seed identification. *Journal of Ornithology* 148: 537–541

LEGAGNEUX, P., BLAIZE, C., LATRAUBE, F., GAUTIER, J. & BRETAGNOLLE, V. 2008. Variation in home-range size and movements of wintering dabbling ducks. *Journal of Ornithology* 150: 183–193.

LEGAGNEUX, P., CLARK, R. G., GUILLEMAIN, M., ERAUD, C., THÉRY, M. & BRETAGNOLLE, V. 2012. Large-scale geographic variation in iridescent structural ornaments of a long-distant migratory bird. *Journal of Avian Biology*, 43: 355–361.

LEHIKOINEN, A. & JAATINEN, K. 2012. Delayed autumn migration in northern European waterfowl. *Journal of Ornithology* 153: 563–570.

LEHMANN, C. 2007. *Update Report on the Use of Non-toxic Shot for Hunting in Wetlands.* UNEP/AEWA Secretariat, accessible from http://unep-aewa.org.

LERCEL, B. A., KAMINSKI, R. M. & COX, R. R. 1999. Mate loss in winter affects reproduction of Mallards. *Journal of Wildlife Management* 63: 621–629.

LI, K. S., GUAN, Y., WANG, J., SMITH, G. J. D., XU, K. M., DUAN, L., RAHARDJO, A. P., PUTHAVATHANA, P., BURANATHAI, C., NGUYEN, T. D., ESTOEPANGESTIE, A. T. S., CHAISINGH, A., AUEWARAKUL, P., LONG, H. T., HANH, N. T. H., WEBBY, R. J., POON, L. L. M., CHEN, H., SHORTRIDGE, K. F., YUEN, K. Y., WEBSTER, R. G. & PEIRIS, J. S. M. 2004. Genesis of a highly pathogenic and potentially pandemic H5N1 influenza virus in eastern Asia. *Nature* 430: 209–213.

LI, Z. W. D. & MUNDKUR, T. 2004. *Numbers and Distribution of Waterbirds and Wetlands in the Asia-Pacific Region. Results of the Asian Waterbird Census: 1997–2001.* Wetlands International, Selangor, Malaysia

LI, Z. W. D., BLOEM, A., DELANY, S., MARTAKIS, G. & QUINTERO, J. O. 2009. *Status of Waterbirds in Asia – Results of the Avian Waterbird Census: 1987–2007.* Wetlands International, Kuala Lumpur, Malaysia.

LIMA, S. L. 1986. Predation risk and unpredictable feeding conditions: determinants of body mass in birds. *Ecology* 67: 377–385.

LIN, B. C., MALANOSKI, A. P., WANG, Z., BLANEY, K. M., LONG, N. C., MEADOR, C. E., METZGAR, D., MYERS, C. A., YINGST, S. L., MONTEVILLE, M. R., SAAD, M. D., SCHNUR, J. M., TIBBETTS, C. & STENGER, D. A. 2009. Universal detection and identification of avian influenza virus by use of resequencing microarrays. *Journal of Clinical Microbiology* 47: 988–993.

LINDBERG, P. 1984. Mercury in feathers of Swedish gyrfalcons, *Falco rusticolus*, in relation to diet. *Bulletin of Environmental Contamination and Toxicology* 32: 453–459.

LINDBERG, P., ODSJÖ, T. & REUTERGÄRDH, L. 1985. Residue levels of polychlorobiphenyls, ΣDDT, and mercury in bird species commonly preyed upon by the Peregrine Falcon (*Falco peregrinus* Tunst.) in Sweden. *Archives of Environmental Contamination and Toxicology* 14: 203–213.

LINKOLA, P. 1962. Notes on the breeding success of ducks in Central Häme. *Suomen Riista* 15 : 157–64. [In Finnish with English summary]

LINNAEI, C. 1758. *Systema naturae per regna tria naturae, secundum classes, ordines, genera, species, cum characteribus, differentiis, synonymis, locis. Tomus I. Edition decima.* Holmie, Impensis Direct, Laurentii Salvii, Stockholm, Sweden.

LIU, J., LUO, X-J., YU, L-H., HE, M-J., CHEN, S-J. & MAI, B-X. 2010. Polybrominated diphenyl ethers (PBDEs), polychlorinated biphenyls (PCBs), hydroxylated and methoxylated-PBDEs, and methylsulfonyl-PCBs in bird serum from South China. *Archives of Environmental Contamination and Toxicology* 59: 492–501.

LLORENTE, G. A., RUIZ, X. & SERRA-COBO, J. 1987. Alimentación otoñal de la cerceta común (*Anas crecca*) en el Delta del Ebro. *Miscellania Zoologica* 11: 319–330. [In Spanish with English summary]

LLORENTE, G. A., FARRAN, A., RUIZ, X. & ALBAIGÉS, J. 1987. Accumulation and distribution of hydrocarbons, polychlorobyphenyls, and DDT in tissues of three species of Anatidae from the Ebro Delta (Spain). *Archives of Environmental Contamination and Toxicology* 16: 563–572.

LONG, P. R., SZÉKELY, T., KERSHAW, M. & O'CONNELLE, M. 2007. Ecological factors and human threats both drive waterfowl population declines. *Animal Conservation* 10: 183–191.

LOOS, E. R. & ROHWER, F. C. 2004. Laying-stage nest attendance and onset of incubation in prairie nesting ducks. *The Auk* 121: 587–599.

LORENZ, K. 1952. Comparative studies on the behaviour of Anatidæ. XIV. The European Teal *Nettion crecca* (L.). *The Avicultural Magazine* 58: 172–175.

LORTET, L. C. E. & GAILLARD, C. 1909. La faune momifiée de l'ancienne Egypte et recherches anthropologiques. *Archives du Muséum d'Histoire Naturelle de Lyon* 10 : 1–336.

LOVVORN, J. R. 1989. Food defendability and antipredator tactics: implications for dominance and pairing in Canvasbacks. *The Condor* 91: 826–836.

LOW, S. H. 1949. Migration of the Green-winged Teal. In Aldrich, J.W. and others (eds), *Migration of some North American Waterfowl. A Progress Report on an Analysis of Banding Records.* U.S. Fish and Wildlife Service Special Scientific Report – Wildlife N0.1. Washington, D.C., USA.

LYDEKKER, R. 1891. On British fossil birds. *Ibis* 33: 381–410.

McCANCH, N. 2012. A pair of Eurasian Teals diving in search of food. *British Birds* 105: 221–223.

McDONALD, M. E. 1969. *Catalogue of Helminths of Waterfowl (Anatidae).* Bureau of Sport Fisheries and Wildlife Special Scientific Report – Wildlife No. 126. Washington, D.C., USA.

McGILVREY, F. B. 1966. Fall food habits of ducks near Santee Refuge, South Carolina. *Journal of Wildlife Management* 30: 577–580.

McILHENNY, E. A. 1940. Sex ratio in wild birds. *The Auk* 57: 85–93.

McKINNEY, F. 1965. The displays of the American Green-Winged Teal. *The Wilson Bulletin* 77: 112–121.

McKINNEY, F. 1985. Primary and secondary male reproductive strategies of dabbling ducks. Pp. 68–82 in Gowaty, P. A. & Mock, D. W. (eds), *Avian Monogamy*. The American Ornithologists' Union (Ornithological Monographs 37), Washington, D.C., USA.

McKINNEY, F. & DERRICKSON, S. 1979. Aerial scratching, leeches and nasal saddles in Green-winged Teal. *Wildfowl* 30: 151–153.

McKINNEY, F. & EVARTS, S. 1998. Sexual coercion in waterfowl and other birds. Pp. 163–195 in Parker, P. G. & Burley, N. T. (eds), *Avian Reproductive Tactics: Female and Male Perspectives*. American Ornithologists' Union (Ornithological Monographs 49), Washington, D.C., USA.

McKINNEY, F. & STOLEN, P. 1982. Extra-pair courtship and forced copulation among captive Green-winged Teal (*Anas crecca carolinensis*). *Animal Behaviour* 30: 461–474.

McKINNEY, F., DERRICKSON, S. R. & MINEAU, P. 1983. Forced copulation in waterfowl. *Behaviour* 86: 250–294.

McKINNON, L., SMITH, P. A., NOL, E., MARTIN, J. L., DOYLE, F. I., ABRAHAM, K. F., GILCHRIST, H. G., MORRISON, R. I. G. & BÊTY, J. 2010. Lower predation risk for migratory birds at high latitudes. *Science* 327: 326–327.

McLANDRESS, M. R. & RAVELING, D. G. 1981. Hyperphagia and social behaviour of Canada geese prior to spring migration. *Wilson Bulletin* 93: 310–324.

McNEIL, R., DRAPEAU, P. & GOSS-CUSTARD, J. D. 1992. The occurrence and adaptive significance of nocturnal habits in waterfowl. *Biological Reviews* 67: 381–419.

MABBOTT, D. C. 1920. *Food Habits of Seven Species of American Shoal-water Ducks*. United States Department of Agriculture Bulletin 862, Washington, D.C., USA.

MACCLUSKIE, M. C. & SEDINGER, J. S. 2000. Nutrient reserves and clutch-size regulation of Northern Shovelers in Alaska. *The Auk* 117: 971–979.

MACKAY, G. H. 1890. Notes on several species of water birds at Muskeget Island, Massachusetts. *The Auk* 7: 294–295.

MADSEN, J. 1988. Autumn feeding ecology of herbivorous wildfowl in the Danish Wadden Sea, and impact of food supplies and shooting on movements. *Danish Review of Game Biology* 13: 1–32.

MADSEN, J. 1998. Experimental refuges for migratory waterfowl in Danish wetlands. II. Tests of hunting disturbance effects. *Journal of Applied Ecology* 35: 398–417.

MADSEN, J. & FOX, A. D. 1995. Impacts of hunting disturbance on waterbirds – a review. *Wildlife Biology* 1: 193–207.

MAGUIRE, B. J. 1963. The passive dispersal of small aquatic organisms and their colonization of isolated bodies of water. *Ecological Monographs* 33: 161–185.

MAINGUY, J., BÊTY, J., GAUTHIER, G. & GIROUX, J. F. 2002. Are body condition and reproductive effort of laying greater snow geese affected by the spring hunt? *The Condor* 104: 156–162.

MARION, W. R. & SHAMIS, J. D. 1977. An annotated bibliography of bird marking techniques. *Bird Banding* 48: 42–61.

MARTIN, A. C. & UHLER, F. M. 1939. *Food of Game Ducks in the United States and Canada*. United States Department of Agriculture Bulletin 634, Washington, D.C., USA.

MARTIN, F. W., POSPAHALA, R. S. & NICHOLS, J. D. 1979. Assessment and population management of North American migratory birds. Pp. 187–239 in Cairns, J. Jr., Patil, G. P. & Waters, W. E. (eds), *Environmental Biomonitoring, Assessment, Prediction,* and Management – Certain Case Studies and Relative Quantitative Issues. International Co-operative Publishing House, Fairland, USA.

MARTIN, G. R. 1986. Total panoramic vision in the mallard duck, Anas platyrhynchos. Vision Research 26: 1303–1306.

MARTÍNEZ-HARO, M., SÁNCHEZ-NAVA, P., SALGADO-MALDONADO, G. & RODRÍGUEZ-ROMERO, F. DE. J. 2012. Gastrointestinal helminth in waterfowl of the upper Lerma river sub-basin, Mexico. Revista Mexicana de Biodiversidad 83: 36–41.

MARTINSON, R. K., ROSASCO, M. E., MARTIN, E. M., SMART, M. G., CARNEY, S. M., KACZYNSKI, C. F. & GEIS, A. D. 1966. *1965 Experimental September Hunting Season on Teal.* U.S. Fish and Wildlife Service Special Scientific Report – Wildlife No. 95. U.S. Department of the Interior, Washington, D.C., USA.

MATEO, R., BONET, A., DOLZ, J. C. & GUITART, R. 2000. Lead shot densities in a site of grit ingestion for Greylag Geese Anser anser in Doñana (Spain). Ecotoxicology and Environmental Restoration 3: 76–80.

MATHEVET, R. & TAMISIER, A. 2002. Creation of a nature reserve, its effects on hunting management and waterfowl distribution in the Camargue (southern France). Biodiversity and Conservation 11: 509–519.

MATTHEWS, G. V. T. 1969. Nacton decoy and its catches. Wildfowl 20: 131–137.

MATTHEWS, G. V. T. 1982. The control of recreational disturbance. Pp. 325–330 in Scott, D. A. (ed), *Managing Wetlands and their Birds – A Manual of Wetland and Waterfowl Management.* IWRB, Slimbridge, UK.

MATTSSON, B. J., RUNGE, M. C., DEVRIES, J. H., BOOMER, G. S., EADIE, J. M., HAUKOS, D. A., FLESKES, J. P., KOONS, D. N., THOGMARTIN, W. E. & CLARK, R. G. 2012. A modeling framework for integrated harvest and habitat management of North American waterfowl: Case-study of northern pintail metapopulation dynamics. Ecological Modelling 225: 146–158.

MAZZUCCHI, L. 1971. Beitrag zur Nahrungsökologie in der Umgebung von Bern überwinternder Krinckenten Anas crecca L. Der Ornithologische Beobachter 68:161–178.

MEIJER, T. & DRENT, R. 1999. Re-examination of the capital and income dichotomy in breeding birds. Ibis 141: 399–414.

MELTOFTE, H. 1982. Shooting disturbance of waterfowl. Dansk Ornitologisk Forenings Tidsskrift 76: 21–35. [In Danish with English summary and figure legends]

MELTOFTE, H. & CLAUSEN, P. 2011. Swimming birds on Tipperne 1929–2007. Dansk Ornitologisk Forenings Tidsskrift 105: 120pp. [In Danish with English summary and figure legends]

MILLER, L. 1925. Food of the Harris Hawk. The Condor 27: 71–72.

MILLER M. R. 1985. Time budgets of northern Pintails wintering in the Sacramento Valley, California. Wildfowl 36: 53–64.

MILLER, M. R. & EADIE, J. McA. 2006. The allometric relationship between resting metabolic rate and body mass in wild waterfowl (Anatidae) and an application to estimation of winter habitat requirements. The Condor 108: 166–177.

MITCHELL, C. 1997. Re-mating in migratory Wigeon Anas penelope. Ardea 85: 275–277.

MITCHELL, C. & OGILVIE, M. 1996. Fifty years of wildfowl ringing by the Wildfowl and Wetlands Trust. Wildfowl 47: 240–247.

MITCHELL, C., FOX, A. D., HARRADINE, J. & CLAUSAGER, I. 2008. Measures of annual breeding success amongst Eurasian Wigeon Anas penelope. Bird Study 55: 43–51.

MITCHELL, T. R. & RIDGWELL, T. 1971. The frequency of Salmonellae in wild ducks. *Journal of Medical Microbiology* 4: 359–361.

MIYABAYASHI, Y. & MUNDKUR, T. 1999. *Atlas of Key Sites for Anatidae in the East Asian Flyway*. Wetlands International, Tokyo, Japan and Kuala Lumpur, Malaysia,

MOISAN, G. 1966. The Green-winged Teal in the Atlantic Flyway. *Naturaliste Canadien* 93: 69–88.

MOISAN, G., SMITH, R. I. & MARTINSON, R. K. 1967. *The Green-winged Teal: its distribution, migration, and population dynamics*. U.S. Fish and Wildlife Service Special Report – Wildlife 100. Washington, D.C., USA.

MOLODOVSKY, A. O. 1971. Feeding of *Anas crecca* L. and *A. querquedula* L. of the Gorky reservoirs. *Biologicheskie Nauki* 11: 20–25.

MONDAIN-MONVAL, J. Y. & GIRARD, O. 2000. Le canard colvert, la sarcelle d'hiver & autres canards de surface. *Faune Sauvage* 251: 124–139.

MONDAIN-MONVAL, J. Y., DESNOUHES, L. & TARIS, J. P. 2002. Lead shot ingestion in waterbirds in the Camargue (France). *Game and Wildlife Science* 19: 237–246.

MONVAL, J. Y. & PIROT, J. Y. 1989. *Results of the IWRB International Waterfowl Census 1967–1986*. IWRB Special Publication 8. IWRB, Slimbridge, UK.

MOOIJ, J. H. 2005. Protection and use of waterbirds in the European Union. *Beiträge zur Jagd und Wildforschung* 30: 49–76.

MOORE, R. L. 1980. Aspects of the ecology and hunting economics of migratory waterfowl on the Texas High Plains. M.Sc. Thesis, Texas Tech University, Lubbock, USA.

MORTON, J. K. & DIETERICH, R. A. 1979. Avian pox infection in an American green-winged teal (*Anas crecca carolinensis*) in Alaska. *Journal of Wildlife Diseases* 15: 451–453.

MOURONVAL, J. B., GUILLEMAIN, M., CANY, A. & POIRIER, F. 2007. Diet of non-breeding wildfowl Anatidae and Coot *Fulica atra* on the Perthois gravel pits, northeast France. *Wildfowl* 57: 68–97.

MUDGE, G. P. 1983. The incidence and significance of ingested lead pellet poisoning in British wildfowl. *Biological Conservation* 27: 333–372.

MULLIÉ, W. C. & MEININGER, P. L. 1983. Waterbird trapping and hunting in Lake Manzala, Egypt, with an outline of its economic significance. *Biological Conservation* 27: 23–43.

MUNRO, J. A. 1939. Food of ducks and coots at Swan Lake, British Columbia. *Canadian Journal of Research* 17: 178–186.

MUNRO, J. A. 1949. Studies of waterfowl in British Columbia – Green-winged Teal. *Canadian Journal of Research* D 27: 149–178.

MUNSTER, V. J., BAAS, C., LEXMOND, P., BESTEBROER, T. M., GULDEMEESTER, J., BEYER, W. E. P., DE WIT, E., SCHUTTEN, M., RIMMELZWAAN, G. F., OSTERHAUS, A. D. M. E. & FOUCHIER, R. A. M. 2009. Practical considerations for high-throughput influenza A virus surveillance studies of wild birds by use of molecular diagnostic tests. *Journal of Clinical Microbiology* 47: 666–673.

MURPHY-KLASSEN, H. M., UNDERWOOD, T. J., SEALY, S. G. & CZYRNYJ, A. A. 2005. Long-term trends in spring arrival dates of migrant birds at Delta Marsh, Manitoba, in relation to climate change. *The Auk* 122: 1130–1148.

NAKAMURA, M. & ATSUMI, T. 2000. Adaptive significance of winter pair bond on male pintail, *Anas acuta*. *Journal of Ethology* 18: 127–131.

NAUMANN, J. A. 1897. *Naturgeschichte der Vögel Mitteleuropas. X. Band (Enten)*. F.E. Köhler, Gera-Untermhaus, Germany.

NECHAEV, V. A. 1991. *Birds of Sakhalin Island.* Amuro-Ussuriiskii tsentr po izucheniiu bioraznoobraziia ptits, Vladivostok, Russia. [In Russian]

NELSON, C. H. 1993. *The Downy Waterfowl of North America.* Delta Station Press, Deerfield, USA.

NEWTON, I. 1989. *Lifetime Reproduction in Birds.* Academic Press, London, UK.

NEWTON, I. 2008. *The Migration Ecology of Birds.* Academic Press, London, UK.

NEWTON, I. 2010. *Bird Migration.* Collins, London, UK.

NEWTON, I. & CAMPBELL, C. R. G. 1975. Breeding of ducks at Loch Leven, Kinross. *Wildfowl* 26: 83–103.

NICHOLS, J. D. & JOHNSON, F. A. 1996. The management of hunting of Anatidae. *Gibier Faune Sauvage / Game & Wildlife* 13: 977–989.

NICHOLS, J. D., CONROY, M. J., ANDERSON, D. R. & BURNHAM, K. P. 1984. Compensatory mortality in waterfowl populations: a review of the evidence and implications for research and management. *Transactions of the North American Wildlife and Natural Resources Conference* 49: 535–554.

NICHOLS, J. D., RUNGE, M. C., JOHNSON, F. A. & WILLIAMS, B. K. 2007. Adaptive harvest management of North American waterfowl populations: a brief history and future prospects. *Journal of Ornithology* 148 (Suppl. 2): S343–S349.

NICHOLS, J. D., BLOHM, R. J., REYNOLDS, R. E., TROST, R. E., HINES, J. E. & BLADEN, J. P. 1991. Band reporting rates for Mallards with reward bands of different dollar values. *Journal of Wildlife Management* 55: 119–126.

NICHOLS, J. D., REYNOLDS, R. E., BLOHM, R. J., TROST, R. E., HINES, J. E. & BLADEN, J. P. 1995. Geographic variation in band reporting rates for Mallards based on reward banding. *Journal of Wildlife Management* 59: 697–708.

NICOLAI, C. A., SEDINGER, J. S., WARD, D. H. & BOYD, W. S. 2012. Mate loss affects survival but not breeding in black brant geese. *Behavioral Ecology* 23: 643–648.

NIEL, C. & LEBRETON, J. D. 2005. Using demographic invariants to detect overharvested bird populations from incomplete data. *Conservation Biology* 19: 826–835.

NIKOLAEV, V. I. 1998. *Birds of the marsh landscapes of "Zavidovo" National Park and Upper Volga River area.* Tver', Russia. [In Russian]

NILSSON, J-Å. & PERSSON, I. 2004. Postnatal effects of incubation length in mallard and pheasant chicks. *Oikos* 105: 588–594.

NORTH AMERICAN WATERFOWL MANAGEMENT PLAN, PLAN COMMITTEE. 2004. *North American Waterfowl Management Plan 2004. Implementation Framework: Strengthening the Biological Foundation.* Canadian Wildlife Service, U.S. Fish and Wildlife Service, Secretaria de Medio Ambiente y Recursos Naturales.

NORTH AMERICAN WATERFOWL MANAGEMENT PLAN COMMITTEE. 2012. *North American Waterfowl Management Plan 2012: People Conserving Waterfowl and Wetlands.* Canadian Wildlife Service, U.S. Fish and Wildlife Service, Secretaria de Medio Ambiente y Recursos Naturales.

NORTON, A. H. 1911. A second European Teal (*Nettion crecca*) in Maine. *The Auk* 28: 255.

NORTON, M. R. & THOMAS, V. G. 1994. Economic analyses of 'crippling losses' of North American waterfowl and their policy implications for management. *Environmental Conservation* 21: 347–353.

NUDDS, T. D. & ANKNEY, C. D. 1982. Ecological correlates of territory and home range size in North American dabbling ducks. *Wildfowl* 33: 58–62.

NUDDS, T. D. & BOWLBY, J. N. 1984. Predator-prey size relationships in North American dabbling ducks. *Canadian Journal of Zoology* 62: 2002–208.

NUDDS, T. D. & COLE, R. W. 1991. Changes in populations and breeding success of boreal forest ducks. *Journal of Wildlife Management* 55: 569–573.

NUDDS, T. D., SJÖBERG, K. & LUNDBERG, P. 1994. Ecomorphological relationships among Palearctic dabbling ducks on Baltic coastal wetlands and a comparison with the Nearctic. *Oikos* 69: 295–303.

NUDDS, T. D., ELMBERG, J., SJÖBERG, K., PÖYSÄ, H. & NUMMI, P. 2000. Ecomorphology in breeding Holarctic dabbling ducks: the importance of lamellar density and body length varies with habitat type. *Oikos* 91: 583–588.

NUMMI, P. 1990. The differences in the diet of flying juvenile teals and mallards. *Suomen Riista* 36: 89–96. [In Finnish with English summary]

NUMMI, P. 1992. The importance of beaver ponds to waterfowl broods: an experiment and natural tests. *Annales Zoologici Fennici* 29: 47–55.

NUMMI, P. 1993. Food niche relations of two sympatric ducks, Mallard and Green-winged Teal. *Canadian Journal of Zoology* 71: 49–55.

NUMMI, P. & HAHTOLA, A. 2008. The beaver as an ecosystem engineer facilitates teal breeding. *Ecography* 31: 519–524.

NUMMI, P. & KUULUVAINEN, T. 2013. Forest disturbance by an ecosystem engineer: beaver in boreal forest landscapes. *Boreal Environment Research* 18 (Suppl. A): 13–24.

NUMMI, P. & PÖYSÄ, H. 1995a. Breeding success of ducks in relation to different habitat factors. *Ibis* 137: 145–150.

NUMMI, P. & PÖYSÄ, H. 1995b. Habitat use by different-aged duck broods and juvenile ducks. *Wildlife Biology* 1: 181–187.

NUMMI, P. & PÖYSÄ, H. 1997. Population and community level responses in *Anas* species to patch disturbance caused by an ecosystem engineer, the beaver. *Ecography* 20: 580–584.

NUMMI, H. & VÄÄNÄNEN, V. M. 2001. High overlap in diets of sympatric dabbling ducks – an effect of food abundance? *Annales Zoologici Fennici* 38: 123–130.

NUMMI, P., PAASIVAARA, A., SUHONEN, S. & PÖYSÄ, H. 2013. Wetland use by brood-rearing female ducks in a boreal landscape: the importance of food and habitat. *Ibis* 155: 68–79.

NUMMI, P., ELMBERG, J., PÖYSÄ, H., GUNNARSSON, G. & SJÖBERG, K. 2005. Breeding success of teals *Anas crecca* varies for different lakes. *Suomen Riista* 51: 27–34. [In Finnish with English summary]

NUMMI, P., PÖYSÄ, H., ELMBERG, J., DESSBORN, L. & SJÖBERG, K. 2010. Importance of food and habitat structure to dabbling duck broods. *Suomen Riista* 56: 16–25. [In Finnish with English summary]

OBENAUER, J. C., DENSON, J., MEHTA, P. K., SU, X., MUKATIRA, S., FINKELSTEIN, D. B., XU, X., WANG, J., MA, J., FAN, Y., RAKESTRAW, K. M., WEBSTER, R. G., HOFFMANN, E., KRAUSS, S., ZHENG, J., ZHANG, Z. & NAEVE, C. W. 2006. Large-scale sequence analysis of avian influenza isolates. *Science* 311: 1576–1580.

OGILVIE, M. A. 1975. *Ducks of Britain and Europe*. T. & A.D. Poyser, Berkhamsted, UK.

OGILVIE, M. A. 1982. Winter 1978/79 hard weather movements and mortality of ducks ringed in the United Kingdom. Pp. 174–180 in Scott, D. A. & Smart, M. (eds), *Proceedings of the Second Technical Meeting on Western Palearctic Migratory Bird Management, Paris 11–13 December 1979*. IWRB, Slimbridge, UK.

OGILVIE, M. A. 1983. A migration study of the Teal (*Anas crecca*) in Europe using ringing recoveries. PhD Thesis, University of Bristol, UK.

OGILVIE, M. 2002. Eurasian Teal (Teal) *Anas crecca* – Green-winged Teal *Anas carolinensis*. Pp. 189–192 in Wernham, C. V., Toms, M. P., Marchant, J. H., Clark, J. A., Siriwardena, G. M. & Baillie, S. R. (eds), *The Migration Atlas: Movements of the Birds of Britain and Ireland*. T. & A.D. Poyser, London, UK.

OHLENDORF, H. M., HOTHEM, R. L., BUNCK, C. M. & MAROIS, K. C. 1990. Bioaccumulation of selenium in birds at Kesterson Reservoir, California. *Archives of Environmental Contamination and Toxicology* 19: 495–507.

OLNEY, P. J. S. 1960. Lead poisoning in wildfowl. *Wildfowl Trust Annual Report* 11: 123–134.

OLNEY, P. J. S. 1963. The food and feeding habits of teal *Anas crecca crecca* L. *Proceedings of the Zoological Society of London* 140: 169–210.

OLNEY, P. J. S. 1965. The autumn and winter feeding biology of certain sympatric ducks. *Transactions of the International Union of Game Biologists Congress* 6: 309–320.

OLSEN, B., MUNSTER, V. J., WALLENSTEN, A., WALDENSTRÖM, J., OSTERHAUS, A. D. M. E. & FOUCHIER, R. A. M. 2006. Global patterns of influenza A virus in wild birds. *Science* 312: 384–388.

OLSON, D. 2013. Survival and recovery rate analysis of Green-winged Teal, 1970–2008. In Fleming, K. & Howell, D. (eds), An assessment of the harvest potential of North American Teal. Unpublished report of the joint USFWS-Flyway-CWS Teal Harvest Potential Working Group.

OLSSON, C. & WIKLUND, J. 1999. *Västerbottens fåglar*. Private publication, Umeå, Sweden.

OMLAND, K. E. 1996. Female mallard mating preferences for multiple male ornaments I. Natural variation. *Behavioral Ecology and Sociobiology* 39: 353–360.

OPDAM, P., THISSEN, J., VERSCHUREN, P. & MÜSKENS, G. 1977. Feeding ecology of a population of Goshawk *Accipiter gentilis*. *Journal of Ornithology* 118: 35–51.

ORING, L. W. & SAYLER, R. D. 1992. The mating system of waterfowl. Pp. 128–189 in Batt, B. D. J., Afton, A. D., Anderson, M. G., Ankney, C. D., Johnson, D. H., Kadlec, J. A. & Krapu, G. L. (eds), *Ecology and Management of Breeding Waterfowl*. University of Minnesota Press, Minneapolis, USA.

OTNES, G. & OTNES, M. 1979. Green-winged Teal *vs*. Peregrine Falcon. *The Loon* 51: 200.

OTTOSSON, U., OTTVALL, R., ELMBERG, J., GREEN, M., GUSTAFSSON, R., HAAS, F., HOLMQVIST, N., LINDSTRÖM, Å., NILSSON, L., SVENSSON, M., SVENSSON, S. & TJERNBERG, M. 2012. *Swedish birds: numbers and distribution*. Sveriges Ornitologiska Förening, Halmstad, Sweden. [In Swedish]

OWEN, M. 1975. An assessment of fecal analysis technique in waterfowl feeding studies. *Journal of Wildlife Management* 39: 271–279.

OWEN, M. 1991. Nocturnal feeding in waterfowl. *Acta XX Congressus Internationalis Ornithologici*: 1105–1112.

OWEN, M. & BLACK, J. M. 1990. *Waterfowl Ecology*. Blackie, Glasgow, UK.

OWEN, M., ATKINSON-WILLES, G. L. & SALMON, D. G. 1986. *Wildfowl in Great Britain* – Second Edition. Cambridge University Press, Cambridge, UK.

PAIN, D. J. 1991a. Why are lead-poisoned waterfowl rarely seen?: the disappearance of waterfowl carcasses in the Camargue, France. *Wildfowl* 42: 118–122.

PAIN, D. J. 1991b. Lead shot densities and settlement rates in Camargue marshes, France. *Biological Conservation* 57: 273–286.

PAIN, D. J. 1996. Lead in waterfowl. Pp. 251–264 in Beyer, W. N., Heinz, G. H. & Redmon-Norwood, A. W. (eds), *Environmental Contaminants in Wildlife – Interpreting Tissue Concentrations*. SETAC Special Publications Series, Lewis Publishers, Boca Raton, Florida, USA.

PAIN, D. & AMIARDTRIQUET, C. 1993. Lead poisoning of raptors in France and elsewhere. *Ecotoxicology and Environmental Safety* 25: 183–192.

PALMER, R. S. 1976. Green-winged Teal *Anas crecca*. Pp. 347–371 in Palmer, R. S. (ed), *Handbook of North American Birds Volume 2 – Waterfowl (first part)*. Yale University Press, London, UK.

PAQUETTE, G. A. & ANKNEY, C. D. 1996. Wetland selection by American Green-winged Teal breeding in British Columbia. *The Condor* 98: 27–33.

PAQUETTE, G. A. & ANKNEY, C. D. 1998. Diurnal time-budgets of American green-winged Teal *Anas crecca* breeding in British Columbia. *Wildfowl* 49:186–194.

PAULUS, G. L. 1988. Time-activity budgets of nonbreeding Anatidae: a review. Pp. 135–152 in Weller, M. W. (ed), *Waterfowl in Winter*. University of Minnsota Press, Minneapolis, USA.

PAULUS, S. L. 1983. Dominance relations, resource use, and pairing chronology of Gadwalls in winter. *The Auk* 100: 947–952.

PEARCE, J. 1995. Radiocesium in migratory bird species in Northern Ireland following the Chernobyl accident. *Bulletin of Environmental Contamination and Toxicology* 54: 805–811.

PEARCE, J. M. 2007. Philopatry: a return to origins. *The Auk* 124: 1085–1087.

PEARSE, A. T. & RATTI, J. T. 2004. Effects of predator removal on Mallard duckling survival. *Journal of Wildlife Management* 68: 342–350.

PEASE, M. L., ROSE, R. K. & BUTLER, M. J. 2005. Effects of human disturbances on the behavior of wintering ducks. *Wildlife Society Bulletin* 33: 103–112.

PEHRSSON, O. 1991. Egg and clutch size in the mallard as related to food quality. *Canadian Journal of Zoology* 69: 156–162.

PENNYCUICK, C. J. 2008. *Modelling the Flying Bird*. Academic Press, Burlington, MA, USA. Flight software can be downloaded from http://www.bio.bristol.ac.uk/people/pennycuick.htm

PERDECK, A. C. & CLASON, C. 1983. Sexual differences in migration and winter quarters of ducks ringed in The Netherlands. *Wildfowl* 34:137–143.

PEREDA-SOLIS, M. E., MARTÍNEZ-GUERRERO, J. H. & TOCA-RAMIREZ, J. A. 2012. Detection of zinc, lead, cadmium and arsenic in dabbling ducks from Durango, Mexico. *Asian Journal of Animal and Veterinary Advances* 7: 761–766.

PÉRON, G. 2013. Compensation and additivity of anthropogenic mortality: life-history effects and review of methods. *Journal of Animal Ecology* 82: 408–417.

PERRY, M. C. & UHLER, F. M. 1981. Asiatic Clam (*Corbicula manilensis*) and other foods used by waterfowl in the James River, Virginia. *Estuaries* 4: 229–233.

PETERS, J. L., McCRACKEN, K. G., PRUETT, C. L., ROHWER, S., DROVETSKI, S. V., ZHURAVLEV, Y. N., KULIKOVA, I., GIBSON, D. D. & WINKER, K. 2012. A parapatric propensity for breeding precludes the completion of speciation in common teal (*Anas crecca*, sensu lato). *Molecular Ecology* 21: 4563–4577.

PETRIDES, G. A. 1944. Sex ratios in ducks. *The Auk* 61: 564–571.

PETRULA, M. J. 1994. Nesting ecology of ducks in interior Alaska. M.Sc. Thesis, University of Alaska Fairbanks, USA.

PHILLIPS, J. C. 1923. *A Natural History of the Ducks. Volume II: The Genus* Anas. Houghton Mifflin Co., Boston, USA.

PILCHER, R. E. M. 1964. Effects of the cold winter of 1962–63 on birds of the north coast of the Wash. *Wildfowl Trust Annual Report* 15: 23–26.

PIRKOLA, M. K. & HÖGMANDER, J. 1974. The age determination of duck breeds in the field. *Suomen Riista* 25: 50–55. [In Finnish with English summary]

PIROT, J. Y. 1981. Partage alimentaire et spatial des zones humides camarguaises par 5 espèces de canards en hivernage et en transit. Thèse de Doctorat, Université Pierre et Marie Curie, Paris, France.

PIROT, J. Y., CHESSEL, D. & TAMISIER, A. 1984. Exploitation alimentaire des zones humides de Camargue par cinq espèces de canards de surface en hivernage et en transit: modélisation spatio-temporelle. *Revue d'Ecologie (Terre et Vie)* 39 : 167–192.

PLATT, M. I. 1951. Diet of Golden Eagle. *The Scottish Naturalist* 63: 67–68.

POLLUX, B. J. A., DE JONG, M., STEEGH, A., OUBORG, N. J., VAN GROENENDAEL, J. M. & KLAASSEN, M. 2006. The effect of seed morphology on the potential dispersal of aquatic macrophytes by the common carp (*Cyprinus carpio*). *Freshwater Biology* 51: 2063–2071.

PÖYSÄ, H. 1984a. Species assembly in the dabbling duck (*Anas* spp.) guild in Finland. *Annales Zoologici Fennici* 21: 451–464.

PÖYSÄ, H. 1984b. Temporal and spatial dynamics of waterfowl populations in a wetland area – a community ecological approach. *Ornis Fennica* 61: 99–108.

PÖYSÄ, H. 1985a. Size and recent changes of breeding waterfowl populations in the Lake Vetsijärvi, Utsjoki area. *Lintumies* 20: 228–233. [In Finnish with English summary]

PÖYSÄ, H. 1985b. Circumstantial evidence of foraging interference between 2 species of dabbling ducks. *Wilson Bulletin* 97: 541–543.

PÖYSÄ, H. 1986a. Species composition and size of dabbling duck (*Anas* spp.) feeding groups: are foraging interactions important determinants? *Ornis Fennica* 63: 33–41.

PÖYSÄ, H. 1986b. Foraging niche shifts in multispecies dabbling duck (*Anas* spp.) feeding groups: harmful and beneficial interactions between species. *Ornis Scandinavica* 17: 333–346.

PÖYSÄ, H. 1987. Feeding-vigilance trade-off in the Teal (*Anas crecca*): effects of feeding method and predation risk. *Behaviour* 103: 108–122.

PÖYSÄ, H. 1989. Foraging patch dynamics in the Teal (*Anas crecca*): effects of sociality and search method switching. *Behaviour* 110: 306–318.

PÖYSÄ, H., ELMBERG, J., NUMMI, P. & SJÖBERG, K. 1994. Species composition of dabbling duck assemblages: ecomorphological patterns compared with null models. *Oecologia* 98: 193–200.

PÖYSÄ, H., ELMBERG, J., NUMMI, P. & SJÖBERG, K. 1996. Are ecomorphological associations among dabbling ducks consistent at different spatial scales? *Oikos* 76: 608–612.

PÖYSÄ, H., RINTALA, J., LEHIKOINEN, A. & VÄISÄNEN, R. A. 2013b. The importance of hunting pressure, habitat preference and life history for population trends of breeding waterbirds in Finland. *European Journal of Wildlife Research* 59: 245–256.

PÖYSÄ, H., ELMBERG, J., GUNNARSSON, G., NUMMI, P., SJÖBERG, G. G. & SJÖBERG, K. 2004. Ecological basis of sustainable harvesting: is the prevailing paradigm of compensatory mortality still valid? *Oikos* 104: 612–615.

PÖYSÄ, H., DESSBORN, L., ELMBERG, J., GUNNARSSON, G., NUMMI, P., SJÖBERG, K., SUHONEN, S. & SÖDERQUIST, P. 2013a. Harvest mortality in North

American mallards: a reply to Sedinger and Herzog. *Journal of Wildlife Management* 77: 653–654.

PRADEL, R. 2009. The stakes of capture-recapture models with state uncertainty. Pp. 781–795 in Thomson, D. L., Cooch, E. G. & Conroy, M. J. (eds), *Modeling Demographic Processes in Marked Populations.* Springer, Berlin, Germany.

PREBLE, E. A. & McATEE, W. L. 1923. A biological survey of the Pribilof Islands, Alaska. *North American Fauna* 46: 1–255.

PREUSS, N. O. 2001. Hans Christian Cornelius Mortensen: aspects of his life and of the history of bird ringing. *Ardea* 89 (Special Issue): 1–6.

PREVOST, M. B., JOHNSON, A. S. & LANDERS, J. L. 1978. Production and utilization of waterfowl foods in brackish impoundments in South Carolina. *Proceedings of the Annual Conference of the Southeastern Association of Fish and Wildlife Agencies* 32: 60–70.

PRICE, R. D. 1971. A review of the Genus *Holomenopon* (Mallophaga: Menoponidae) from the Anseriformes. *Annals of the Entomological Society of America* 64: 633–646.

PRIKLONSKI, S. G. & SAPETINA, I. M. 1990. Game statistics in the USSR. Pp. 113–114 in Matthews, G. V. T. (ed), *Managing Waterfowl Populations. Proceedings of an IWRB symposium, Astrakhan, USSR, 2–5 October 1989.* IWRB Special Publication 12, IWRB, Slimbridge, UK.

PROMISLOW, D., MONTGOMERIE, R. & MARTIN, T. E. 1994. Sexual selection and survival in North American waterfowl. *Evolution* 48: 2045–2050.

PYBUS, M. J. 2011. Avian botulism in Alberta: a history. Pp. 121–135 in Bollinger, T. K., Evelsizer, D. D., Dufour, K. W., Soos, C., Clark, R. G., Wobeser, G., Kehoe, F. P., Guyn, K. L. & Pybus, M. J. (eds), *Ecology and Management of Avian Botulism on the Canadian Prairies.* Report to the Prairie Habitat Joint Venture, Canada.

PYLE, P. 2008. *Identification Guide to North American Birds. Part II. Anatidae to Alcidae.* Slate Creek Press, Point Reyes Station, California, USA.

QUAY, T. L. & CRITCHER, T. S. 1962. Food habits of waterfowl in Currituck Sound, North Carolina. *Proceedings of the Annual Conference of the Southeastern Association of Fish and Wildlife Agencies* 16: 200–209

QUINLAN, E. E. & BALDASSARRE, G. A. 1984. Activity budgets of nonbreeding Green-winged Teal on playa lakes in Texas. *Journal of Wildlife Management* 48: 838–845.

RAFTOVICH, R. V., WILKINS, K. A., WILLIAMS, S. S. & SPRIGGS, H. L. 2012. *Migratory Bird Hunting Activity and Harvest during the 2010 and 2011 Hunting Seasons.* U.S. Fish and Wildlife Service, Laurel, Maryland, USA.

RAVE, D. P. & BALDASSARRE, G. A. 1989. Activity budget of Green-winged Teal wintering in coastal wetlands of Louisiana. *Journal of Wildlife Management* 53: 753–759.

RAVE, D. P. & BALDASSARRE, G. A. 1991. Carcass mass and composition of Green-winged Teal wintering in Louisiana and Texas. *Journal of Wildlife Management* 55: 457–461.

RAY, J. & WILLUGHBY, F. 1678. *The Ornithology of Francis Willughby.* A.C. for John Martyn, London, UK.

REICHHOLF, J. 1970. Der Einfluß von Störungen durch Angler auf den Entenbrutbestand auf den Altwässern am Unteren Inn. *Vogelwelt* 91: 68–72.

REMPEL, R. S., ABRAHAM, K. F., GADAWSKI, T. R., GABOR, S. & ROSS, R. K. 1997. A simple wetland habitat classification for boreal forest waterfowl. *Journal of Wildlife Management* 61: 746–757.

RENDÓN, M. A., GREEN, A. J., AGUILERA, E. & ALMARAZ, P. 2008. Status, distribution and long-term changes in the waterbird community wintering in Doñana, south-west Spain. *Biological Conservation* 141: 1371–1388.

RHODES, O. E. JR., DEVAUILT, T. L. & SMITH, L. M. 2006. Seasonal variation in carcass composition of American Wigeon wintering in the Southern High Plains. *Journal of Field Ornithology* 77: 220–228.

RIDGILL, F. C. & FOX, A. D. 1990. *Cold Weather Movements of Waterfowl in Western Europe*. IWRB Special Publication 13, Slimbridge, UK.

ROBERTSON, G. J. & COOKE, F. 1999. Winter philopatry in migratory waterfowl. *The Auk* 116: 20–34.

ROBINSON, J. A. & POLLITT, M. S. 2002. Sources and extent of human disturbance to waterbirds in the UK: an analysis of Wetland Bird Survey data, 1995/96 to 1998/99. *Bird Study* 49: 205–211.

ROBINSON, J. A., CULZAC, L. G. & ALDRIDGE, N. S. 2002. Age-related changes in the habitat use and behaviour of Mallard *Anas platyrhynchos* broods at artificially created lakes in southern Britain. *Wildfowl* 53: 107–118.

ROCKE, T. E. 2006. The global importance of avian botulism. Pp. 422–426 in Boere, G., Galbraith, C. & Stroud, D. (eds), *Waterbirds around the World*. The Stationary Office, Edinburgh, UK.

ROCKE, T. E. & BOLLINGER, T. K. 2007. Avian botulism. pp. 377–416 in Thomas, N. J., Hunter, D. B. & Atkinson, C. T. (eds), *Infectious Diseases of Wild Birds*. Blackwell Publishing, Oxford, UK.

ROCKE, T. E., SAMUEL, M. D., SWIFT, P. K. & YARRIS, G. S. 2000. Efficacy of a type C botulism vaccine in Green-winged Teal. *Journal of Wildlife Diseases* 36: 489–493.

RODRIGUES, D., FIGUEIREDO, M. & FABIAO, A. 2002. Mallard (*Anas platyrhynchos*) summer diet in central Portugal ricefields. *Game and Wildlife Science* 19: 55–62.

RODRIGUES, D., FIGUEIREDO, M. E., FABIÃO, A. & ENCARNAÇAO, V. 2006. Ducks and the risk of avian influenza in Portugal. *Airo* 16: 69–74.

RODRIGUES, D. J. C., FABIAO, A. M. D. & FIGUEIREDO, M. E. M. A. 2001. The use of nasal markers for monitoring Mallard populations. Pp. 316–318 in Field, R., Warren, R. J., Okarma, H. & Sievert, P. R. (eds), *Wildlife, Land, and People: Priorities for the 21st Century. Proceedings of the Second International Wildlife Management Congress*. The Wildlife Society, Bethesda, Maryland, USA.

RODWAY, M. S. 2007. Timing of pairing in waterfowl I: reviewing the data and extending the theory. *Waterbirds* 30: 488–505.

ROGERON, G. 1903. *Les canards considérés à l'état sauvage et comme oiseaux d'agrément en domesticité*. Librairie J.B. Baillière et Fils, Paris, France.

ROGERS, J. P., NICHOLS, J. D., MARTIN, F. W., KIMBALL, C. F. & POSPAHALA, R. S. 1979. Can ducks be managed by regulation? An examination of harvest and survival rates of ducks in relation to hunting. *Transactions of the North American Wildlife and Natural Resources Conference* 44: 114–126.

ROHWER, F. C. 1984. Patterns of egg laying in prairie ducks. *The Auk* 101: 603–605.

ROHWER, F. C. & ANDERSON, M. G. 1988. Female-biased philopatry, monogamy, and the timing of pair formation in migratory waterfowl. In Johnson, R. F. (ed), *Current Ornithology* Volume 5: 187–221. Plenum Press, New York, USA.

ROLLO, J. D. & BOLEN, E. G. 1969. Ecological relationships of Blue and Green-winged Teal on the High Plains of Texas in early fall. *The Southwestern Naturalist* 14: 171–188.

ROSENBERG, K. V. OHMART, R. D., HUNTER, W. C. & ANDERSON, B. W. 1991. *Birds of the Lower Colorado River Valley*. University of Arizona Press, Tucson, USA.

ROSS, R. K., ABRAHAM, K. F., GADAWSKI, T. R., REMPEL, R. S., GABOR, T. S. & MAHER, R. 2002. Abundance and distribution of breeding waterfowl in the Great Clay Belt of Northern Ontario. *The Canadian Field-Naturalist* 116: 42–50.

ROTHSCHILD, R. F. N. & DUFFY, L. K. 2005. Mercury concentrations in muscle, brain and bone of Western Alaskan waterfowl. *Science of the Total Environment* 349: 277–283.

ROUSSELOT, J. C. & TROLLIET, B. 1991. *Critères de détermination du sexe et de l'âge des canards*. Office National de la Chasse, Paris, France.

ROUX F., MAHÉO R. & TAMISIER A. 1978. L'exploitation de la basse vallée du Sénégal (quartier d'hiver tropical) par trois espèces de canards paléarctiques et éthiopien. *Revue d'Ecologie (Terre et Vie)* 32: 387–416.

ROWE, M., CZIRJÁK, G. Á., McGRAW, K. J. & GIRAUDEAU, M. 2011. Sexual ornamentation reflects antibacterial activity of ejaculates in mallards. *Biology Letters* 7: 740–742.

ROWLEY, I. 1983. Re-mating in birds. Pp. 331–360 in Bateson, P. (ed), *Mate Choice*. Cambridge University Press, Cambridge, UK.

RÜGER, A., PRENTICE, C. & OWEN, M. 1986. *Results of the IWRB International Waterfowl Census 1967–1983*. IWRB Special Publication 6. IWRB, Slimbridge, UK.

RUNSTADLER, J. A., HAPP, G. M., SLEMONS, R. D., SHENG, Z. M., GUNDLACH, N., PETRULA, M., SENNE, D., NOLTING, J., EVERS, D. L., MODRELL, A., HUSON, H., HILLS, S., ROTHE, T., MARR, T. & TAUBENBERGER, J. K. 2007. Using RRT-PCR analysis and virus isolation to determine the prevalence of avian influenza virus infections in ducks at Minto Flats State Game Refuge, Alaska, during August 2005. *Archives of Virology* 152: 1901–1910.

SAAD, M. D., AHMED, L. S., GAMAL-ELDEIN, M. A., FOUDA, M. K., KHALIL, F. M., YINGST, S. L., PARKER, M. A. & MONTEVILLEL, M. R. 2007. Possible avian influenza (H5N1) from migratory bird, Egypt. *Emerging Infectious Diseases* 13: 1120–1121.

SAGE, B. L. 1958. On the avian hosts of the leech *Theromyzon* (*Protoclepsis*) *tessellata* (O.F. Muller). *Bulletin of the British Ornithologists' Club* 78: 113–114.

SAGE, B. L. 1960. Notes on some Pintail x Teal hybrids. *Bulletin of the British Ornithologists' Club* 80: 80–86.

SAINT-GERAND, T. 1982. Les stationnements d'Anatidés en France en janvier 1979. Pp. 170–173 in Scott, D. A. & Smart, M. (eds), *Proceedings of the Second Technical Meeting on Western Palearctic Migratory Bird Management, Paris 11–13 December 1979*. IWRB, Slimbridge, UK.

SALMON, D. G. & FOX, A. D. 1994. Changes in the wildfowl populations wintering on the Severn Estuary. *Biological Journal of the Linnean Society* 51: 229–236.

SALOMONSEN, F. 1968. The moult migration. *Wildfowl* 19: 5–24.

SAMUEL, M. D. & BOWERS, E. F. 2000. Lead exposure in American black ducks after implementation of non-toxic shot. *Journal of Wildlife Management* 64: 947–953.

SANGSTER, G., COLLINSON, M., HELBIG, A. J., KNOX, A. G., PARKIN, D. T. & PRATER, T. 2001. The taxonomic status of Green-winged Teal *Anas carolinensis*. *British Birds* 94: 218–226.

SARGEANT, A. B. 1972. Red fox spatial characteristics in relation to waterfowl predation. *Journal of Wildlife Management* 36: 225–230.

SARGEANT, A. B. & RAVELING, D. G. 1992. Mortality during the breeding season. Pp. 396–422 in Batt, B. D. J., Afton, A. D., Anderson, M. G., Ankney, C. D., Johnson, D. H., Kadlec, J. A. & Krapu, G. L. (eds), *Ecology and Management of Breeding Waterfowl*. University of Minnesota Press, Minneapolis, USA.

SARGEANT, A. B., ALLEN, S. H. & EBERHARDT, R. T. 1984. Red fox predation on breeding ducks in midcontinent North America. *Wildlife Monograph* 89: 1–41.

SCHILLER, E. L. 1953. Studies of the Helminth fauna of Alaska. XIV. Some cestode parasites of the Aleutian Teal (*Anas crecca* L.) with the description of *Diorchis longiovum* n. sp. *Proceedings of the Helminthological Society of Washington* 20: 7–12.

SCHRICKE, V. 1983. Distribution spatio-temporelle des populations d'anatidés en transit et en hivernage en baie du Mont Saint-Michel, en relation avec les activités humaines. PhD Thesis, University of Rennes I, France.

SCHUMMER, M. L., KAMINSKI, R. M., RAEDEKE, A. H. & GRABER, D. A. 2010. Weather-related indices of autumn–winter dabbling duck abundance in Middle North America. *Journal of Wildlife Management* 74: 94–101.

SCOTT, B. & DICKSON, W. 2001. Taxonomic changes. *British Birds* 93: 462–465.

SCOTT, D. A. & ROSE, P. M. 1996. *Atlas of Anatidae Populations in Africa and Western Eurasia*. Wetlands International Publication 41, Wageningen, The Netherlands.

SEDINGER, J. S. 1992. Ecology of prefledging waterfowl. Pp. 109–127 in Batt, B. D. J., Afton, A. D., Anderson, M. G., Ankney, C. D., Johnson, D. H., Kadlec, J. A. & Krapu, G. L. (eds), *Ecology and Management of Breeding Waterfowl*. University of Minnesota Press, Minneapolis, USA.

SEDINGER, J. S. & HERZOG, M. P. 2012. Harvest and dynamics of duck populations. *Journal of Wildilfe Management* 76: 1108–1116.

SELL, D. L. 1979. Fall foods of Teal on the Texas High Plains. *Southwestern Naturalist* 24: 373–375.

SHAW, M. G. & KOCAN, A. A. 1980. Helminth fauna of waterfowl in central Oklahoma. *Journal of Wildlife Diseases* 16: 59–64.

SHCHELKANOV, M., ANAN'EV, V., L'VOV, D. K., KIREEV, D. E., GUR'EV, E. L., AKANINA, D. S., GALKINA, I. V., ARISTOVA, V. A., MOSKVINA, T. M., CHUMAKOV, V. M., BARANOV, N. I., GORELIKOV, V. N., USACHEV, E. V., AL'KHOVSKĬ, S. V., LIAPINA, O. V., POGLAZOV, A. B., SHLIAPNIKOVA, O. V., BURUKHINA, E. G., BORISOVA, O. N., FEDIAKINA, I. T., BURTSEVA, E. I., MOROZOVA, T. N., GRENKOVA, E. P., GREBENNIKOVA, T. V., PRILIPOV, A. G., SAMOKHVALOV, E. I., SABEREZHNYĬ, A. D., KOLOMEETS, S. A., MIROSHNIKOV, V. A., OROPAĬ, P. L., GAPONOV, V. V., SEMENOV, V. I., SUSLOV, I. O., VOLKOV, V. A., IAMNIKOVA, S. S., ALIPER, T. I., DUNAEV, V. G., GROMASHEVSKĬ, V. L., MASLOV, D. V., NOVIKOV, F. T., VLASOV, N. A., DERIABIN, P. G., NEPOKLONOV, E. A., ZLOBIN, V. I. & L'VOV, D. K. 2007. Complex environmental and virological monitoring in the Primorye Territory in 2003–2006. *Voprosy Virusologii* 52: 37–48.

SHEPPARD, J. L., CLARK, R. G., DEVRIES, J. H. & BRASHER, M. G. 2013. Reproductive effort and success of wild female mallards: Does male quality matter? *Behavioural Processes* 100: 82–90.

SHUTLER, D., GLOUTNEY, M. L. & CLARK, R. G. 1998. Body mass, energetic constraints, and duck nesting ecology. *Canadian Journal of Zoology* 76: 1805–1814.

SIMPSON, J. W., YERKES, T. J., SMITH, B. D. & NUDDS, T. D. 2005. Mallard duckling survival in the Great Lakes region. *The Condor* 107: 898–909.

SJÖBERG, K. 1988. The flightless period of free-living male Teal *Anas crecca* in Northern Sweden. *Ibis* 130: 164–171.

SJÖBERG, K. & DANELL, K. 1982. Feeding activity of ducks in relation to diel emergence of chironomids. *Canadian Journal of Zoology* 60: 1383–1387.

SJÖBERG, K. & DANELL, K. 1983. Changes in the abundance of invertebrates and ducks after flooding of a wetland area in the boreal forest region (N. Sweden). Pp.

921–930 in Hell, P. (ed), *Proceedings from XVI. Congress of the International Union of Game Biologists*. Vysoké Tatry, Štrbské Pleso, ČSSR.

SMART, G. 1965. Body weights of newly hatched Anatidae. *The Auk* 82: 645–648.

SMIETANKA, K., MINTA, Z., WLODARCZYK, R., WYROSTEK, K., JÓZWIAK, M., OLSZEWKA, M., MINIAS, P., KACZMAREK, K., JANISZEWSKI, T. & KLESZCZ, A. 2012. Avian influenza viruses in wild birds at the Jeziorsko reservoir in Poland in 2008–2010. *Polish Journal of Veterinary Sciences* 15: 323–328.

SOLMAN, V. E. F. 1945. The ecological relations of Pike, *Esox lucius* L., and waterfowl. *Ecology* 26: 157–170.

SOONS, M. B., VAN DER VLUGT, C., VAN LITH, B., HEIL, G. W. & KLAASSEN, M. 2008. Small seed size increases the potential for dispersal of wetland plants by ducks. *Journal of Ecology* 96: 619–627.

SOTNIKOV, V. N. 1999. *Birds of the Kirov region and adjacent territories. Vol. 1. Non-Passerines*. Triada-S Press, Kirov, Russia. [In Russian]

SOUCHAY, G., GAUTHIER, G. & PRADEL, R. 2013. Temporal variation of juvenile survival in a long-lived species: the role of parasites and body condition. *Oecologia* 173: 151–160.

SOWLS, L. K. 1955. *Prairie Ducks. A Study of their Behavior, Ecology and Management*. The Stackpole Company, Harrisburg, Pennsylvania and the Wildlife Management Institute, Washington, D.C., USA.

SPACKMAN, E., STALLKNECHT, D. E., SLEMONS, R. D., WINKER, K., SUAREZ, D. L., SCOTT, M. & SWAYNE, D. E. 2005. Phylogenetic analyses of type A influenza genes in natural reservoir species in North America reveals genetic variation. *Virus Research* 114: 89–100.

SPÄRCK, R. 1947. Ten years of wildlife biological studies in Denmark. *Svensk Jakt* 85: 287–292. [In Swedish]

SPÄRCK, R. 1957. An investigation of the food of swans and ducks in Denmark. *Danish Review of Game Biology* 3: 45–47.

SPARLING, D. W., VANN, S. & GROVE, R. A. 1998. Blood changes in Mallards exposed to white phosphorus. *Environmental Toxicology and Chemistry* 17: 2521–2529.

SPINA, F. & VOLPONI, S. 2008. *Atlas of bird migration in Italy. 1. Non-passerines*. Ministero dell' Ambiente e della Tutela del Territorio e del Mare, Instituto Superiore per la Protezione e la Ricerca Ambientale (ISPRA). Tipografia CSR, Roma, Italia. [In Italian]

SREBRODOL'SKAYA, N. I. & PAVLUK, R. S. 1976. Nutrition of the Mallard *Anas platyrhynchos* in the western part of the Ukraine Polesye, USSR. *Vestnik Zoologii* 2: 78–80.

STAAV, R. 1998. Longevity list of birds ringed in Europe. *EURING Newsletter* 2: 9–18.

STAFFORD, J. D. & PEARSE, A. T. 2007. Survival of radio-marked mallard ducklings in South Dakota. *The Wilson Journal of Ornithology* 119: 585–591.

STAFFORD, J. D., FLAKE, L. D. & MAMMENGA, P. W. 2002. Survival of mallard broods and ducklings departing overwater nesting structures in eastern South Dakota. *Wildlife Society Bulletin* 30: 327–336.

STALLKNECHT, D. E., NAGY, E., HUNTER, D. B. & SLEMONS, R. D. 2007. Avian Influenza. Pp. 108–130 in Thomas, N. J., Hunter, D. B. & Atkinson, C. T. (eds), *Infectious Diseases of Wild Birds*. Blackwell Publishing Ltd., Oxford, UK.

STALLKNECHT, D. E., SENNE, D. A., ZWANK, P. J., SHANE, S. M. & KEARNEY, M. T. 1991. Avian paramyxoviruses from migrating and resident ducks in coastal Louisiana. *Journal of Wildlife Diseases* 27: 123–128.

STEELE, B. B., REITSMA, L. R., RACINE, C. H., BURSON, S. L. III., STUART, R. & THEBERGE, R. 1997. Different susceptibilities to white phosphorus poisoning among five species of ducks. *Environmental Toxicology and Chemistry* 16: 2275–2282.

STOTT, T. & MITCHELL, C. 1991. Orielton duck decoy – the story of its decline. *Field Studies* 7: 759–769.

STOUT, I. J. & CORNWELL, G. W. 1976. Nonhunting mortality of fledged North American waterfowl. *Journal of Wildlife Management* 40: 681–693.

STERBERTZ, I. 1969. Investigations on wild-ducks in the inundation area of the River Tisza. *Aquila* 76–77: 141–143.

STREET, M. 1977. The food of Mallard ducklings in a wet gravel quarry, and its relation to duckling survival. *Wildfowl* 28: 113–125.

STRUBBE, E. T. 1967. Blue-winged Teal X Green-winged Teal hybrid. *The Loon* 39: 59.

STRYAN, T. W. 1891. On the birds of the Lower Yangtze Basin, Part II. *Ibis* 33: 481–510.

SUAREZ-R, C. & URIOS, V. 1999. La contaminación por saturnismo en las aves acuáticas del Parque Natural de El Hondo y su relación con los hábitos alimenticios. *Humedales Mediterráneos* 1: 83–90.

SUGDEN, L. G. 1973. Feeding ecology of pintail, gadwall, American wigeon and lesser scaup ducklings in Southern Alberta. *Canadian Wildlife Service Report Series* 24.

SUGDEN, L. G. & POSTON, H. J. 1968. A nasal marker for ducks. *Journal of Wildlife Management* 32: 984–986.

SUHONEN, S., NUMMI, P. & PÖYSÄ, H. 2011. Long term stability of boreal lake habitats and use by breeding ducks. *Boreal Environment Research* 16 (Suppl. B): 71–80.

SULGOSTOWSKA, T. 2007. Intestinal digeneans of birds (superfamily Diplostomoidea) of the Masurain Lakes. *Wiadomosci Parazytologiczne* 53: 117–128.

SUTHERLAND, W. J. 2004. Diet and foraging behavior. Pp. 233–250 in Sutherland, W. J., Newton, I. & Green, R. E. (eds), *Bird Ecology and Conservation – A Handbook of Techniques*. Oxford University Press, Oxford, UK.

SVAZAS, S. 1994. The pattern of diurnal and nocturnal migratory activity of autumnal bird migrants in the inland part of Lithuania. *The Ring* 16: 48–54.

SWANSON G. A. & BARTONEK, J. C. 1970. Bias associated with food analysis in gizzards of blue-winged teal. *Journal of Wildlife Management* 34: 739–746.

SWANSON, G. A. & SARGEANT, A. B. 1972. Observation of night time feeding behavior of ducks. *Journal of Wildlife Management* 36: 959–961.

SWANSON, G. A., KRAPU, G. L. & SERIE, J. R. 1979. Foods of laying female dabbling ducks on the breeding grounds. Pp. 47–55 in Bookhout, T. A. (ed), *Waterfowl and Wetlands – An Integrated Review. Proceedings of the 1977 Symposium, Madison*. Wildlife Society, Washington, D.C., USA.

SWANSON, G. A., KRAPU, G. L., BARTONEK, J. C., SERIE, J. R. & JOHNSON, D. H. 1974. Advantages in mathematically weighing waterfowl food habits data. *Journal of Wildlife Management* 38: 302–307.

SWARTH, H. S. 1924. Birds and mammals of the Skeena river region of northern British Columbia. *University of California Publications in Zoology* 24: 315–394.

SWIDEREK, P. K., JOHNSON, A. S., HALE, P. E. & JOYNER, R. L. 1988. Production, management, and waterfowl use of Sea Purslane, Gulf Coast Muskgrass, and Widgeongrass in brackish impoundments. Pp. 441–457 in Weller, M. W. (ed), *Waterfowl in Winter*. University of Minnesota Press, Minneapolis, USA.

SZIJJ, J. 1965. ökologische Untersuchungen an Entenvögeln (*Anatidae*) des Ermatinger Beckens (Bodensee). *Die Vogelwarte* 23: 24–71.

SZYMANSKI, M. L., JOHNSON, M. A. & GROVIJAHN, M. 2013. Effects of hunting pressure and collection method bias on body mass of drake mallards. *Journal of Wildlife Management* 77: 235–242.

SZYMCZAK, M. R. & RINGELMAN, J. K. 1986. Differential habitat use of patagial-tagged female Mallards. *Journal of Field Ornithology* 57: 230–232.

TAMISIER, A. 1966. Dispersion crépusculaire des sarcelles d'hiver *Anas crecca crecca* L. en recherche de nourriture. *Revue d'Ecologie (Terre et Vie)* 3: 316–337.

TAMISIER, A. 1970. Chasse et mortalité chez les sarcelles d'hiver *Anas crecca crecca* L. baguées en Camargue. Unpublished note from presentation at the IVth Entretiens de Chizé, France.

TAMISIER, A. 1971. Régime alimentaire des sarcelles d'hiver *Anas crecca* L. en Camargue. *Alauda* 39 : 261–311.

TAMISIER, A. 1972. Etho-écologie des Sarcelles d'hiver *Anas crecca* L. pendant leur hivernage en Camargue. PhD Thesis, University of Montpellier, France.

TAMISIER, A. 1976. Diurnal activities of Green-winged Teal and Pintail wintering in Louisiana. *Wildfowl* 27: 19–32.

TAMISIER, A. 1978. The functional units of dabbling ducks: a spatial integration of their comfort and feeding requirements. *Verhandlungen der Ornithologischen Gesellschaft in Bayern* 23: 229–238.

TAMISIER, A. 1995. Hunting as a key environmental parameter for the Western Palearctic duck populations. *Wildfowl* 36: 95–103.

TAMISIER, A. & DEHORTER, O. 1999. *Camargue, canards et foulques – fonctionnement et devenir d'un prestigieux quartier d'hiver*. Centre Ornithologique du Gard, Nîmes, France.

TAMISIER, A. & TAMISIER, M. C. 1981. L'existence d'unités fonctionnelles démontrée chez les sarcelles d'hiver en Camargue par la biotélémétrie. *Revue d'Ecologie (Terre et Vie)* 35: 563–579.

TAMISIER, A., ALLOUCHE, L., AUBRY, F. & DEHORTER, O. 1995. Wintering strategies and breeding success: hypothesis for a trade-off in some waterfowl species. *Wildfowl* 46: 76–88.

TAMISIER, A., BECHET, A., JARRY, G., LEFEUVRE, J. C. & LE MAHO, Y. 2003. Effets du dérangement par la chasse sur les oiseaux d'eau. Revue de littérature. *Revue d'Ecologie (Terre & Vie)* 58 : 435–449.

TERNIER, L. 1922. *La Sauvagine en France. Chasse, description et histoire naturelle de toutes les espèces visitant nos contrées*. Emonet, Dupuy & Cie, Paris, France.

TERREGINO, C., DE NARDI, R., GUBERTI, V., SCREMIN, M., RAFFINI, E., MARTIN, A. M., CATTOLI, G., BONFANTI, L. & CAPUA, I. 2007. Active surveillance for avian influenza viruses in wild birds and backyard flocks in Northern Italy during 2004 to 2006. *Avian Pathology* 36: 337–344.

TERRILL, S. B. & ABLE, K. P. 1988. Bird migration terminology. *The Auk* 105: 205–206.

TESKY, J. L. 1993. *Anas crecca*. In: Fire Effects Information System, U.S. Department of Agriculture, Forest Service, Rocky Mountain Research Station, Fire Sciences Laboratory (Producer). Accessed on 8 May 2012 at: http://www.fs.fed.us/database/feis/

THERKILDSEN, O. R. & BREGNBALLE, T. 2006. The importance of salt-marsh wetness for seed exploitation by dabbling ducks *Anas* sp. *Journal of Ornithology* 147: 591–598.

THOMAS, G. J. 1975. Ingested lead pellets in waterfowl at the Ouse Washes, England, 1968–73. *Wildfowl* 26: 43–48.

THOMAS, G. J. 1981. Field feeding by dabbling ducks around the Ouse Washes, England. *Wildfowl* 32: 69–78.

THOMAS, G. J. 1982. Autumn and winter feeding ecology of waterfowl at the Ouse Washes, England. *Journal of Zoology (London)* 197: 131–172.

THOMPSON, A. L. 1931. On 'abmigration' among the ducks, an anomaly shown by the results of bird-marking. *Proceedings of the International Ornithological Congress* 7: 382–388.

THOMPSON, D. R. 1996. Mercury in birds and terrestrial mammals. Pp. 341–356 in Beyer, W. N., Heinz, G. H. & Redmon-Norwood, A. W. (eds), *Environmental Contaminants in Wildlife – Interpreting Tissue Concentrations*. SETAC Special Publications Series, Lewis Publishers, Boca Raton, Florida, USA.

TOFT, C. A., TRAUGER, D. L. & MURDY, H. W. 1982. Tests for species interactions: breeding phenology and habitat use in subarctic ducks. *The American Naturalist* 120: 586–613.

TOFT, C. A., TRAUGER, D. L. & MURDY, H. W. 1984. Seasonal decline in brood sizes of sympatric waterfowl (*Anas* and *Aythya*, Anatidae) and a proposed evolutionary explanation. *Journal of Animal Ecology* 53: 75–92.

TOLKAMP, C. R. 1993. Filter-feeding efficiencies of dabbling ducks (*Anas* spp.) in relation to microhabitat use and lamellar spacing. Master's Thesis, University of Guelph, Ontario, Canada.

TOWEILL, D. E. 1979. Eurasian Green-winged Teal observed at Lakeview, Oregon. *The Murrelet* 60: 77–79.

TOWNSHEND, D. J. & O'CONNOR, D. A. 1993. Some effects of disturbance to waterfowl from bait-digging and wildfowling at Lindisfarne National Nature Reserve, North-East England. In Davidson, N. & Rothwell, P. (eds), Disturbance to Waterfowl on estuaries. *Wader Study Group Bulletin* 68 Special Issue: 47–52.

TRAIL, P. W. 2006. Avian mortality at oil pits in the United States: a review of the problem and efforts for its solution. *Environmental Management* 38: 532–544.

TREASURER, J. W. 1980. The occurrence of duck chicks in the diet of pike. *North East Scotland Bird Report 1979*, University of Aberdeen, Scotland.

TREMBLAY, S., & COUTURE, R. 1986. Morphologie bucco-linguale d'une guilde de canards barboteurs. *Canadian Journal of Zoology* 64: 2176–2180.

TUBBS, C. R. & TUBBS, J. M. 1983. The distribution of zostera and its exploitation by wildfowl in the Solent, southern England. *Aquatic Botany* 15: 223–239.

TUCK, L. M. 1968. Recent Newfoundland bird records. *The Auk* 85: 304–311.

TUITE, C. H., HANSON, P. R. & OWEN, M. 1984. Some ecological factors affecting winter wildfowl distribution on inland waters in England and Wales, and the influence of water-based recreation. *Journal of Applied Ecology* 32: 41–62.

TUITE, C. H., OWEN, M. & PAYNTER, D. 1983. Interaction between wildfowl and recreation at Llangorse Lake and Talybont Reservoir, South Wales. *Wildfowl* 34: 48–63.

TURNER, B. C. & THRELFALL, W. 1975. The metazoan parasites of Green-winged Teal (*Anas crecca* L.) and Blue-winged Teal (*Anas discors* L.) from Eastern Canada. *Proceedings of the Helminthological Society of Washington* 42: 157–169.

TWEDT, D. J. & NELMS, C. O. 1999. Waterfowl density on agricultural fields managed to retain water in winter. *Wildlife Society Bulletin* 27: 924–930.

UNEP / AEWA SECRETARIAT. 2013. *Proceedings of the Fifth Session of the Meeting of the Parties to the Agreement on the Conservation of African-Eurasian Migratory Waterbirds.* La Rochelle, France, 14–18 May 2012. AEWA Secretariat, Bonn, Germany.

USGS. 2011. *Summaries of Banding and Encounter Data*. Accessed on 22 November 2011 at: http://www.pwrc.usgs.gov/bbl/homepage/start.cfm

USGS. 2013. *Longevity Records of North American Birds (Current Through January 2013)*. Accessed on 17 May 2013 at: http://www.pwrc.usgs.gov/bbl/longevity/Longevity_main.cfm

USGS BIRD BANDING LABORATORY. 2011. *Unpublished data retrieved 01 December*. Patuxent Wildlife Research Center, Laurel, MD, USA.

U.S. FISH AND WILDLIFE SERVICE. 2012. *Waterfowl Population Status, 2012*. U.S. Department of the Interior, Washington, D.C., USA.

VÄÄNÄNEN, V. M. 2001. Hunting disturbance and the timing of autumn migration in *Anas* species. *Wildlife Biology* 7: 3–9.

VÄÄNÄNEN, V. M. & NUMMI, P. 2003. Diet of sympatric ducks in eutrophic wetlands. *Suomen Riista* 49: 7–16. [In Finnish with English summary]

VÄHÄTALO, A. V., RAINIO, K., LEHIKOINEN, A. & LEHIKOINEN, E. 2004. Spring arrival of birds depends on the North Atlantic Oscillation. *Journal of Avian Biology* 35: 210–216.

VÄISÄNEN, R. A. 1974. Timing of waterfowl breeding on the Krunnit Islands, Gulf of Bothnia. *Ornis Fennica* 51: 61–84.

VÄISÄNEN, R. A., HARIO, M. & SAUROLA, P. 2011. Population estimates of Finnish birds. In Valkama, J., Vepsäläinen, V. & Lehikoinen, A. (2011) *The Third Finnish Breeding Bird Atlas*. Finnish Museum of Natural History and Ministry of Environment. http://atlas3.1intuatlas.fi/english

VÄISÄNEN, R. A., LAMMI, E. & KOSKIMIES, P. 1998. *Second Finnish Bird Atlas*. Otava, Helsinki, Finland. [In Finnish]

VAN DEN AKKER, J. B. & WILSON, V. T. 1949. Twenty years of bird banding at Bear River migratory bird refuge, Utah. *Journal of Wildlife Management* 13: 359–373.

VAN EERDEN, M. R. 1984. Waterfowl movements in relation to food stocks. Pp. 84–100 in Evans, P. R., Goss-Custard, J. D. & Hale, W. G. (eds), *Coastal Waders and Wildfowl in Winter*. Cambridge University Press, Cambridge, UK.

VAN EERDEN, M. R. & MUNSTERMAN, M. J. 1997. Patch use upon touch: filter-feeding European Teal *Anas crecca* have environmentally and socially determined foraging goals. Pp. 165–185 in Van Eerden, M. R. (ed), Patchwork – Patch use, habitat exploitation and carrying capacity for water birds in Dutch freshwater wetlands. PhD Thesis, Groningen University, The Netherlands.

VAN GILS, J. A., DE ROOIJ, S. R., VAN BELLE, J., VAN DER MEER, J., DEKINGA, A., PIERSMA, T. & DRENT, R. 2005. Digestive bottleneck affects foraging decisions in red knots *Calidris canutus*. I. Prey choice. *Journal of Animal Ecology* 74: 105–119.

VAN LEEUWEN, C. H. A., VAN DER VELDE, G., VAN GROENENDAEL, J. M. & KLAASSEN, M. 2012. Gut travellers: internal dispersal of aquatic organisms by waterfowl. *Journal of Biogeography* 39: 2031–2040.

VANSCHOENWINKEL, B. WATERKEYN, A., VANDECAETSBEEK, T., PINEAU, O., GRILLAS, P. & BRENDONCK, L. 2008. Dispersal of freshwater invertebrates by large terrestrial mammals: a case study with wild boar (*Sus scrofa*) in Mediterranean wetlands. *Freshwater Biology* 53: 2264–2273.

VARTAPETOV, L. G. 1998. *Birds of northern taiga of West Siberian plain* Nauka Sibirskoe Predpriiatie Ran, Novosibirsk, Russia. [In Russian]

VAUGHAN, R. E. & JONES, K. H. 1913. The birds of Hong Kong, Macao, and the West River or Si Kiang in South-eastern China, with special reference to their nidification and seasonal movements. *Ibis* 1: 163–201.

VESELOVSKÝ, Z. 1952. Postembryonic development of ducks. *Sylvia* 14: 36–73. [In Czech]

VEST, J. L. & CONOVER, M. R. 2011. Food habits of wintering waterfowl on the Great Salt Lake, Utah. *Waterbirds* 34: 40–50.

VEST, J. L., CONOVER, M. R., PERSCHON, C., LUFT, J. & HALL, J. O. 2009. Trace element concentrations in wintering waterfowl from the Great Salt Lake, Utah. *Archives of Environmental Contamination and Toxicology* 56: 302–316.

VIANA, D. S., SANTAMARÍA, L., MICHOT, T. C. & FIGUEROLA, J. 2013. Migratory strategies of waterbirds shape the continental-scale dispersal of aquatic organisms. *Ecography* 36: 430–438.

VIETTI, F., FROMENTIN, M. & COLLOT, A. 2010. La sarcelle d'hiver *Anas crecca*. *Aviornis International*, 213: 50–56.

VINICOMBE, K. E. 1994. Common Teals showing mixed characters of Eurasian and North American races. *British Birds* 87: 88–89.

VIKSNE, J., SVAZAS, S., CZAJKOWSKI, A., JANAUS, M., MISCHENKO, A., KOZULIN, A., KURESOO, A. & SEREBRYAKOV, V. 2010. *Atlas of Duck Populations in Eastern Europe*. Akstis, Vilnius, Lithuania.

VRTISKA, M. P., GAMMONLEY, J. H., NAYLOR, L. W. & RAEDEKE, A. H. 2013. Economic and conservation ramifications from the decline of waterfowl hunters. *Wildlife Society Bulletin* 37: 380–388.

VYAZOVICH, Y. 1996. Dynamics of the radionuclide contamination and ecology of wild Anatidae species in Belarus after the Chernobyl nuclear accident. *Gibier Faune Sauvage / Game & Wildlife* 13: 723–736.

WALDRUP, K. A. & KOCAN, A. A. 1985. Screening of Free-Ranging Waterfowl in Oklahoma for *Salmonella*. *Journal of Wildlife Diseases* 21: 435–437.

WAINWRIGHT, M. G. C. B. 1967. Results of wildfowl ringing at Abberton Reservoir, Essex 1949 to 1966. *Wildfowl Trust Annual Report* 18: 28–35.

WARE, L. L, PETRIE, S.A., BADZINSKI, S. S. & BAILEY, R. C. 2011. Selenium concentrations in greater scaup and Dreissenid mussels during winter on Western Lake Ontario. *Archives of Environmental Contamination and Toxicology* 61: 292–299.

WEATHERHEAD, P. J. & ROBERTSON, R. J. 1979. Offspring quality and the polygyny threshold: "The sexy son hypothesis". *American Naturalist* 113: 201–208.

WEBSTER, R. G., BEAN, W. J., GORMAN, O. T., CHAMBERS, T. M. & KAWAOKA, Y. 1992. Evolution and ecology of influenza A viruses. *Microbiological Reviews* 56: 152–179.

WERSAL, R. M., MCMILLAN, B. R. & MADSEN, J. D. 2005. Food habits of dabbling ducks during fall migration in a prairie pothole system, Heron Lake, Minnesota. *The Canadian Field Naturalist* 119: 546–550.

WETLANDS INTERNATIONAL. 2012. *Waterbird population estimates*. Accessed on 01 August 2012 at: http://wpe.wetlands.org

WHITE, D. H. & CROMARTIE, E. 1985. Bird use and heavy metal accumulation in waterbirds at dredge disposal impoundments, Corpus Christi, Texas. *Bulletin of Environmental Contamination and Toxicology* 34: 295–300.

WILLIAMS, B. K. 1990. Population levels in North American waterfowl. An assessment of recent trends. Pp. 90–96 in Matthews, G. V. T. (ed), *Managing Waterfowl Populations. Proceedings of an IWRB symposium, Astrakhan, USSR, 2–5 October 1989.* IWRB Special Publication 12, IWRB, Slimbridge, UK.

WINGER, P. V., LASIER, P. J., WHITE, D. H. & SEGINAK, J. T. 2000. Effects of contaminants in dredge material from the Lower Savannah River. *Archives of Environmental Contamination and Toxicology* 38: 128–136.

WINGS OVER WETLANDS PROJECT. 2010. *Critical Site Network Tool*. UNEP-WCMC, BirdLife International and Wetlands International, Cambridge, UK and Wageningen, The Netherlands. Accessed on 26 September 2012 at: http://dev.unep-wcmc.org/csn/default.html#state=home

WISHART, R. A. 1983. Pairing chronology and mate selection in the American wigeon (*Anas americana*). *Canadian Journal of Zoology* 61: 1733–1743.

WITTER, M. S. & CUTHILL, I. C. 1993. The ecological costs of avian fat storage. *Philosophical Transactions of the Royal Society of London B* 340: 73–92.

WOBESER, G. A. 1997a. *Diseases of Wild Waterfowl – Second Edition*. Plenum Press, New York, USA.

WOBESER, G. A. 1997b. Avian botulism – another perspective. *Journal of Wildlife Diseases* 33: 181–186.

WOBESER, G. A., RAINNIE, D. J., SMITH-WINDSOR, T. B. & BOGDAN, G. 1983. Avian botulism during late autumn and early spring in Saskatchewan. *Journal of Wildlife Diseases* 19: 90–94.

WOLFF, W. J. 1966. Migration of Teal ringed in the Netherlands. *Ardea* 54: 230–270.

WORLD HEALTH ORGANIZATION GLOBAL INFLUENZA PROGRAM SURVEILLANCE NETWORK. 2005. Evolution of H5N1 avian influenza viruses in Asia. *Emerging Infectious Diseases* 11: 1515–1521.

YEH, J. Y., PARK, J. Y. & OSTLUND, E. N. 2011. Serologic evidence of West Nile Virus in wild ducks captured in major inland resting sites for migratory waterfowl in South Korea. *Veterinary Microbiology* 154: 96–103.

YOCOM, C. F. 1951. *Waterfowl and their Food Plants in Washington*. University of Washington Press, Seattle, USA.

YOCOM, C. F. & KELLER, M. 1961. Correlation of food habits and abundance of waterfowl, Humboldt Bay, California. *California Fish and Game* 47: 41–53.

YURLOV, A. K., YANOVSKJI, A. P. & CHERNYSOV, V. M. 1990. Structure of the waterfowl bags in Baraba forest-steppe (West Siberia). P. 117 in Matthews, G. V. T. (ed), *Managing Waterfowl Populations. Proceedings of an IWRB symposium, Astrakhan, USSR, 2–5 October 1989*. IWRB Special Publication 12, IWRB, Slimbridge, UK.

ZAMANI-AHMADMAHMOODI, R., ESMAILI-SARI, A., GHASEMPOURI, S. M. & SAVABIEASFAHANI, M. 2009. Mercury in wetland birds of Iran and Iraq: contrasting resident Moorhen, *Gallinula chloropus*, and migratory Common Teal, *Anas crecca*, life strategies. *Bulletin of Environmental Contamination and Toxicology* 82: 450–453.

ZARUDNYY, N. A. 2003. The birds of Pskov Province. *Russkiy Ornitologicheskiy Zhurnal* 236: 1011–1021. [In Russian]

ZHANG, X-L., LUO, X-J., LIU, J., LUO, Y, CHEN, S-J. & MAI, B-X. 2011. Polychlorinated biphenyls and organochlorinated pesticides in birds from a contaminated region in South China: association with trophic level, tissue distribution and risk assessment. *Environmental Science and Pollution Research* 18: 556–565.

ZIMIN, V. B., SAZONOV, S. V., LAPSHIN, N. V., KHOKHLOVA, T. Y., ARTEMYEV, A. V., ANNENKOV, V. G. & YAKOVLEVA, M. V. 1993. *Bird Fauna of Karelia*. Institute of Biology of the Karelina Branch of the Russian Academy of Sciences, Petrozavodsk, Russia. [In Russian]

ZIMMER, C., BOOS, M., PETIT, O. & ROBIN, J. P. 2010. Body mass variations in disturbed mallards *Anas platyrhynchos* fit to the mass-dependent starvation-predation risk trade-off. *Journal of Avian Biology* 41: 637–644.

ZIMMER, C., BOOS, M., POULIN, N., GOSLER, A., PETIT, O. & ROBIN, J. P. 2011. Evidence of the trade-off between starvation and predation risks in ducks. *PLoS ONE* 6: e22352.

ZIMPFER, N. L., RHODES, W. E., SILVERMAN, E. D., ZIMMERMAN, G. S. & RICHKUS, K. D. 2011. *Trends in Duck Breeding Populations, 1955–2011*. U.S. Fish & Wildlife Service, Patuxent Wildlife Research Center, Laurel, MD.

ZIMPFER, N. L., RHODES, W. E., SILVERMAN, E. D., ZIMMERMAN, G. S. & RICHKUS, K. D. 2013. *Trends in Duck Breeding Populations, 1955–2013*. U.S. Fish & Wildlife Service, Patuxent Wildlife Research Center, Laurel, MD.

ZINK, R. M., ROHWER, S., ANDREEV, A. V. & DITTMANN, D. L. 1995. Trans-Beringia comparisons of mitochondrial DNA differentiation in birds. *The Condor* 97: 639–649.

ZINKL, J. G., DEY, N., HYLAND, J. M., HURT, J. J. & HEDDLESTON, K. L. 1977. An epornitic of Avian Cholera in waterfowl and Common Crows in Phelps County, Nebraska, in the spring, 1975. *Journal of Wildlife Diseases* 13: 194–198.

ZUBEROGOITIA, I., MARTÍNEZ, J. E., GONZÁLEZ-OREJA, J. A., CALVO, J. F. & ZABALA, J. 2013. The relationship between brood size and prey selection in a Peregrine Falcon population located in a strategic region on the Western European flyway. *Journal of Ornithology* 154: 73–82.

ZUUR, B. J., SUTER, W. & KRÄMER, A. 1983. Zur Nahrungsökologie auf dem Ermatinger Becken (Bodensee) überwinternder Wasservögel. *Der Ornithologische Beobachter* 80: 247–262.

ZWARTS, L. 1976. Density-related processes in feeding dispersion and feeding activity of Teal (*Anas crecca*). *Ardea* 64: 192–209.

Index

A
Abberton Reservoir, England 60
abmigration 51, 62, 196
AEWA (African-Eurasian Waterbird Agreement) 40, 47
Afghanistan 51
Africa 21, 38, 40, 42
   north east African population 45, 47, 213, 218
age determination 30–5
   tail feathers 31–2
agricultural development 38, 48, 183, 185, 201, 210, 212
Aix sponsa 73
Alaska, USA 38, 39, 41, 51, 57, 127, 132, 138, 139, 186
Alberta, Canada 39, 128, 157
albinism 30
Aleutian Islands 138
amphipods 104
Anas 14, 21, 99
   acuta 21
   americana 98
   carolinensis 22, 23, 35, 128, 142, 205
   clypeata 30
   crecca crecca 21, 35, 37, 128, 142, 205
   crecca nimia 25, 29, 37, 38, 138, 142
   cyanoptera 29
   discors 23
   falcata 29
   flavirostris flavirostris 23
   flavirostris oxyptera 23
   formosa 29
   penelope 60, 98
   platyrhynchos 21
   querquedula 21–2
   rubripes 73
   strepera 29
Anatidae 21
annual cycle 49–50
Anser rossii 125
artistic representations 198–9
Asia 38, 40, 42, 167, 213, 218
   central Asia 14
   eastern Asia 14, 59
   south east Asian population 45, 46
   south west Asian population 45, 46, 47
Asian Waterbird Census 40
avian botulism 156–7, 162, 174
avian influenza viruses (AIV) 153, 154
Aythya 21
   affinis 161
   americana 21
   collaris 21
   ferina 21
   fuligula 21
   marila 161
   valisineria 117

B
bacteria 118, 156–7
Badger, American 138
Baltic 39, 50, 54, 74
bands *see* ringing
beaver ponds 69, 71, 72–3, 185, 212
Beaver, American 69, 72
   European 69, 72
Belgium 50
Bering Sea 25, 51
Bern Convention on the Conservation of European Wildlife and Natural Habitats 47
bill colour 118
bill functioning 98
   lamellae 85, 99–101
BirdLife International 46
'black boxes' 214
   linking the seasons 215–16
   moulting 215
   nesting 215
   spring migration 214–15
Black Sea 39, 40, 41, 45, 46, 47
Boar, Wild 167
body mass 35, 83–4
   measurements 36–7
   migration 49, 60–1, 63–4
   Resting Metabolic Rate (RMR) 82–3
Bonn Convention 46
boreal zone 55, 69, 76, 138, 185
   forest 15, 16, 48, 50, 71, 76
   lakes 133, 145
Brant, Black 116
Branta bernicla nigricans 116
   leucopsis 195
breeding density 70–1, 73
breeding habitats 69–72, 144–5
   testing breeding habitat selection 149–52
breeding propensity 194
breeding success 56, 117, 127–8, 179, 182, 183, 194, 214, 216, 218
British Columbia, Canada 39, 51, 52, 71, 87, 145
brood amalgamation 149
brood density 71–2, 73, 139
brood size 131, 139, 147–8
brood success 127, 147
bulrushes (*Scirpus* spp.) 104

C
caddisflies 95
California, USA 51, 52, 79, 104, 171, 203
Camargue, France 41, 42, 52–4, 56, 60, 61, 62, 67, 79, 81, 84–5, 104, 113, 116, 119, 120, 122,

149–50, 151, 160, 164, 165, 166, 177, 178, 179, 181, 182–3, 187, 190, 193, 196, 204, 208–9, 210
Canada 38, 41, 47, 51, 52, 70, 71, 127, 170, 207
Canis latrans 200
Canvasback 117, 122, 165
capital breeders 75, 85, 215
captivity 204–5
carry-over effects 181, 214, 215–16
Castor canadensis 69, 72
   fiber 69, 72
censuses 40, 42
Characeae 104, 114
Chen caerulescens atlantica 162
   caerulescens caerulescens 125
chironomids 95–6, 101, 103, 104, 113, 138, 145
Circus 78
   aeruginosus 78
   cyaneus 167
CITES (Convention on International Trade in Endangered Species of Wild Fauna and Flora) 46
Cladocera 114
climate change 182–3, 212, 214, 217, 218
cloaca 124
cloacal examination 33–4, 154
   bursa of Fabricius 34
Cloacotaenia megalops 160, 161
Clostridium botulinum 156
clutches 129–30
   clutch predation 131, 135
   clutch size 127, 131, 133–4, 139, 148, 166
   second clutch 125, 140
CMS (Convention on Migratory Species) 46
communal swims 119
commuting flights 77, 79
conservation 14, 218
conservation status 48
   Teal conservation status by listing authorities 46–7
copulation 121–2, 124
   extra-pair copulation 125–6, 129
   forced copulation 116, 124–6
corn (*Zea mays*) 83, 104
Corvus corax 139
   monedula 139
counts 41–2, 172, 173, 185, 196, 213, 217
courtship 118–19
   courtship behaviour 123–4
   timing of pair formation 119–23
coverts 29, 30, 32
Coyote 200
crustaceans 101, 103, 104, 105, 106
Czech Republic 170

D
Dakota, USA 39, 127, 129, 130, 132, 134, 137, 138, 139
Delta Marsh, Manitoba 55, 183
demographic heterogeneity 196, 197, 210
Denmark 42, 53, 57, 74, 87, 170, 171, 178, 203, 216
density-dependence 129–30, 197
diet 101
   general food habits over the year 101–3
   methods to assess diet 105–8
   preferred foods 103–4
   Teal and Mallard strategies during food depletion 110–12
   Teal in the dabbling duck community 108–10
   see also foraging

digestive system
   assessment of digestive tract contents 106–7
   methods and problems in the presentation of gut content analysis results 107–8
   oesophagus 101, 104, 106
   proventriculus 106
Diptera 91, 95, 145
disease 126, 153–4
   avian botulism 156–7, 162, 174
   avian influenza viruses (AIV) 153, 154
   parasites 157–61
   Summary table of diseases known to occur in Common and Green-winged Teal 155
dissection 34
   digestive tract 106–7
distribution 43–5
   Mean teal numbers in European countries 44
   Teal, Common 39–41, 43
   Teal, Green-winged 38–9, 43
DNA 24–6
Duck, American Black 73
   Black 84, 118, 122–3
   Falcated 29
   Ring-necked 21
   Tufted 21, 128
   Wood 73
duck decoys 175, 200–1, 202, 205
ducklings 69, 138–9
   dependence on mother 140
   diet 102–3, 145
   duckling description and ageing of Teal broods 141–4
   duckling plumage 142–4
   duckling weight 140
   first hours and days 140
   foraging methods 95–7, 145–6
   growth 146
   habitat use by mother 144–5
   mortality 72–3, 145–6, 147
   predation 148–9
   survival 146–9
   yolk sac absorption 140
Ducks Unlimited 199, 212

E
Eagle, Bald 167
eggs 115
   appearance 134
   clutch predation 131, 135
   clutch size 127, 131, 133–4, 139, 148, 166
   egg formation 85
   egg-laying 131–4
   second clutch 125, 140
Egypt, ancient 198, 204
Eider, Common 118
energy requirements 82–5, 102–3
England 61, 195, 204
Ephemeroptera 95
Esox lucius 69
Eurasia 14, 35, 38, 51
   pro-active approach to Common Teal population management in Eurasia 217–18
Europe 15, 38
   hunting bags 79, 170–2
European Directive 2009/147/CE ('Birds Directive') 47
eutrophic wetlands 74, 75, 112

F

faecal analysis 105
Falco peregrinus 78
fasting ability 64, 84
feathers *see* plumage
fecundity and age structure 193–5
Federal List of Endangered and Threatened Wildlife and Plants, USA 47
Fennoscandia 50, 51, 55, 70, 73, 127
Finland 15, 40, 50, 70, 71, 73, 111, 127, 129, 132, 138, 139, 147, 148, 149, 170, 171, 176, 181, 183, 190, 213, 216, 217
flash marks 29, 30
flight altitude 60
flight speed 14, 16, 56–7
Florida, USA 177
flyways 41, 52, 53, 55, 56, 62, 212, 214, 218
    Europe 15, 75, 149–52, 172, 190, 213, 216
    Himalayas 59
    North America 39, 41, 51, 171, 195, 205
food calorific density 102
food competition 109, 111–12
food depletion 110–12
food habits 101–3
    preferred foods 103–4
food intake rate 98, 112
foraging 83
    bill functioning 98–101
    diurnal foraging 80, 87–8
    diurnal foraging in winter 86–9
    ducklings 95–7
    foraging during incubation 136–7
    foraging methods 90–1
    foraging methods and the structure of foraging behaviour 97–8
    foraging time 82–3
    nocturnal foraging 77–81, 83, 85–6
    spring and summer patterns 89–90
    see also diet
foraging depth 91–2
    foraging depth over time 94–7
    trading off foraging depth and predation risk 92–4, 97–8
Fox, Red 137, 138, 148, 167
France 15, 42, 50, 57, 61, 65, 67, 77, 78, 85, 87, 111, 170, 171, 178, 183, 185, 190, 195, 203, 204, 205, 207

G

Gadwall 29, 117, 122, 136, 150
Gallinula chloropus 139
Garganey 21–2, 29, 128, 142
gastroliths 112–13, 164
geese 21, 62, 65, 89, 198
gene flow 25, 26
generation time 195, 197
genetics 23, 24–6, 37, 41, 51, 126, 133, 198
genitalia 33–4
Germany 50, 67, 74, 210
gizzard 101, 106, 107, 112–13, 164–5, 166, 207
Goose, Barnacle 195
    Lesser Snow 125
    Ross's 125
    Greater Snow 162
GPS tags 67, 68, 216
Great Britain 50, 61, 116, 164, 200, 201, 204
Great Lakes 148

grit 112–13, 164
Gull, Yellow-legged 78
gulls 167

H

habitats
    agricultural habitats 76, 86, 104, 151
    breeding season 69–72, 144–5
    daily habitat use strategies by wintering Teal 78–81
    functional units 78–80
    habitat loss 183–5
    habitat management 185, 207–12, 218
    habitat selection 74, 127, 130, 196
    intertidal areas 74, 87, 177
    migrations 74–7
    winter areas 77–81
Haliaeetus leucocephalus 167
Harrier, Marsh 78, 167
    Northern 167
harriers 78
harvest 170, 171–2
    commercial harvest 175, 199–204, 217
    demographic features and resilience to harvest 197
    deriving harvest rate from ring recoveries 172–4
    harvest management 14, 170, 205–7, 217, 218
hatching 135, 138–40
    hatching dates 138
    hatching success 139
hawks 78
heavy metals 163–4
heterogeneity 196, 197, 210
homing rate 62
hormones 119, 162
human disturbance 176–9
hunting 47, 103
    commercial hunting 205
    crippling loss 174–5
    deriving harvest rate from ring recoveries 172–4
    hunting bags in North America and in Europe 79, 170–2
    hunting disturbance 177–8
    hunting pressure 57, 172
    mortality 175–6
    special hunting season 205
    spring 103, 106
hybrids 30, 205
hyperphagia 89

I

Iberian Peninsula 39, 50, 181
Iceland 51, 52, 71, 119, 120, 122, 127, 129, 132, 133, 138, 139, 140, 145, 147
income breeders 75, 85, 89, 128, 137, 214–15
income migrants 85, 89
incubation 115–16, 125, 135–8
    incubation length 137–8
    incubation recesses 136
India 40, 45, 50, 59–60, 203, 204
Indonesia 40, 45
influenza 154
insects 98, 101, 103, 104, 113, 145
    emergence 85, 133
International Waterbird Census 42
International Waterbird Counts 172
interspecific competition 108–10, 111–12

invertebrates 101–3
Iran 51, 204
Ireland, Northern 162, 204
isotopic signature 105–6
Italy 50, 85
IUCN (International Union for Conservation of Nature) 46

J
Jackdaw, Eurasian 139
Japan 45, 51

K
Kamchatka, Russia 51
Kanin Peninsula, Russia 74
kill rate 175, 197
kriging 52–3

L
Labrador, Canada 51
Larus michahellis 78
lead poisoning 113, 164–6
life cycle 14, 187
    life cycle and population trends 195–6
life expectancy 116–17, 170, 192–3, 194
lifetime reproductive success 117, 193
lipid fat reserves 64, 84
loop migration 50, 53
lores 29, 30
Louisiana, USA 51, 57, 78, 82, 83, 87, 120, 121, 171

M
Maldives 40
Mallard 14, 21, 22, 25, 29, 41, 51, 72, 75, 77, 157
    artistic representations 198
    ducklings 140, 141, 144, 145, 146, 147, 148, 149
    energy requirements 82
    foraging 92, 100–1, 109
    hunting 170, 173, 201, 202–3
    lead poisoning 165
    mating 117, 118, 122, 124–5
    nesting 131, 133, 136, 138, 149, 150, 151
    Teal and Mallard strategies during food depletion 110–12
mammals 92, 138
mandibles 99, 100
Manitoba, Canada 39, 55
marking techniques 64–9
mating system 115–17
    courtship behaviour 123–4
    forced copulations 116, 124–6
    mate guarding 116, 117, 125, 126
    mate selection 117–19
    timing of pair formation 119–23
mayflies 95
measurements 35–7
    body mass 36–7
Mediterranean 39, 40, 41, 45, 46, 47, 50, 62, 149–50, 213, 218
Mephitis mephitis 138
metabolic rate 82–3, 131, 179
Mexico 47, 57, 170
Michigan, USA 38, 39
midges 71, 90, 95, 133, 145
migration 38, 49
    abmigration 51, 62, 196
    annual movements of teal wintering in the Camargue, southern France 52–4
    autumn migration 55, 56, 62, 74
    breeding habitat selection 149–52
    cold weather movements 62–4
    differential migration speed 60–1
    loop migration 50, 53
    marking techniques 64–9
    migration distance 57–60
    migration flight 56–60
    migration routes 50–1
    migration timing 54–6
    spring migration 55–6, 62, 74–7, 103, 214–15
    stopover areas 74–5, 85, 212, 217
    transatlantic flights 51–2
    winter site fidelity 62
Migratory Bird Treaty Act (USA) 47, 203
Migratory Birds Convention Act (Canada) 47
mink 139, 148, 167
    American 138
Minnesota, USA 38, 39
Mississippi, USA 38, 39, 51, 171, 195
molluscs 101, 103, 105
monogamy 115, 125
Moorhen 139
mortality 117, 153
    additive mortality 175–7
    compensatory mortality 176
moulting 30–1, 215
    moulting sites 54–5, 74

N
nasal disks 65
nasal saddles 62, 65–7, 116–17, 121, 209, 216
neck collars 65
Neovison vison 138
nesting 71–3, 127, 215
    egg-laying 131–4
    hatching 138–40
    incubation 135–8
    nest defence 137
    nest density 127, 129–30
    nest description 130
    nest establishment 128–31
    nest parasitism 133
    nest predation 131, 135–6, 137, 138–9, 151–2
    nest success 138
    pre-nesting period 127–8
    semi-natural nests 150–1
Netherlands 42, 50, 74, 87, 119, 120, 170, 200, 201, 203
Newfoundland, Canada 52
niche partitioning 108–9
Nile Valley 40
Norma Oficial Mexicana 47
North America 14, 29–30, 38
    hunting bags 79, 170–2
North Carolina, USA 119, 122, 192
Norway 172

O
Ob River, Russia 74
Oklahoma, USA 38
oligotrophic wetlands 50, 70, 71, 112
Ontario, Canada 39, 51, 71, 212
Ostracoda 104, 114, 160
oviduct 33, 34, 132

P
pair formation 115–17
    courtship behaviour 123–4
    forced copulations 116, 124–6

mate selection 117–19
    timing of pair formation 119–23
Pakistan 40
parasites 157–61
patagial tags 65
PCB 162, 163
penis 33–4, 122, 124
Peregrine 78, 167, 168
pesticides 161, 163
philopatry 62, 115–16, 128
Phoebe, Eastern 62, 64
phylogeny 131
    Phylogeny of the tribe Anatini (dabbling ducks) 23
Pike 69, 149, 167–8
Pintail, Northern 21, 23, 25, 29, 101, 116, 165, 198, 205
plumage 27–30, 118
    alternate plumage 30, 118
    breeding 27–9, 118
    ducklings 142–4
    eclipse (basic) 30, 31, 118
    females 27, 29
    plumage care 33
    rachis 31
Pochard, Common 21, 60, 128
pollution 161–4
polygyny 116
poplar (*Populus* spp.) 212
population 186
    demographic features and resilience to harvest 197
    fecundity and age structure 193–5
    general numbers and trends 213–14
    life cycle and population trends 195–6
    limiting factors for Teal populations 216–17
    population estimates 38
    population growth rate 196
    population sizes and trends 45–6
    pro-active approach to Common Teal population management in Eurasia 217–18
    teal survival rates 186–93
Portugal 57, 65, 67
potatoes (*Solanum tuberosum*) 98
potholes 69, 129, 139
prairies 38, 41, 48, 71, 127, 129, 138, 139, 185
predation 76, 166–70
    duckling predation 148–9
    lead poisoning 165–6
    nest predation 131, 135–6, 137, 138–9, 151–2
    predation risk 78, 152
    Reported Teal predators in the Holarctic 169–70
    trading off foraging depth and predation risk 92–4, 97–8
preening 33, 124
primaries 30, 32, 143
Procyon lotor 138
prolactin 140
propagules 113–14
punt guns 175, 200

Q
Québec, 51, 71, 170, 186

R
Raccoon 138
radar tracking 56, 60

radio tracking 67–9
radio transmitters (tags) 67, 147
Ramsar Convention 48
raptors 65, 78, 92, 93, 94, 106, 149, 167
Raven, Common 139
Redhead 21
refuges 177–8, 217
remiges 143
reproductive value 196
rice (*Oryza sativa*) 81, 104
ringing 50–1, 66
    development of ringing 64–5
    reporting rate 65, 172–4
    reward rings 173–4
    ring recoveries 52–4, 57–9, 62, 63, 172–4
Rome, ancient 198
roosts 76, 77, 167, 209
    day-roosts 78–9, 92, 114, 209–10
Russia 40, 41, 50, 54, 56, 69, 70, 71, 74, 132, 134, 138, 139, 147, 171, 172, 182

S
Sahara 40, 59
Salix spp. 130, 209
Saskatchewan, Canada 39, 168
satellite geologgers 67–9
Sayornis phoebe 62
scapulars 27, 31
Scaup, Greater 161
    Lesser 161
Scotland 139, 204
secondaries 29, 31
sedges (*Carex* spp.) 104, 144
seeds 101, 102, 113, 114
    Cyperaceae 104
    smartweed (*Polygonum* spp.) 103
selenium 161
sex determination 30–5
sex ratio 116
sexual characters 118
sexual competition 116, 123, 126
sexual dimorphism 123
sexual selection 29
Shoveler, Northern 30, 92, 101, 128, 136
Siberia 47, 51, 138
Skunk, Striped 138, 139, 167
Slovakia 170
Slovenia 170
smartweed (*Polygonum* spp.) 103
snakes 138
Solanum tuberosum (potatoes) 98
Somateria mollissima 118
Spain 67, 113, 165, 203, 213, 217
spatial heterogeneity 196, 197
speculum 29, 31
sperm competition 125–6
Spermophilus spp. 138
Squirrel, Ground 138
Sri Lanka 40
stable isotope analysis 105–6
stoneworts (*Chara* spp.) 114
Sudan 40
surveys 38, 41–2
    history of waterfowl surveys 41–2
survival rates 186–93
    survival of juvenile Teal during first months after fledging 187–90

Sus scrofa 167
swans 21, 62, 65
Sweden 15, 67, 70, 74, 90, 145, 150, 151, 183
Switzerland 172
syrinx morphology 34–5
systematics 21

T

Tamarix spp. 209
Taxidea taxus 138
taxonomy 23–6
Teal, Aleutian 25, 29, 49, 52, 81, 167
   population estimates 38, 45–6
Teal, Baikal 29, 30
   Blue-winged 23, 29, 104, 128, 136, 137, 142, 175, 205
   Cinnamon 29, 142, 205
Teal, Common 14–15, 16–17
   annual movements of teal wintering in the Camargue, southern France 52–4
   common names 21–2, 23
   distribution 39–41
   Geographic ranges 39
   methods to assess diet 105–8
   population estimates 38, 45
   pro-active approach to Common Teal population management in Eurasia 217–18
   Summary table of diseases known to occur in Common and Green-winged Teal 155
   taxonomy 24–6
   Teal and Mallard strategies during food depletion 110–12
   Teal in the dabbling duck community 108–10
   transatlantic flights 51–2
Teal, Green-winged 14, 16–17
   common names 22–3
   distribution 38–9
   Geographic ranges 39
   methods to assess diet 105–8
   population estimates 38, 46
   Summary table of diseases known to occur in Common and Green-winged Teal 155
   taxonomy 24–6
   Teal and Mallard strategies during food depletion 110–12
   transatlantic flights 51–2
Teal, Sharp-winged Speckled 23
   Speckled 23
   Yellow-billed 23
tealeries 203, 204

tertials 29, 31, 32
testosterone 119, 125
Texas, USA 51, 57, 84, 103, 120, 160, 209
thermoregulation 37, 63, 87, 89, 102, 122
thyroid 31
tidal cycle 87
time budgets 85, 87, 93, 132, 176
Tour du Valat, France 42
Trichoptera 95, 97
trophic status 112
Turkey 74

U

United Kingdom (UK) 38, 49, 177, 178, 179, 185, 193, 201, 204
United States (USA) 23, 38, 41, 47, 51, 52, 170, 205, 210
Ural Mountains 45, 56, 60
uropygial gland 33

V

vagrancy 51, 52
VHF tags 67–9, 79
vigilance 92, 93–4, 97–8, 110, 111
visual fields 92
Volga Delta, Russia 54, 55
Vulpes vulpes 137

W

Wadden Sea 74
Wales 132, 139, 144, 147, 204
water depth 90–8
weather 179–83
   cold weather movements 62–4
wheat (*Triticum aestivum*) 104
Wigeon 92, 98, 133
   American 83, 118–19
   Eurasian 60, 65, 77, 117, 190, 201
willow (*Salix* spp.) 130
wind tunnels 56
wing length 35
Wisconsin, USA 38, 39

Y

Yukon, USA 39

Z

zoochory 113–14